STUDY GUIDE FOR

MICROBIOLOGY

AN INTRODUCTION

ELEVENTH EDITION

Berdell R. Funke
NORTH DAKOTA STATE UNIVERSITY

PEARSON

Boston   Columbus   Indianapolis   New York   San Francisco   Upper Saddle River
Amsterdam   Cape Town   Dubai   London   Madrid   Milan   Munich   Paris   Montréal   Toronto
Delhi   Mexico City   São Paulo   Sydney   Hong Kong   Seoul   Singapore   Taipei   Tokyo

Acquisitions Editor: Kelsey Churchman
Managing Editor: Deborah Cogan
Production Project Manager and Manufacturing Buyer: Dorothy Cox
Production Management and Composition: Element LLC
Main Text Cover Design: Riezebos Holzbaur Design Group
Supplement Cover Design: 17th Street Studios
Executive Marketing Manager: Neena Bali
Cover Photo Credit: Alfred Pasieka/Photo Researchers Inc.

www.pearsonhighered.com

ISBN 10: 0-321-80299-3
ISBN 13: 978-0-321-80299-6

1 2 3 4 5 6 7 8 9 10—EBM—16 15 14 13 12

Contents

Preface

TO THE STUDENT

Welcome to microbiology. You will encounter few subjects in college that touch on daily living in so many ways. But as with any subject you'll find some parts more interesting than others. On the one hand, you'll find out things that are quite certain to interest you: how you become immune to diseases, how penicillin works, why oranges get moldy, and other "facts of life." On the other hand, now and then you may feel like the little girl thanking her aunt for a book on penguins. "Thank you very much," she wrote, "for the book on penguins. It told me a lot about penguins, even more than I wanted to know."

THE IMPORTANCE OF TUTORING YOURSELF

I have taught microbiology regularly to students required to take it for nursing, pharmacy, home economics, agriculture, civil engineering, and a number of other fields, including mortuary science. Once a mortuary science student attempted the course and failed miserably. In none of three exams did he reach a score of 50%. At the end of the following term he told me, woefully, that he had just repeated the course and had missed a grade of C by two points. To transfer the credit to the school of mortuary science he needed at least a C.

It happens that our school has a policy of providing tutors for those who need them. I turned his case over to a microbiology graduate student who tutored in his spare time. On the first exam this formerly hapless undergraduate scored a 92 … and never looked back. Had three terms of exposure to my lectures finally enlightened him? Probably not; the tutor reported that in preparing for the first exam, the student denied having ever heard of the prokaryotic cell.

What, then, made the difference? The tutor did, of course. But a tutor can't really put anything into your head that you can't put there all by yourself. What a tutor does is force you to think about the material instead of just staring at it. The tutor exposes your areas of ignorance but, instead of letting it go at that, forces you to pave over these areas with information.

You can do the same thing the tutor does—tutor yourself. The suggestions that follow and the use of this study guide will help you.

LECTURES AND NOTE TAKING

Instructors vary in their approaches to microbiology courses. Some give lectures that largely paraphrase the text, and exams are based mainly on the text. Others scarcely use the book at all in their lectures and base the exams solely on the lectures; successful students must be careful note takers. (I have even heard of some cases in which there seems to be no apparent relationship between the information available to the student and the exams. These cases, of course, can only be left to the chaplain.)

It is important to take notes in an organized fashion such as this:

Staining of Bacteria
  Gram Stain
    Insert all material about Gram staining.
    Take down all terms likely to be associated with Gram staining.
  Acid-Fast Stain
    Insert all material about the acid-fast stain.
    Take down all terms likely to be associated with acid-fast staining.

If you look back at notes like this, you can see that you were hearing about the staining of bacteria and that there were basically two types of bacterial stains discussed. Note that you also clearly kept the information about the Gram stain separate from information about the acid-fast stain. Never let your notes run together so that one topic merges with another. On objective tests, this failure to separate discussions can be fatal. Make sure you have definitions straight. Never leave your notes or your mind a blank on something as obvious as the definition of a term.

STUDYING

I always invite students with grade problems in the course to come see me. I ask them to bring their notes and a copy of the offending exam, corrected from a key. Very often certain patterns show up. One is that the notes have no "study marks," that is, no underlining or circling. Another is that the student often misses a cluster of questions pertaining to one group of topics—for example, the characteristics of different antibiotics. In other words, the student has encountered a list of antibiotics and failed to differentiate one from another. This is why it is important to organize information by general headings and a system of indentation as I have described. I also notice occasionally that a term, such as the word *autotroph*, stands alone in the notes without definition or explanation. Surely the student could have filled in such an omission by consulting the book or the instructor. In my presentation of the material I am not trying to keep things secret, nor, presumably, is your instructor.

My recommendations are these: First, *organize your notes* so that it is obvious where a discussion of a topic, such as the disease diphtheria, starts and where it ends. The most obvious way of doing this is to underline the main topic and insert all the information about that topic. (If you have trouble writing rapidly enough, or you have trouble with English as a language, ask the instructor if you can record the lectures.)

Second, *never confuse staring at a page with studying a page.* As you read, do your eyes ever travel down half a page while your mind fails to take in a single word written on it? A person has a limited capability for concentration, and by staying up all night before an exam, he or she is likely to exceed it. Do your studying in short bursts within the limits of your full attention span; start early. To ensure concentration, go through the notes or the text with a pencil or highlighting marker. Look deliberately at the material, and try to think like the person who goes through it to create an exam. You will find that your mind and the instructor's mind will often travel the same path. If the topic is penicillin, for example, mark the important facts: that it is produced by a mold, that its mode of action affects cell wall synthesis, that it has a β-lactam ring in its structure, that it affects mainly gram-positive bacteria, and so on. When you have done this, you have actually *thought* about the material instead of just staring at it. This method will also teach you how to take notes. You'll begin to recognize material that is likely to be on an exam and get it down, leaving to casual memory the instructor's anecdote about his memories of childhood chickenpox.

STUDY GUIDE CONTENTS

This study guide provides a chapter-by-chapter summary that contains the important terms and concepts likely to be on examinations. This chapter summary is usually organized by the headings used in the text. Important terms are printed in boldface and defined, and important figures and tables from the text are included. Following each chapter synopsis is a sample exam, which is an extensive self-testing section containing matching, fill-in-the-blanks, critical-thinking, and label-the-art questions. An answer key is provided. (You may wish to supplement these questions with the Study Questions at the end of each chapter in the textbook.) Obviously, not all questions can be anticipated, and a good instructor will usually introduce material other than that found in the text. But going through these sample exams will work as a self-tutoring device to direct attention to any areas of ignorance.

If you are going to improve your time in running the mile or the amount of weights you can lift, you must do more than read a book on techniques. You must apply them to build up your endurance or your muscles. The same applies to study habits. A guide cannot do your work for you; it can only serve as a tutor. You must do the work yourself.

1 The Microbial World and You

MICROBES IN OUR LIVES

Most **microorganisms,** or **microbes,** are not harmful and indeed play a vital role in maintaining our global environment. Only a minority are **pathogenic** (disease-producing). Microbes are part of the food chain in oceans, lakes, and rivers; they break down wastes, incorporate nitrogen gas from the air into organic compounds, and participate in *photosynthesis*, which generates food and oxygen.

NAMING AND CLASSIFYING MICROORGANISMS

The system of naming **(nomenclature)** we now use was established by **Carolus Linnaeus.** Scientific nomenclature assigns each organism two names. The **genus** (plural: *genera*), the first name, is always capitalized, and the **specific epithet (species)**, which follows, is not capitalized. The scientific names of organisms are always either underlined or italicized. At first, organisms were grouped into either the animal kingdom or the plant kingdom; however, in 1978 Carl Woese devised a system of classification that groups all organisms into three domains: bacteria, archaea, and eukarya.

TYPES OF MICROORGANISMS

Bacteria and Archaea

Bacteria and archaea are simple, one-celled organisms whose genetic material is not enclosed in a special nuclear membrane. For this reason, bacteria and archaea are called **prokaryotes** (from Greek words meaning prenucleus). (*Note:* The word **Bacteria,** when capitalized, refers to the domain. When not capitalized, it usually describes any prokaryotic cell.) Bacterial cells generally have one of three shapes: **bacillus** (rodlike), **coccus** (spherical or ovoid), or **spiral** (curved or corkscrew). Individual bacteria may form pairs, chains, or other groupings, which are usually the same within a species. Bacteria (singular: *bacterium*) are enclosed in cell walls largely made of a carbohydrate and protein complex called *peptidoglycan* (cellulose is the main substance of plant cell walls). Bacteria generally reproduce by *binary fission* into two equal daughter cells. Many move by appendages called flagella, and although most use organic material for nutrition, some use inorganic substances or carry out photosynthesis. Microorganisms in the domain **Archaea** are often found in extreme environments. Their cell walls lack peptidoglycan. Groups included are *methanogens*, *extreme halophiles*, and *extreme thermophiles*. They are not known to cause disease in humans.

Fungi

Fungi (singular: *fungus*) are **eukaryotes;** they contain DNA within a distinct nucleus surrounded by a nuclear membrane. They may be unicellular or multicellular. Their cell walls are composed primarily of *chitin*. **Yeasts** are unicellular nonfilamentous fungi larger than bacteria. **Molds** form mycelia of long filaments (*hyphae*).

Protozoa

Protozoa (singular: *protozoan*) are unicellular, eukaryotic microbes, members of the kingdom Protista. They are classified by their means of locomotion, which include *pseudopods* (false feet), *cilia*, or *flagella*.

Algae

Algae (singular: *alga*) are photosynthetic eukaryotes, mostly of the kingdom Protista, and are usually unicellular. They need light for growth and produce oxygen and carbohydrates by photosynthesis.

Viruses

Viruses are very small and are not cellular. They have a core of either DNA or RNA, surrounded by a protein coat. They may have a lipid envelope layer as well. They reproduce only inside the cells of a host organism.

Multicellular Animal Parasites

Flatworms and **roundworms**, collectively called **helminths**, are not strictly microorganisms. A part of their life cycles involves microscopic forms, however, and identifying them requires many of the techniques used in identifying traditional microorganisms.

A BRIEF HISTORY OF MICROBIOLOGY

The First Observations

Anton van Leeuwenhoek was the first to report observing microorganisms with the use of magnifying lenses, beginning in 1674. He made detailed drawings of "animalcules" that have since been identified as bacteria and protozoa. About this time, **Robert Hooke** observed with a microscope the boxlike openings in slices of plants. He called them "cells." His discovery, reported in 1665, marked the beginning of the **cell theory**—that all living things are composed of cells.

The Debate over Spontaneous Generation

Until the second half of the nineteenth century, it was generally believed that life could arise spontaneously from nonliving matter, a process known as **spontaneous generation**. An early opponent of spontaneous generation, **Francesco Redi**, demonstrated in 1668 that maggots, the larvae of flies, do not arise spontaneously from decaying meat.

Many, however, still believed that the simpler organisms observed by Leeuwenhoek might undergo spontaneous generation. In 1745, **John Needham** found that heated nutrient fluids poured into covered flasks were soon teeming with microorganisms. He took this as evidence of spontaneous generation. Twenty years later, **Lazzaro Spallanzani** showed that Needham's microorganisms had entered the fluid after boiling. Heating such fluids in a sealed flask, he showed, prevented the growth Needham had observed. Objections still remained; Needham thought that the heating had destroyed some vital force necessary for spontaneous generation. The concept of **biogenesis**, that living cells can arise only from other living cells, was introduced in 1858 by **Rudolf Virchow**.

In 1861, **Louis Pasteur** designed the experiments that finally ended the debate about spontaneous generation. He showed that flasks left open to the air after boiling would soon be contaminated, but if they were sealed, they remained free of microorganisms. He also used flasks whose long necks he bent into S-shaped curves. Air, with its presumed vital force, could enter these flasks, but airborne microorganisms were trapped in the tubes. The flask contents remained sterile. Pasteur showed that microorganisms are present throughout the environment and that they can be destroyed. He also devised methods of blocking the access of airborne microorganisms to nutrient environments; these methods were the basis of **aseptic** (germ-free) **techniques**, which are among the first things that a beginning microbiologist learns.

THE GOLDEN AGE OF MICROBIOLOGY

Fermentation and Pasteurization

At this time Pasteur was asked to investigate the problem of spoilage of beer and wine. Pasteur showed that, contrary to the belief that air acted on the sugars to convert them to alcohol, microorganisms called yeasts were responsible—and in the absence of air. This process is called **fermentation** and is used in making wine and beer. Spoilage occurs later when bacteria, in the presence of air, change the alcoholic beverage into acetic acid (vinegar). Pasteur prevented spoilage by heating the wine or beer just enough to kill such bacteria, a process that came to be known as **pasteurization**.

The Germ Theory of Disease

This association of yeasts with fermentation was the first concept to link a microorganism's activity to physical and chemical changes in organic materials. It suggested the possibility that microorganisms might be able to cause diseases as well—the **germ theory of disease**. In 1835, **Agostino Bassi** made the first association between a microorganism and a disease by proving that a fungus was the cause of a silkworm disease. In 1865, Pasteur found that another silkworm disease was caused by a protozoan. Also in the 1860s, **Joseph Lister** applied the germ theory to medicine. He used carbolic acid (phenol) on surgical dressings and wounds, and the number of infections and deaths greatly declined. In the 1840s, **Ignaz Semmelweis** demonstrated that chemically disinfecting the hands of physicians minimized infections of obstetrical patients. In 1876, **Robert Koch** demonstrated that certain bacteria in the blood of cattle that had died of anthrax were the cause of death. He showed that these bacteria could be isolated and grown in pure culture, be injected into healthy animals, and cause their death by anthrax. The same bacteria could then be isolated from the dead animals. This demonstration, which proved that a specific microbe is the cause of a specific disease, followed a set of criteria known today as **Koch's postulates**.

Vaccination

In 1798, **Edward Jenner** showed that the mild disease, cowpox, gave immunity to smallpox. He inoculated people with cowpox material by scratching their arm with a cowpox-infected needle. This process became known as **vaccination** (*vacca* is the Latin word for cow). The protection from disease that vaccination provides is called **immunity**. Years later, around 1880, Pasteur showed why vaccinations work. He found that the bacterium for fowl cholera lost its **virulence** (ability to cause disease), or became **avirulent**, after it was grown for long periods in the laboratory. However, he showed that the weakened bacteria, which he called a **vaccine**, still retained their ability to induce immunity. Apparently the cowpox virus is related closely enough to smallpox to induce effective immunity.

THE BIRTH OF MODERN CHEMOTHERAPY: DREAMS OF A "MAGIC BULLET"

The treatment of disease by chemical substances is called **chemotherapy**. When prepared from chemicals in the laboratory, these substances are called **synthetic drugs**, and when produced naturally by bacteria and fungi, they are called **antibiotics**. **Paul Ehrlich** speculated about a "magic bullet" that would destroy a pathogen without harming the infected host. In 1910, he found *salvarsan*, an arsenic derivative, that was effective against syphilis. *Quinine*, an extract of South American tree bark, had until then been the only other such chemical available. Spanish conquistadors used it to treat malaria. In the late 1930s, a survey of dye derivatives led to the development of antibacterial *sulfa drugs*. The first antibiotic was discovered by **Alexander Fleming**, who observed the inhibition of bacterial growth by the mold, *Penicillium notatum* (later renamed *P. chrysogenum*). The inhibitor, which he called penicillin, was mass-produced and clinically tested in the 1940s. Since then, many antibiotics have been discovered.

MODERN DEVELOPMENTS IN MICROBIOLOGY

Bacteriology is the study of bacteria. **Mycology** is the study of fungi. **Parasitology** is the study of protozoa and parasitic worms. **Genomics** is the study of all of an organism's genes.

Immunology, the study of immunity, expanded rapidly in the twentieth century. Smallpox was eliminated, and many new vaccines became available. A major challenge now will be to defeat the AIDS virus, which attacks the immune system. In 1933, **Rebecca Lancefield** proposed an immunologically based classification system for streptococci bacteria, classifying them as serotypes (variants within a species).

Virology, the study of viruses, really began in 1892, when **Dmitri Iwanowski** demonstrated that tobacco plant pathogens would pass through filters too fine for known bacteria. Much later, **Wendell Stanley** showed that the organism, called the tobacco mosaic virus, was so simple and homogeneous it could be crystallized.

Recombinant DNA Technology

Recombinant DNA technology, or **genetic engineering**, had its origin in **microbial genetics** (how microbes inherit traits) and **molecular biology** (how genetic information is carried in DNA, which is then used to direct synthesis of proteins). Because of their simplicity and rapid reproduction rate, bacteria are the preferred organisms in this field. Beginning in the early 1940s, **George W. Beadle** and **Edward L. Tatum** demonstrated the relationship between genes and enzymes. DNA was established as the hereditary material by **Oswald Avery**, **Colin MacLeod**, and **Maclyn McCarty**. **Joshua Lederberg** and **Edward L. Tatum** discovered bacterial genetic transfer by conjugation. In 1958, **James Watson** and **Francis Crick** proposed the structure of DNA. In the 1960s, **François Jacob** and **Jacques Monod** discovered messenger RNA, important in protein synthesis, and later made major discoveries about the regulation of gene function in bacteria.

MICROBES AND HUMAN WELFARE

Recycling Vital Elements

Microbes recycle vital elements such as nitrogen, carbon, oxygen, sulfur, and phosphorus. *Cyanobacteria* and certain soil bacteria may use atmospheric nitrogen directly (through a process called *nitrogen fixation*). The nitrogen fixed from the air is incorporated into living organisms and eventually returned as gaseous nitrogen, making up the **nitrogen cycle**. **Microbial ecology**, the study of the relationship between microbes and their environment, originated with the work of **Martinus Beijerinck** and **Sergei Winogradsky**. They first showed how bacteria helped to recycle vital elements. In the **carbon cycle,** plants and algae remove carbon dioxide from the air, converting it to food. In the **oxygen cycle**, oxygen is recycled to the air during photosynthesis. There will probably be increasing use of microbes to produce methane and ethanol as alternatives to fossil fuels. Especially promising is the use of bacterial enzymes to break down plant cellulose to sugars that yeasts can metabolize into ethanol. Microbes are used in treatment of sewage. Microbes are useful in treating oil spills, toxic waste sites, and so on, a process called **bioremediation**. Bacteria such as *Bacillus thuringiensis* are used in control of insect pests.

Modern Biotechnology and Recombinant DNA Technology

Practical applications of microbiology are called **biotechnology**. The use of recombinant DNA technology has led to genetic engineering, which now produces insulin, interferon, clotting substances, vaccines, and other substances. Eventually it may become common to replace missing or defective genes in human cells, a process called **gene therapy**. Agricultural applications, including resistance to drought, insects, and microbial diseases, may also result from genetic engineering.

MICROBES AND HUMAN DISEASE

In nature, microorganisms may exist as single cells that exist independently, or they may attach to each other and/or some usually solid surface. This latter mode of behavior is called a **biofilm**. Biofilms can be beneficial, for example, by protecting mucous membranes. However, they also can clog water pipes and cause other similar problems. Most important, biofilms protect pathogens from the action of antimicrobials.

The relationship between microbes and disease will remain of great interest to us all. Indeed, we are seeing many **emerging infectious diseases**, including outbreaks of influenza by the H1N1 virus (swine flu), West Nile encephalitis (WNE), prion-caused bovine spongiform encephalopathy (BSE), a new variant of Creutzfeldt-Jakob disease (CJD), *E. coli* O157:H7, Ebola and Marburg hemorrhagic fevers, the parasitic disease cryptosporidiosis, and, of course, the human immunodeficiency virus (HIV) that is the cause of AIDS.

These diseases are caused by viruses, bacteria, protozoa, and prions. This course in microbiology will give you a greater familiarity with these organisms and the procedures used to study them.

SELF-TESTS

In the matching section, there is only one answer to each question; however, the lettered options (a, b, c, etc.) may be used more than once or not at all.

I. Matching

_____ 1. In 1668, demonstrated that maggots appeared only in decaying meat that had been exposed to flies.

_____ 2. Introduced the concept that living cells arise from other living cells.

_____ 3. Introduced the technique of vaccination for smallpox.

_____ 4. First to use the microscope to observe "cells."

_____ 5. Made an association between silkworm disease and a fungus.

_____ 6. A surgeon who used carbolic acid to control wound infections.

_____ 7. First to speculate about the possibility of a "magic bullet" that would destroy a pathogen without harming the host.

_____ 8. Discovered penicillin.

_____ 9. Using anthrax as a model, demonstrated that a specific microorganism is the cause of a specific disease.

_____ 10. Originated our system of scientific nomenclature.

a. Anton van Leeuwenhoek ✓

b. John Needham

c. Lazzaro Spallanzani

d. Louis Pasteur ∨

e. Francesco Redi

f. Agostino Bassi

g. Joseph Lister

h. Robert Koch ↲

i. Paul Ehrlich

j. Alexander Fleming

k. Edward Jenner

l. Carolus Linnaeus

m. Robert Hooke

n. Rudolph Virchow

II. Matching

_____ 1. Assigned a microbial cause to fermentation.

_____ 2. First to crystallize a virus.

_____ 3. Showed that mild heating of spirits kills spoilage bacteria without damage to the beverage.

_____ 4. Devised a classification system for the streptococci based on an immunological system of serotypes.

_____ 5. Demonstrated that infections in obstetrical wards could be minimized by disinfecting the hands of physicians.

_____ 6. Participated in determining the structure of DNA.

_____ 7. First demonstrated that genetic information could be exchanged between bacteria by conjugation.

a. Louis Pasteur

b. Wendell Stanley

c. Francis Crick

d. Robert Koch

e. Ignaz Semmelweis

f. Rebecca Lancefield

g. George Beadle

h. Joshua Lederberg

III. Matching

____ 1. Prokaryotes.

____ 2. Noncellular; reproduce only inside cells of host organism.

____ 3. Helminths.

____ 4. Yeasts.

____ 5. An infectious protein.

____ 6. Unicellular eukaryotic microorganisms; members of Kingdom Protista.

a. Protozoa

b. Elephants

c. Fungi

d. Bacteria

e. Viruses

f. Multicellular animal parasites

g. Prion

IV. Matching

____ 1. Protection from a disease that is provided by vaccination.

____ 2. The use of chemical substances to treat a disease.

____ 3. The use of microbes to clean up, for example, an oil spill.

____ 4. The process by which yeasts change sugars into alcohol.

____ 5. Techniques that keep areas free of unwanted microorganisms.

a. Bioremediation

b. Chemotherapy

c. Fermentation

d. Aseptic

e. Immunity

V. Matching

____ 1. Photosynthetic bacteria; may fix nitrogen from air.

____ 2. Photosynthetic eukaryotes.

____ 3. Eukaryotes classified primarily by their means of locomotion.

____ 4. General name for a rod-shaped bacterium.

____ 5. General name for a spherical bacterium.

____ 6. Prokaryotes whose cell walls lack peptidoglycan and are often found in extreme environments.

a. Cyanobacteria

b. Coccus

c. Bacillus

d. Protozoa

e. Archaea

f. Algae

Fill in the Blanks

1. Bacteria generally reproduce by a process called _____ into two equal daughter cells.

2. The set of criteria that prove that a specific microorganism is the cause of a specific disease is known today as _____ .

3. The concept that living cells can arise only from other living cells is called _____ .

4. Responding to experiments in which nutrient fluids were heated in sealed containers, proponents of spontaneous generation objected that heating destroyed some _____ in the air.

5. According to the rules applied to the scientific naming of a biological organism, the _____ name is always capitalized.

6. Paul Ehrlich discovered an arsenic derivative, _____ , that was effective against syphilis.

7. Antimicrobial chemicals produced naturally by bacteria and fungi are called _____ .

8. The penicillin-producing mold that Fleming discovered was named *Penicillium* _____ ; later it was renamed *P. chrysogenum*.

9. Bacteria usually exist in nature not as single cells but as aggregations of cells called a(n) _____ .

Critical Thinking

1. What are the advantages of using microorganisms to control insect pests?

2. Discuss the similarities and differences between the syphilis epidemic of the 1940s and the current AIDS epidemic.

3. List three characteristics unique to prokaryotes.

   peptidoglycan cell walls
   no nucleus
   70s ribosomes

4. How did Bassi's and Pasteur's work on silkworm disease help to reinforce the germ theory of disease?

ANSWERS

Matching
   I.   1. e   2. n   3. k   4. m   5. f   6. g   7. i   8. j   9. h   10. l
   II.   1. a   2. b   3. a   4. f   5. e   6. c   7. h
   III.   1. d   2. e   3. f   4. c   5. g   6. a
   IV.   1. e   2. b   3. a   4. c   5. d
   V.   1. a   2. f   3. d   4. c   5. b   6. e

Fill in the Blanks
1. binary fission   2. Koch's postulates   3. biogenesis   4. vital force
5. genus   6. salvarsan   7. antibiotics   8. *notatum*   9. biofilm

Critical Thinking

1. Biological control agents, such as *Bacillus thuringiensis*, are specific for insects, so they don't pose a threat to humans and other animals. Biological control agents do not remain in the soil as toxic pollutants that enter and concentrate in the food chain as do many chemical agents, such as DDT.

2. Similarities:

    1. Both diseases are transmitted sexually.

    2. No vaccine is available for either disease.

    3. There is (or was) no effective drug available.

    4. Control through behavior played an important role in lowering the incidence of syphilis and is also emphasized in efforts to check the spread of AIDS.

   Differences:

    1. AIDS may also be transmitted by sharing of needles, in breast milk, and via blood.

    2. Syphilis is effectively treated with antibiotics, but several difficulties complicate the development of antiviral drugs to treat AIDS.

    3. Prokaryotes have a single, circular chromosome that is not enclosed by a nuclear envelope. In contrast, eukaryotic organisms have a true membrane-bound nucleus.

       Prokaryotes lack true (membrane-bound) organelles.

       Most prokaryotes (members of Archaea excepted) have peptidoglycan in their cell walls.

       Binary fission (rather than mitosis) is a common form of reproduction in prokaryotic organisms.

    4. Although Pasteur disproved spontaneous generation and established that yeasts play a role in fermentation, the germ theory was still difficult for people to accept. Many people still believed that disease was either God's punishment for unacceptable behavior or was caused by foul odors from sewage or swamps. They couldn't accept that tiny microbes could travel through the air or be transmitted by bedding or clothing from person to person.

       Bassi and Pasteur both worked on silkworm diseases. Bassi (1835) demonstrated that a microscopic microbe, a fungus, caused disease in silkworms. Several years later (1865), Pasteur established that yet another microbe, this time a protozoan, also caused infections in silkworms. Both of these discoveries established a link between the "dubious" microbes and disease.

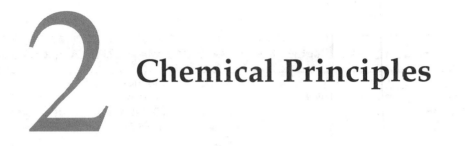

2 Chemical Principles

THE STRUCTURE OF ATOMS

All matter is made up of small units called **atoms**. Atoms in certain combinations form **molecules.** All atoms have a centrally located **nucleus** and particles called **electrons**, which have a negative (−) charge, moving around the nucleus. The nucleus of an atom is made up of positively (+) charged particles called **protons** and uncharged particles called **neutrons**. Both neutrons and protons have about the same weight, 1840 times that of an electron. Because the total positive charge on the nucleus equals the total negative charge of the electrons, each atom is an electrically neutral unit. Different kinds of atoms are listed by **atomic number**, the number of protons in the nucleus. The **atomic weight** is the total number of protons and neutrons in an atom.

Chemical Elements

All atoms with the same atomic number are classified as the same **chemical element**. Examples of elements are hydrogen (H), carbon (C), sodium (Na), nitrogen (N), and sulfur (S). About 26 of the 92 naturally occurring elements are commonly found in living things. Most elements have **isotopes**, or forms of the element with the same number of protons in the nucleus but different numbers of neutrons (and therefore different atomic weights). For example, oxygen isotopes have the same atomic number, 8, but different atomic weights: $^{16}_{8}O$, $^{17}_{8}O$, $^{18}_{8}O$.

Electronic Configurations

Electrons are arranged in **electron shells**, which are regions corresponding to different **energy levels** (Table 2.1). The arrangement is called an **electronic configuration**. The chemical properties of atoms are largely a function of the number of electrons in the outermost electron shell. When the outer shell of an atom is filled, as in helium, it is stable, or inert; it does not react with other atoms. Partially filled electron shells make for atoms that tend to react with other atoms to become more stable.

HOW ATOMS FORM MOLECULES: CHEMICAL BONDS

The **valence**, or combining capacity of an atom, depends on the number of extra electrons or missing electrons in the atom's outermost electron shell. Hydrogen has a valence of 1, and carbon has a valence of 4. This means that hydrogen can form only one chemical bond and carbon can form four. When atoms gain stability by completing the full complement of electrons, they form **molecules**. A molecule containing at least two kinds of atoms, such as water (H_2O), is a **compound**.

Ionic and Covalent Bonds

The attractive forces holding molecules together are **chemical bonds**. In general, atoms form bonds in two ways. One way is by gaining or losing electrons from the outer electron shell to make atoms that are positively or negatively charged. The *attraction* between such atoms forms an **ionic bond**. (An **ion** is a negatively or positively charged atom or group of atoms.) Sodium chloride (NaCl) is an example of a substance formed by ionic bonding. The other way is by *sharing* outer electrons to form **covalent bonds**. Sodium (Na) has a single electron in the outer shell, which it tends to lose; it is an *electron donor*. The sodium ion is thus positively charged.

Table 2.1 **Electronic Configurations for the Atoms of Some Elements Found in Living Organisms**

Element	First Electron Shell (2)*	Second Electron Shell (8)*	Third Electron Shell (8)*	Diagram	Number of Valence (Outermost) Shell Electrons	Number of Unfilled Spaces	Maximum Number of Bonds Formed
Hydrogen	1	—	—		1	1	1
Carbon	2	4	—		4	4	4
Nitrogen	2	5	—		5	3	5
Oxygen	2	6	—		6	2	2
Magnesium	2	8	2		2	6	2
Phosphorus	2	8	5		5	3	5
Sulfur	2	8	6		6	2	6

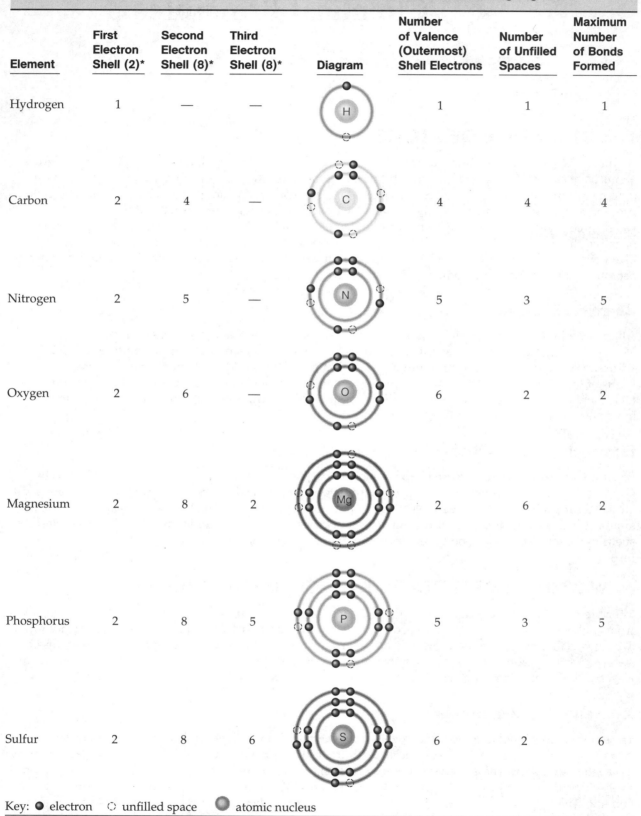

Key: ● electron   ○ unfilled space   ⬤ atomic nucleus

*Numbers in parentheses indicate the maximum number of electrons in their respective shells.

This ion reacts strongly with chlorine, which lacks one electron of outer-shell capacity. Chlorine has a negative charge because it usually fills this vacancy with electrons from other atoms; that is, it is an *electron acceptor*. A **cation**, such as potassium (K^+) or sodium (Na^+), is a positively charged ion; its outer electron shell is less than half filled, and it loses electrons. An **anion** is a negatively charged ion; with its outer shell *more* than half filled, it tends to gain electrons. Examples of anions are the iodide ion (I^-), chloride ion (Cl^-), and sulfide ion (S^{2-}). Covalent bonds are stronger and more common in organisms than are ionic bonds.

Hydrogen Bonds

The **hydrogen bond** consists of a hydrogen atom covalently bonded to one oxygen or nitrogen atom (these being the most commonly involved elements). In water, for example, the electrons are all closer to the oxygen nucleus than to the hydrogen nucleus. As a result, the oxygen portion of the molecule has a slight negative charge, and the hydrogen portion a slight positive charge. The hydrogen bond is a result of the attraction of the water molecule's slightly positive polar charge to the negative end of the other molecules. Although these bonds are relatively weak, large molecules may have hundreds of them.

Molecular Weight and Moles

The **molecular weight** of a molecule is the sum of the atomic weights of all its atoms. A **mole** is the number of grams equal to the molecular weight.

CHEMICAL REACTIONS

Chemical reactions involve the making or breaking of bonds between atoms. After a chemical reaction, the total number of atoms remains the same, but there are new molecules with new properties because the atoms have been rearranged.

Energy in Chemical Reactions

All chemical bonds require energy when they are broken and release chemical energy when they are formed. An **endergonic reaction** (*endo* = within) is one that absorbs more energy than it releases. An **exergonic reaction** (*exo* = out) releases more energy than it absorbs.

Synthesis Reactions

In a **synthesis reaction**, two or more atoms, ions, or molecules combine to form new and larger molecules. A synthesis reaction forms new bonds. The combining substances are called the *reactants*; the substance formed by the combination is the *product*. Pathways of synthesis reactions in living organisms are collectively called anabolic reactions, or **anabolism**.

Decomposition Reactions

A **decomposition reaction** is the reverse of a synthesis reaction. In a decomposition reaction, bonds are broken. Decomposition reactions that occur in living organisms are collectively called catabolic reactions, or simply **catabolism**.

Exchange Reactions

Many reactions, such as **exchange reactions**, are actually part synthesis and part decomposition. Bonds are broken in a decomposition process, and then new bonds are formed in a synthesis process.

The Reversibility of Chemical Reactions

All chemical reactions are, theoretically, reversible, which means they can occur in either direction. In practice, some reactions do this more easily than others. A **reversible reaction**, in which the end product can revert to the original molecules, is indicated by two arrows.

__ IMPORTANT BIOLOGICAL MOLECULES __

INORGANIC COMPOUNDS

Water

Inorganic compounds, such as water, are essential for living cells. **Water**, a **polar molecule**, has an unequal charge distribution that gives it four important characteristics: (1) a high boiling point and, incidentally, a solid form (ice) that is less dense than the liquid, a characteristic that allows it to float; (2) excellent **solvent** properties (that is, it is a good dissolving medium). Polar substances undergo **dissociation** into individual molecules in water; they dissolve because the negative part of the water molecule is attracted to the positive part of the molecules in the **solute**, or dissolving substance; (3) an important role as a reactant or product in many chemical reactions; and (4) an excellent temperature buffer capability that helps protect the cell from changes in environmental temperature.

Acids, Bases, and Salts

When inorganic salts such as NaCl are dissolved in water, they dissociate, or undergo **ionization**. An **acid** dissociates into one or more hydrogen ions (or protons H^+) and one or more negative ions. In other words, acids are proton (H^+) donors; HCl, for example, yields H^+ and Cl^-. A **base**, such as NaOH, dissociates into Na^+ and the hydroxide ion (OH^-). A **salt** is a substance that dissociates into ions that are neither H^+ nor OH^-. NaCl is an example; it dissociates into Na^+ and Cl^-.

Acid–Base Balance: The Concept of pH

The **pH** of a solution is the negative logarithm to the base 10 of the hydrogen ion concentration in moles per liter. Acidic solutions contain more H^+ ions than OH^- ions and have a pH lower than 7. Alkaline, or basic, solutions have more OH^- ions than H^+ ions. A pH of 7, with equal concentrations of H^+ and OH^-, is neutral. Each change of number on the logarithmic pH scale represents a tenfold change in concentration. **Buffers** are compounds that keep the pH from changing drastically.

ORGANIC COMPOUNDS

All **organic compounds** contain carbon, whose four covalent bonds allow it to combine with neighboring or other atoms to form large structures. Hydrogen fills most of the free bonds in these structures. There are many other **functional groups**, such as the **hydroxyl group** (—OH) (not to be confused with the hydroxide ion (OH^-) of bases). Such fundamental groups can be added to **carbon skeletons** to make alcohols or amino acids. When single bonds hold a carbon atom to four other atoms to form a skeleton of three-dimensional shape, it is called a **tetrahedron**. Small organic molecules can be combined into very large molecules called **macromolecules**. Generally, these are **polymers**, which are formed by repeating small molecules called **monomers**. When two monomers join together, they usually release a molecule of water, a reaction called **condensation reaction** or **dehydration synthesis**.

Carbohydrates

Carbohydrates have the general formula CH_2O, with a hydrogen-to-oxygen ratio of 2:1. Carbohydrates form **monosaccharides**, simple sugars of three to seven carbon atoms; **disaccharides**, two monosaccharides bonded together (sucrose, for example, is made up of glucose and fructose); and **polysaccharides**, eight or more monosaccharides (such as glucose) joined by dehydration. Familiar examples are *cellulose*, *starch*, and *chitin*. The latter is found in the cell walls of fungi and the exoskeletons of lobsters, crabs, and some insects. Two molecules with the same chemical formula but different structures are called **isomers**.

Lipids

Lipids also are composed of atoms of carbon, hydrogen, and oxygen, but they do not have the 2:1 hydrogen-to-oxygen ratio. They are diverse in structure but share the property of being soluble in nonpolar solvents such as ether and alcohol, but not in water. **Simple lipids** are **fats**, or *triglycerides*, which are formed from *glycerol* and *fatty acids* joined by an *ester link*. **Saturated fats** have no double bonds; **unsaturated fats** have several. Important in the membranes of cells are **phospholipids**, which also contain phosphorus, nitrogen, and sulfur. **Complex lipids** called **sterols** (members of the **steroids**) are important constituents in the plasma membranes of cells.

Proteins

Amino acids are the subunits of **proteins**, which are organic molecules containing carbon, hydrogen, oxygen, and nitrogen, as well as some sulfur. Amino acids have at least one *carboxyl* (—COOH) *group* and one *amino* (—NH_2) *group* attached to the same carbon atom, the alpha carbon—hence the name **alpha-amino acid**. Amino acids exist in configurations called stereoisomers, which are designated D or L. (Only L-isomers are found in biological proteins, except for the relatively few D-isomers found in some bacterial cell walls and in some antibiotics.) Only 20 different kinds of amino acids occur naturally in proteins. Amino acids in proteins are connected by **peptide bonds**.

Proteins have four levels of organization: **primary**, the order in which amino acids are linked (their sequence); **secondary**, localized, repetitious twisting or folding of the polypeptide chain that forms helixes and pleated sheets, held together by hydrogen bonding; **tertiary**, three-dimensional shapes in which *sulfhydryl groups* (—SH) can form *covalent disulfide links* (—S—S—), hydrogen bonds providing the holding force; and a **quaternary** structure, an aggregation of two or more individual polypeptide units that operate as a single functional unit. A protein that loses its characteristic shape has undergone **denaturation** and is not functional. **Conjugated proteins** are formed by combinations of amino acids and other organic or inorganic components. **Glycoproteins**, for example, contain sugars; **nucleoproteins** contain nucleic acids; and **lipoproteins** contain lipids.

Nucleic Acids

There are two principal nucleic acids: **deoxyribonucleic acid (DNA)** and **ribonucleic acid (RNA)**. The basic units of **nucleic acids** are **nucleotides**. Each nucleotide unit of DNA contains three parts: a nitrogen-containing base, a five-carbon sugar (**deoxyribose** or **ribose**), and a phosphoric acid molecule (Figure 2.1). The nitrogen-containing base is either **adenine**, **guanine**, **cytosine**, or **thymine**. Adenine and guanine are double-ring structures, **purines**. Thymine and cytosine are single-ring structures, **pyrimidines**. Note in the figure how adenine always pairs with thymine and cytosine with guanine. This **complementary pairing** allows one strand to be reproduced from the structure of another. RNA differs from DNA in several ways. It is usually single stranded, the five-carbon sugar is ribose, and uracil replaces thymine. **Nucleosides** have a purine or pyrimidine attached to a pentose sugar but have no phosphate group.

Adenosine Triphosphate (ATP)

The principal energy-carrying molecule in all cells is **adenosine triphosphate (ATP).** It consists of an adenosine unit of adenine and ribose joined to three phosphate groups (Figure 2.2). It releases much energy when converted to **adenosine diphosphate (ADP)** by the addition of a water molecule, which separates the terminal phosphate group. ATP is generated by energy-yielding reactions in the cell, such as the decomposition of glucose. The energy carried in ATP can be used to perform synthesis.

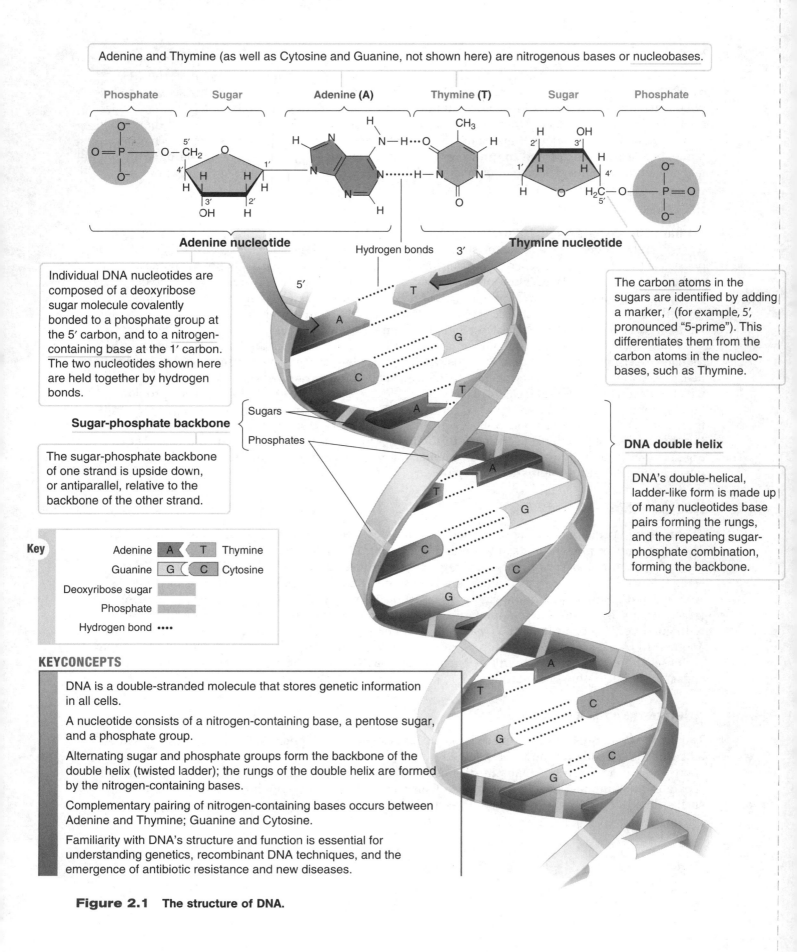

Adenine and Thymine (as well as Cytosine and Guanine, not shown here) are nitrogenous bases or nucleobases.

Phosphate · Sugar · Adenine (A) · Thymine (T) · Sugar · Phosphate

Adenine nucleotide · Hydrogen bonds · **Thymine nucleotide**

Individual DNA nucleotides are composed of a deoxyribose sugar molecule covalently bonded to a phosphate group at the 5′ carbon, and to a nitrogen-containing base at the 1′ carbon. The two nucleotides shown here are held together by hydrogen bonds.

The carbon atoms in the sugars are identified by adding a marker, ′ (for example, 5′, pronounced "5-prime"). This differentiates them from the carbon atoms in the nucleo-bases, such as Thymine.

Sugar-phosphate backbone

Sugars

Phosphates

The sugar-phosphate backbone of one strand is upside down, or antiparallel, relative to the backbone of the other strand.

DNA double helix

DNA's double-helical, ladder-like form is made up of many nucleotides base pairs forming the rungs, and the repeating sugar-phosphate combination, forming the backbone.

Key

Adenine A — T Thymine
Guanine G — C Cytosine
Deoxyribose sugar
Phosphate
Hydrogen bond ••••

KEYCONCEPTS

DNA is a double-stranded molecule that stores genetic information in all cells.

A nucleotide consists of a nitrogen-containing base, a pentose sugar, and a phosphate group.

Alternating sugar and phosphate groups form the backbone of the double helix (twisted ladder); the rungs of the double helix are formed by the nitrogen-containing bases.

Complementary pairing of nitrogen-containing bases occurs between Adenine and Thymine; Guanine and Cytosine.

Familiarity with DNA's structure and function is essential for understanding genetics, recombinant DNA techniques, and the emergence of antibiotic resistance and new diseases.

Figure 2.1   The structure of DNA.

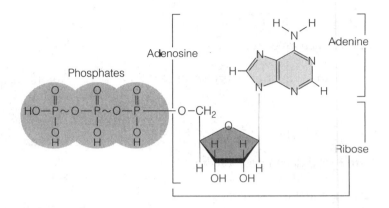

Figure 2.2 **The structure of ATP.** High-energy phosphate bonds are indicated by wavy lines. When ATP breaks down to ADP and inorganic phosphate, a large amount of chemical energy is released for use in other chemical reactions.

SELF-TESTS

In the matching section, there is only one answer to each question; however, the lettered options (a, b, c, etc.) may be used more than once or not at all.

I. Matching

___ 1. The strongest of the three chemical bonds.

___ 2. An uncharged particle in the atomic nucleus.

___ 3. A hydrogen ion.

___ 4. The number of protons in the nucleus.

___ 5. Particles with a negative charge that move in shells around the nucleus.

___ 6. A bond formed by sharing electrons in the outermost shell.

___ 7. A weak bond formed, for example, by the slight positive charge at the hydrogen end of the water molecule reacting with the negative end of other molecules.

___ 8. A bond formed by the gain or loss of electrons from the outer electron shell.

a. Proton

b. Atomic weight

c. Electron

d. Nucleus

e. Atomic number

f. Neutron

g. Ionic bond

h. Covalent bond

i. Hydrogen bond

II. Matching

____ 1. The substance upon which an enzyme acts.

____ 2. A protein that lowers the activation energy required for a reaction.

____ 3. The sum of the atomic weights of a molecule's atoms.

____ 4. The collective term for all decomposition reactions.

____ 5. The number of grams equal to molecular weight.

____ 6. The collective term for all synthesis reactions.

____ 7. The combining capacity of an atom.

____ 8. An ion with a positive charge.

____ 9. One of two molecules with the same chemical formula but different structures.

a. Cation

b. Anion

c. Substrate

d. Valence

e. Mole

f. Molecular weight

g. Anabolism

h. Catabolism

i. Enzyme

j. Isomer

III. Matching

____ 1. Prevents drastic change in pH.

____ 2. Substance that dissociates into ions that are neither OH^- nor H^+.

____ 3. A proton donor.

____ 4. Dissociates into one or more negative hydroxide ions, such as OH^-.

____ 5. Combinations of atoms that have gained stability by completing the full complement of electrons in the outermost shell.

a. Compound

b. Molecule

c. Acid

d. Salt

e. Base

f. Hydroxide ion

g. Buffer

h. Hydroxyl group

IV. Matching

____ 1. Eight or more glucose molecules in a chain.

____ 2. Sterol.

____ 3. Fat.

____ 4. Production of a molecule of water during synthesis.

____ 5. Formed from chains of amino acids.

____ 6. Lipoprotein.

a. Condensation

b. Monosaccharide

c. Disaccharide

d. Polysaccharide

e. Complex lipid

f. Protein

___ 7. Results from the release of energy by separation of the terminal phosphate group.

___ 8. DNA.

___ 9. Triglycerides.

g. Conjugated proteins

h. Nucleic acid

i. Adenosine triphosphate

j. Adenosine diphosphate

k. Simple lipids

V. Matching

___ 1. Same number of protons in the nucleus but different weights.

___ 2. A molecule containing at least two kinds of atoms, such as water (H_2O).

___ 3. An atom that is stable because it has its full complement of electrons.

___ 4. The D and L forms of an amino acid.

___ 5. The substance formed by a reaction.

a. Molecule

b. Product

c. Isotope

d. Compound

e. Stereoisomer

VI. Matching

___ 1. Sucrose.

___ 2. Soluble in solvents such as ether or alcohol, but not in water.

___ 3. Purines or pyrimidines attached to a pentose sugar, but without a phosphate group.

___ 4. Have a hydrogen-to-oxygen ratio of 2:1, a general formula of CH_2O.

a. Lipids

b. Nucleotide

c. Nucleoside

d. Carbohydrate

e. Disaccharide

VII. Matching

___ 1. In DNA, it will pair with guanine.

___ 2. In RNA, replaces thymine.

___ 3. The five-carbon sugar in DNA.

___ 4. Bonds between amino acids in proteins.

a. Peptide

b. Deoxyribose

c. Ribose

d. Uracil

e. Cytosine

Fill in the Blanks

1. All atoms with the same atomic number are classified as the same _____.

2. When discussing synthesis, the combining substances are called _____, and the substance formed is the product.

3. Carbon has a valence of _____.

4. The principal energy-carrying molecule in all cells is _____.

5. RNA differs from DNA in being usually _____ stranded.

6. In a protein, the order of the amino acid sequence is the _____ level of organization.

7. Thymine and cytosine are single-ring structures called _____.

8. An example of a nitrogen-containing base in a nucleotide is _____. (More than one answer is acceptable.)

9. The _____ level of protein organization provides it with a three-dimensional shape.

10. About _____ different kinds of amino acids occur naturally in proteins.

11. Some important characteristics of water are its high _____ and its capacity as a temperature _____.

12. Cations are positively charged ions; their outer electron shell is _____ than half filled, and they lose electrons. (less, more)

13. The minimum collision energy required for a chemical reaction to occur is its _____.

14. Neutrons and protons have a weight about 1840 times that of _____.

15. Decomposition yields energy, which is called an _____ reaction.

Critical Thinking

1. Why is carbon so important in the chemistry of life?

2. Why are lipids important components of living cells?

3. Explain the role of pH buffers in the growth of bacterial cultures in laboratory media.

4. Why is ATP indispensable to cells?

ANSWERS

Matching

I.	1. h	2. f	3. a	4. e	5. c	6. h	7. i	8. g	
II.	1. c	2. i	3. f	4. h	5. e	6. g	7. d	8. a	9. j
III.	1. g	2. d	3. c	4. e	5. b				
IV.	1. d	2. e	3. e	4. a	5. f	6. g	7. j	8. h	9. k
V.	1. c	2. d	3. a	4. e	5. b				
VI.	1. e	2. a	3. c	4. d					
VII.	1. e	2. d	3. b	4. a					

Fill in the Blanks

1. chemical element   2. reactants   3. four   4. adenosine triphosphate (ATP)   5. single   6. primary
7. pyrimidines   8. adenine, guanine, cytosine, thymine, uracil   9. tertiary   10. 20
11. boiling point; buffer   12. less   13. activation energy   14. electrons   15. exergonic

Critical Thinking

1. Carbon is important because of the condition of its outer electron shell. In this shell there are four electrons and four vacancies. This characteristic allows carbon to react with many other atoms and with other carbon atoms to form a variety of chains and rings. These chains and rings are the basis of many organic compounds in living cells.

2. Lipids are a major component of membranes, in terms of both structure and function. Membranes serve as a boundary between the cell and the environment. Lipids function in energy storage and are also a component of some bacterial cell walls.

3. Waste products excreted by bacteria alter the pH of the medium. pH buffers—such as potassium phosphate, monobasic—are added to media to prevent drastic changes in pH. Without the addition of buffers, the media would become acidic enough to inactivate bacterial enzymes and kill the organism.

4. ATP is important because it is the principal energy-carrying molecule of the cell. It stores the chemical energy released by some reactions and releases energy for reactions that require energy. ATP releases large amounts of usable energy when it loses its terminal phosphate group in the following reaction:

$$ATP \rightarrow ADP + P + energy$$

The energy released from decomposition reactions can be stored by attaching a terminal phosphate to ADP, the reverse of the previous reaction.

3 Observing Microorganisms through a Microscope

UNITS OF MEASUREMENT

Microorganisms are measured by metric units unfamiliar to many of us. The **micrometer (µm),** formerly known as the **micron,** is equal to 0.000001 (10^{-6}) meter. The prefix *micro* indicates that the unit following should be divided by 1 million. A **nanometer (nm),** formerly known as a **millimicron (mµ)** is equal to 0.000000001 (10^{-9}) meter. *Nano* tells us that the unit should be divided by 1 billion. An **angstrom (Å)** is equal to 0.0000000001 (10^{-10}) meter.

MICROSCOPY: THE INSTRUMENTS

Compound Light Microscopy

The **compound light microscope** has two sets of lenses: the **objective** and the **ocular.** Specimens magnified by the objective lens—magnified 100 times, for example—are remagnified by the ocular, usually 10 times. Thus, the total magnification is 1000 times. Most microscopes provide magnifications of 100, 400, and 1000. A magnification of 2000 times is about the highest obtainable. The specimen is illuminated by visible light from the light source—the **illuminator**—that is passed through a **condenser,** which directs the light rays through the specimen (Figure 3.1). **Resolution,** or **resolving power,** is the ability of a microscope to distinguish two points. The shorter the wavelength of the illumination, the better the resolution. The white light used in a compound light microscope limits resolving power to about 0.2 µm.

The lenses closest to the specimen are the **objective lenses.** For the highest magnification, it is necessary to use **oil immersion objectives.** Immersion oil has the same **refractive index** as glass; that is, the relative velocity of light passing through it is the same. Without immersion oil filling the space between the slide bearing the specimen and the objective, the image will be fuzzy, with poor resolution (Figure 3.3 in the text).

Darkfield Microscopy

Some microorganisms, such as the very thin spirochete *Treponema pallidum,* which causes syphilis, are best seen with **darkfield microscopy.** In the darkfield microscope, an opaque disk blocks light from entering the objective directly. The light hits only the sides of the specimen, and scattered light enters the objective and reaches the eyes. The specimen appears white against a black background.

Phase-Contrast Microscopy

Living microorganisms do not show up well in the ordinary compound light microscope. The **phase-contrast microscope** takes advantage of subtle differences in the refractive index of different parts of the living cell and its surrounding medium. As light is slowed down in portions of differing density, it travels slightly different pathways. When recombined for viewing, the 'phase differences' are seen as areas of differing brightness. The microorganism (and many of its internal structures) is seen in its natural state, alive and unstained.

Differential Interference Contrast (DIC) Microscopy

Differential interference contrast microscopy is similar to phase-contrast microscopy. It uses differences in refractive indexes but uses two beams of light instead of one. Prisms split each light beam, adding contrasting colors. Compared to standard phase-contrast microscopes, DIC yields images that are more brightly colored, are nearly three-dimensional in appearance, and have higher resolution.

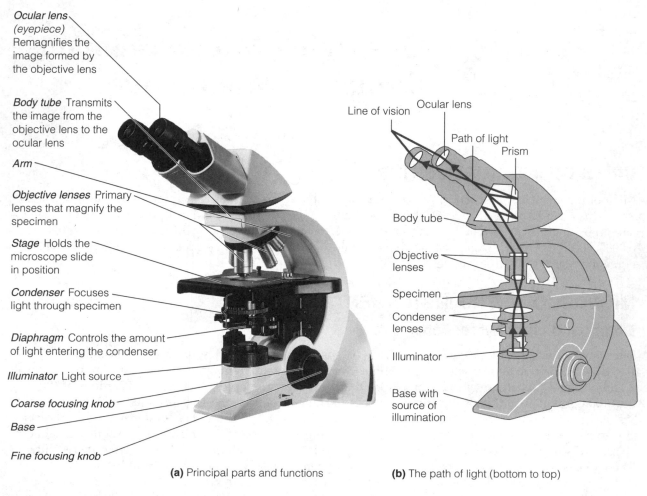

Figure 3.1   The compound light microscope.

Fluorescence Microscopy

Certain *fluorochrome dyes*, which glow with visible light—yellow or green, for example—when illuminated by ultraviolet light, can be used to view and identify microorganisms. This is **fluorescence;** certain substances, when illuminated by a short wavelength, emit light of a longer wavelength. **Fluorescence microscopy** techniques use a special microscope with ultraviolet light illumination; this light illuminates the specimen but is not permitted to reach the eye. The stained microorganism is highly visible against a dark background in such a microscope. However, the principal use of these dyes and microscopes is in the **fluorescent-antibody** (FA) technique, or **immunofluorescence.** In this technique, the organisms are allowed to react on a slide with antibodies (highly specific proteins produced by the body's defense system). A fluorescent dye is attached to the **antibody.** The combination of the antibody, the attached dye, and the microorganism for which the antibody is specific (called an **antigen;** it stimulates the body to produce these antibodies) allows the microorganism's presence to be detected (Figure 3.2). Because the antibody is specific for a particular microorganism, this is a very useful diagnostic technique. It is often used for diagnosis of syphilis and rabies.

Confocal Microscopy

In **confocal microscopy,** as in fluorescent microscopy, specimens are stained with fluorochromes that emit light when illuminated with a short-wavelength (blue) light. Specifically, one plane of a specimen is illuminated with a laser, and emitted light is returned through a pinhole aperture. Each plane corresponds to an image of a fine slice that has been cut from a specimen, and successive planes and regions

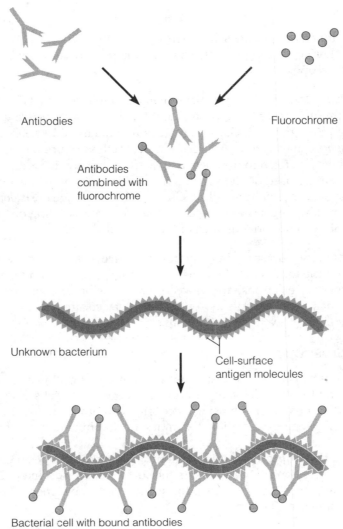

Antibodies

Fluorochrome

Antibodies
combined with
fluorochrome

Unknown bacterium

Cell-surface
antigen molecules

Bacterial cell with bound antibodies
combined with fluorochrome

Figure 3.2   The principle of immunofluorescence. A type of fluorochrome is combined with antibodies against a specific type of bacterium. When the preparation is added to bacterial cells on a microscope slide, the antibodies attach to the bacterial cells, and the cells fluoresce when illuminated with ultraviolet light.

are scanned. Most confocal microscopes are used in conjunction with computers to construct a series of successive focal-plane images that can be assembled into a three-dimensional image.

Two-Photon Microscopy

In **two-photon microscopy (TPM),** as in confocal microscopy, specimens are stained with a fluorochrome. However, TPM uses light with a long-wavelength (red, relatively lower energy). Therefore, two photons, instead of one, are needed to excite the fluorochrome to emit light. TPM images cells to a depth of 1 mm, considerably greater than confocal microscopy. It also causes less cellular damage and can track activity in real time.

Scanning Acoustic Microscopy

Scanning acoustic microscopy (SAM) consists of interpreting the action of a sound wave sent through a specimen. The sound wave, of a specific frequency, is reflected back every time it hits an interface within the material. SAM is useful in the study of living cells attached to another surface; examples are cancer cells, artery plaque, and bacterial biofilms.

Electron Microscopy

Electrons travel in waves much as light does, but their wavelengths are only about 1/100,000 as long as those of visible light and therefore have much better resolving power. They can be focused by magnetic lenses in an **electron microscope.**

Transmission Electron Microscopy.   In **transmission electron microscopy (TEM),** a beam of electrons is passed through ultrathin sections of the specimen and focused on a fluorescent screen, where it is visible to the eye and can be photographed. Objects are generally magnified 10,000 times to 100,000 times, and structures, called *artifacts,* may appear as a result of the method of preparation.

Salts of heavy metals may be fixed to the specimen *(positive staining)* to increase the density and make a darker image. If the metals are used on the surrounding field, the technique is termed *negative staining,* which is useful for viewing exceptionally small specimens. The technique of *shadow casting* produces a three-dimensional effect by spraying a heavy metal at an angle, accumulating on one side and leaving a clear area on the other. This also provides an idea of the size and shape.

Scanning Electron Microscopy.   In **scanning electron microscopy (SEM),** the electron beam is directed at the intact specimen from the top, rather than passing through a section, and electrons leaving the surface of the specimen (secondary electrons) are viewed on a television-like screen. Spectacular pictures of seemingly three-dimensional, intact organisms are possible. Objects are generally magnified 1000 times to 10,000 times with a resolving power of about 10 nm.

Scanned-Probe Microscopy

Scanned-probe microscopes can be used to map atomic and molecular shapes, to characterize magnetic and chemical properties, and to determine temperature variations inside cells. They use various kinds of probes to examine the surface of a specimen using electronic current which does not modify the specimen or expose it to damaging, high-energy radiation. Examples are scanning tunneling microscopes and atomic force microscopes.

Scanning tunneling microscopy uses a thin metal probe that scans a specimen and produces an image of the bumps and depressions of the atoms on the surface of the specimen. In **atomic force micros-copy,** a metal-and-diamond probe is moved along the surface of the specimen. The recorded movements yield a three-dimensional image of the surface.

PREPARATION OF SPECIMENS FOR LIGHT MICROSCOPY

Preparing Smears and Staining

Most microorganisms are viewed in stained preparations; that is, they are colored with a dye to make them visible or to emphasize certain structures. A thin film of a microbial suspension, called a **smear,** is spread on the surface of a slide. Flaming the air-dried smear coagulates the microbial proteins and **fixes** the micro-organisms to the slide so they do not wash off. The smear can then be stained. **Basic dyes** have a colored ion that is positive, helping them adhere to bacteria, which are slightly negative. Examples of basic dyes are crystal violet, methylene blue, and safranin. **Acidic dyes,** having a negative color ion, are more attracted to the background than to the negatively charged bacteria; thus, a field of colorless bacteria is presented against a stained background. This is an example of **negative staining.** An example of an acidic dye is eosin.

Simple Stains

To visualize shapes and arrangements of cells, a **simple stain** is usually sufficient. A chemical called a **mor-dant** may be added to make the microorganism stain more intensely or increase its size to enhance visibility.

Differential Stains

Gram Stain.   The most useful **differential stain** is the **Gram stain,** developed by Hans Christian Gram. It divides bacteria into two large groups: **gram-positive** and **gram-negative.** To prepare a Gram stain, (1) a purple dye, crystal violet, is applied to a heat-fixed smear. This stains all the cells and is called the

primary stain. After a water rinse, (2) an iodine mordant is added. When a smear stained in this manner is (3) washed with ethanol or an ethanol-acetone solution, some species of bacteria are decolorized and others are not. If the smear retains the purple dye, the organism is gram-positive. If the alcohol removes the dye, the colorless microorganisms are no longer visible. (4) Safranin, a red dye, is then applied, and the decolorized, or gram-negative, bacteria appear pink. Safranin is used here as a **counterstain.** The Gram stain reflects a basic difference in the cell wall structure of bacteria. It is a first step in identification, and the susceptibility of microorganisms to antibiotics is often related to the Gram reaction.

Acid-Fast Stain.   Members of the genera *Mycobacterium* (which includes the microorganisms that cause tuberculosis and leprosy) and *Nocardia* possess a cell wall with waxy components. The red dye carbolfuchsin is more soluble in these waxes than in acid-alcohol and is retained by the cell. Therefore, the **acid-fast stain,** in which carbolfuchsin is applied and gently steamed for several minutes, will stain them red. This dye is held so firmly that the cells are not decolorized by acid-alcohol, which does remove the dye from bacteria that are not acid-fast. A methylene blue counterstain will produce a slide in which acid-fast organisms are red and others are blue. The acid-fast stain is an invaluable aid in diagnosing tuberculosis and leprosy.

Special Stains

A colloidal suspension of dark particles such as India ink or nigrosin can be used as a **capsule stain.** The capsule will appear around each bacterial cell as a halo from which the India ink carbon particles are excluded. Endospores do not stain by ordinary methods, but the **Schaeffer-Fulton endospore stain,** which uses malachite green as a primary stain and safranin as a counterstain, shows endospores as green within red or pink cells. Flagella are too small to be resolved by light microscopes. In a **flagella stain,** a mordant can be used to increase the diameter of the flagella until they are visible in a light microscope.

SELF-TESTS

In the matching section, there is only one answer to each question; however, the lettered options (a, b, c, etc.) may be used more than once or not at all.

I. Matching

_____ 1. The electrons pass through a thin section of the specimen.

_____ 2. Visible light passes through the specimen; uses separate objective and ocular lenses.

_____ 3. Details become visible because of differences in the refractive index of different parts of the cell.

_____ 4. Visible light is scattered after striking the specimen, and the specimen is visible against a darkened background.

_____ 5. A special microscope using ultraviolet illumination.

_____ 6. The electrons strike the surface of the specimen, and secondary electrons leaving the surface are viewed on a television-like screen.

_____ 7. Makes use of relatively low-energy red light to excite fluorochromes; can track cellular activity in real time.

a. Compound light microscope

b. Scanning electron microscope

c. Phase-contrast microscope

d. Transmission electron microscope

e. Fluorescence microscope

f. Darkfield microscope

g. Two-photon microscopy

h. Confocal microscopy

II. Matching

___ 1. Pertaining to the relative velocities of light through a substance.

___ 2. Involves the use of antibodies and ultraviolet light.

___ 3. One-millionth of a meter.

___ 4. One ten-billionth of a meter.

___ 5. The ability to separate two points in a microscope field.

a. Micrometer

b. Nanometer

c. Ångstrom

d. Resolving power

e. Refractive index

f. Immersion oil

g. Immunofluorescence

III. Matching

___ 1. Adhere(s) best to bacteria, which have a negative charge, because the color molecule has a positive charge.

___ 2. Used in diagnosis of tuberculosis.

___ 3. Involve(s) the use of a negative stain made from India ink particles.

___ 4. Schaeffer-Fulton stain.

___ 5. Use(s) carbolfuchsin dye.

___ 6. Use(s) malachite green.

___ 7. Reflect(s) a basic difference between microbial cell walls; ethanol will not remove stain from bacteria.

a. Basic dyes

b. Acidic dyes

c. Gram stain

d. Acid-fast stain

e. Capsule stain

f. Endospore stain

IV. Matching

___ 1. A microscope that uses laser illumination.

___ 2. Extremely thin microbes, for example, the spirochete *Treponema pallidum*, are best seen with this type of light microscope.

___ 3. This type of electron microscope yields images with seemingly three-dimensional views of the specimen.

___ 4. Light rays that pass through different portions of the specimen reach the eye with their wave peaks reinforced or cancelled, making structures of the specimen relatively light or dark.

___ 5. A microscope that uses sound waves to form an image.

a. Confocal

b. Phase-contrast

c. Darkfield

d. Transmission

e. Scanning

f. Scanning acoustic

V. Matching

___ 1. Formerly known as a micron.

___ 2. Formerly known as a millimicron.

___ 3. This is 10^{-10} of a meter.

___ 4. A billionth of a meter.

a. Micrometer

b. Nanometer

c. Ångstrom

d. Millimeter

Fill in the Blanks

1. About the highest magnification possible in a compound light microscope is

   _____ .

2. Immersion oil has about the same refractive index as _____ .

3. Fluorochrome dyes glow with visible light when illuminated by _____ light.

4. Electron wavelengths are only about 1/100,000 as long as visible light and therefore have much

   _____ resolving power. (better, poorer)

5. Bacteria tend to have a slightly _____ electrical charge. (positive, negative)

6. The thin film of a microbial suspension spread on the surface of a slide is called a(n)

   _____.

7. Flaming the slide before applying the stain is called _____.

8. Transmission electron microscopy permits magnifications as high as about 10,000 times to

   _____.

9. In the flagella stain, a _____ is used to increase the diameter of the flagella.

10. Two bacterial genera that are acid-fast are _____ and

    _____.

11. A disease for which the acid-fast stain is useful in diagnosis is _____.

12. To see shapes and arrangements of cells, a(n) _____ stain is usually sufficient.

13. _____ dyes have a negative color ion. (acidic, basic)

Label the Art

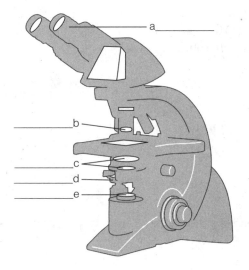

Critical Thinking

1. The equation that describes the resolving power of a microscope is:
   Resolving power = Wavelength of illumination/2 × Numerical aperture
   (The numerical aperture of an oil immersion objective is usually 1.30.)
   If the wavelength of light is 0.52 µm, what is the resolving power of this objective?

2. What type of microscopy would be most appropriate for viewing the following specimens or for the following situations?

   a. To identify pathogenic bacteria in clinical specimens.

   b. To view objects smaller than 0.2 µm, such as viruses.

   c. To view heat-fixed, stained bacterial cells.

   d. To view microorganisms that can't be stained by standard methods, such as *Treponema pallidum*.

   e. To view the internal structure of living microorganisms.

3. For each of the following specimens or situations, indicate which stain(s) or technique would be most appropriate.

   a. To detect bacterial capsules and evaluate an organism's virulence.

   b. To provide the necessary contrast for viewing specimens with a compound light microscope.

   c. To diagnose infections of *Mycobacterium* or *Nocardia*.

   d. To help determine what antibiotic will be most effective against a certain disease organism.

4. Why do gram-positive cells retain the crystal violet through the alcohol wash of Gram staining, whereas gram-negative cells do not?

ANSWERS

Matching

I.   1. d   2. a   3. c   4. f   5. e   6. b   7. g
II. 1. e   2. g   3. a   4. c   5. d
III. 1. a   2. d   3. e   4. f   5. d   6. f   7. c
IV. 1. a   2. c   3. e   4. b   5. f
V.   1. a   2. b   3. c   4. b

Fill in the Blanks

1. 2000   2. glass   3. ultraviolet   4. better   5. negative   6. smear   7. fixing   8. 100,000×
9. mordant   10. *Mycobacterium; Nocardia*   11. tuberculosis or leprosy   12. simple   13. Acidic

Label the Art

a. ocular lens   b. objective lenses   c. condenser lenses   d. diaphragm   e. illuminator

Critical Thinking

1. It would be about 0.2 µm. (The figure for the wavelength of light is for green light, changed from 520 nm.)

2. a. Fluorescence microscopy

   b. Electron microscopy

   c. Brightfield microscopy

   d. Darkfield microscopy

   e. Phase-contrast microscopy

3. a. A negative stain using India ink or nigrosin. The India ink (or nigrosin) stains the background but doesn't penetrate the capsule. The capsule shows up as a halo surrounding the cell against a dark background.

   b. For this purpose, a simple stain such as safranin or methylene blue will work fine.

   c. Acid-fast staining would be appropriate to diagnose infections of *Mycobacterium* and *Nocardia*. The red dye, carbolfuchsin, binds strongly to a waxy substance in the cell wall of these organisms but not to other, non–acid-fast bacteria.

   d. The Gram staining reaction is helpful information when choosing an antibiotic, which often shows specificity for either gram-positive or gram-negative bacteria.

4. When iodine is added to a smear after previous staining with crystal violet, they combine to form a complex (CV-I complex) that is larger than the crystal violet molecule that initially entered the cells. The CV-I complex is too large to be washed out of the intact peptidoglycan layer of gram-positive cells. When decolorizing gram-negative cells, the alcohol washes away the outer lipoprotein layer and the crystal violet from the thin layer of peptidoglycan.

4 Functional Anatomy of Prokaryotic and Eukaryotic Cells

COMPARING PROKARYOTIC AND EUKARYOTIC CELLS: AN OVERVIEW

The chief distinguishing characteristics of **prokaryotes** (from the Greek words meaning prenucleus) are as follows:

1. Their DNA is not enclosed within a membrane and is usually a singular circularly arranged chromosome. (Some bacteria, such as *Vibrio cholerae* have two chromosomes, and some bacteria have a linearly arranged chromosome.)
2. Their DNA is not associated with histones (special chromosomal proteins found in eukaryotes); other proteins are associated with the DNA.
3. They lack membrane-enclosed organelles.
4. Their cell walls almost always contain the complex polysaccharide peptidoglycan.
5. They usually divide by **binary fission.** During this process, the DNA is copied, and the cell splits into two cells. Binary fission involves fewer structures and processes than eukaryotic cell division.

Eukaryotes (from the Greek words meaning true nucleus) have the following distinguishing characteristics:

1. Their DNA is found in the cell's nucleus, which is separated from the cytoplasm by a nuclear membrane, and the DNA is found in multiple chromosomes.
2. Their DNA is consistently associated with chromosomal proteins called histones and with nonhistones.
3. They have a number of membrane-enclosed organelles, including mitochondria, endoplasmic reticulum, Golgi complex, lysosomes, and sometimes chloroplasts.
4. Their cell walls, when present, are chemically simple.
5. Cell division usually involves mitosis, in which chromosomes replicate and an identical set is distributed into each of two nuclei. This process is guided by the mitotic spindle, a football-shaped assembly of microtubules. Division of the cytoplasm and other organelles follows so that the two cells produced are identical to each other.

———— THE PROKARYOTIC CELL ————

THE SIZE, SHAPE, AND ARRANGEMENT OF BACTERIAL CELLS

Most bacteria range in size from 0.2 to 2.0 μm in diameter and 2 to 8 μm in length. Basic bacterial shapes are the spherical **coccus** (meaning berry), the rod-shaped **bacillus** (meaning little staff), and the **spiral. Diplococci** form pairs; **streptococci** form chains; **tetrads** divide in two planes, forming groups of four; **sarcinae** divide in three regular planes and form cubelike packets; **staphylococci** divide in irregular, random planes and form grapelike clusters. Most bacilli are single rods, but they can appear in pairs—**diplobacilli**—or in chains—**streptobacilli. Coccobacilli** are ovals. **Vibrios**—slightly curved, commalike rods—are also included among spiral bacteria. **Spirilla** have a helical, corkscrew shape and are motile by means of flagella. **Spirochetes** are shaped like spirilla but have axial filaments for motility. **Pleomorphic** bacteria have an irregular morphology; if they maintain a single shape, they are **monomorphic.**

STRUCTURES EXTERNAL TO THE CELL WALL

Glycocalyx

The general term for substances surrounding bacterial cells is **glycocalyx,** which is usually a polysaccharide, polypeptide, or both. If organized and tightly attached, it is called a **capsule** (Figure 4.1). If unorganized and loosely attached, the glycocalyx is called a **slime layer.** A glycocalyx that helps cells in a biofilm attach to their target environment and to each other is called an **extracellular polymeric substance (EPS)**. The glycocalyx aids in attachment to surfaces; capsules contribute to pathogenicity by protecting from phagocytosis, an important part of the body's defenses.

Flagella

Flagellar filaments are composed of a protein, **flagellin.** The base of the flagellar filament widens to a hook. Attached to the hook is a **basal body** (a rod with rings), which anchors the flagellum to the cell wall and plasma membrane (Figure 4.2). The basal body of gram-negative bacteria is anchored to the cell wall and plasma membrane; in gram-positive bacteria, it is anchored only at the plasma membrane. Flagella, when present, are arranged in certain ways: **peritrichous** (distributed over the entire cell) or **polar**

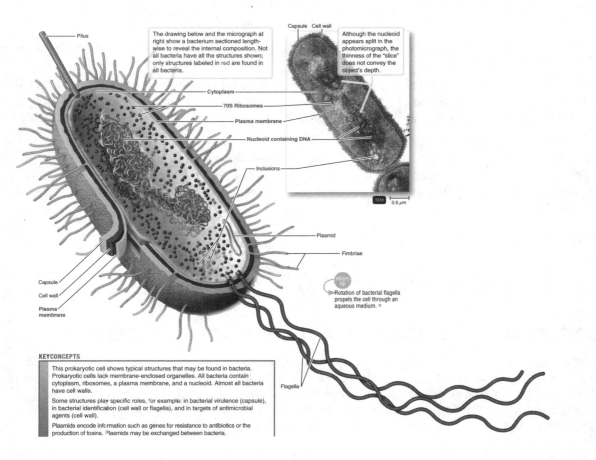

Figure 4.1   **Structure of a typical prokaryotic (bacterial) cell.**

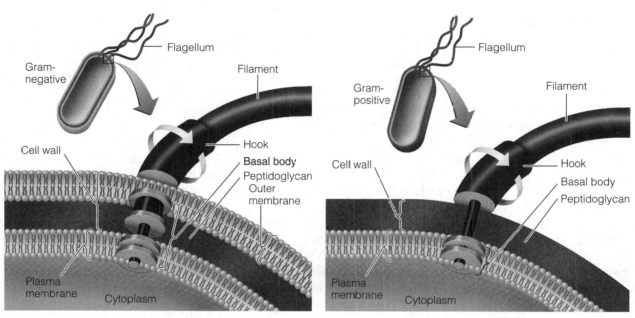

(a) Parts and attachment of a flagellum of a gram-negative bacterium

(b) Parts and attachment of a flagellum of a gram-positive bacterium

Figure 4.2   The structure of a prokaryotic flagellum. The parts and attachment of a flagellum of a gram-negative bacterium and gram-positive bacterium are shown in these highly schematic diagrams.

(at one or both poles of a cell). Polar flagella may be **monotrichous** (a single flagellum at one pole), **lophotrichous** (a tuft of flagella at one pole), or **amphitrichous** (flagella at both poles). Bacteria with no flagella are **atrichous.** Flagella may spin clockwise or counterclockwise, producing directional movement (a "run" or "swim") or random changes in direction ("tumbles"). Movement to or from a stimulus is called **taxis;** the stimulus may be chemicals **(chemotaxis)** or light **(phototaxis).** A flagellar protein, **H antigen**, is useful for helping to distinguish **serovars,** or variation within a species.

Axial Filaments

Spirochetes move by means of **axial filaments (endoflagella),** bundles of fibrils that arise near cell poles beneath an outer sheath and wrap in spiral fashion around the cell. These can cause the spirochetes to move in a corkscrew manner.

Fimbriae and Pili

Many bacterial cells have numerous hairlike appendages called **fimbriae** that are shorter than flagella (see Figure 4.11 in the text) and consist of a protein, **pilin.** They help the cell adhere to surfaces such as mucous membranes—often a factor in pathogenicity. **Pili** are longer than fimbriae and number only one or two per cell. These are sometimes called **sex pili** because they can function to transfer DNA from one cell to another, called **conjugation.**

There are several types of motility. In **twitching motility,** also called the *grappling hook model,* a pilus extends by addition of subunits of pilin and contacts another cell, causing movement by retraction as subunits of pilin are removed). **Gliding motility** is used by myxobacteria; the exact mechanism is unknown.

THE CELL WALL

The bacterial **cell wall** is a complex, semirigid structure responsible for the characteristic shape of the cell.

Composition and Characteristics

The cell wall of gram-positive bacteria is composed of **peptidoglycan,** which consists of two sugars, N-acetylglucosamine and N-acetylmuramic acid (*murein* from *murus*, meaning wall), and also chains of amino acids. The two sugars alternate with each other, forming a carbohydrate (glycan) backbone. Peptide side chains of four amino acids attached to the N-acetylmuramic acid are cross-linked to form the macromolecule of the cell wall. Many gram-positive bacteria also contain polysaccharides called *teichoic acids*. The cell wall of acid-fast bacteria (otherwise considered gram-positive) consists of peptidoglycan and a waxy lipid, mycolic acid.

The cell wall of gram-negative bacteria also contains peptidoglycans, but only thin layers. These cells have a lipoprotein, lipopolysaccharide (**LPS**), and a phospholipid **outer membrane** surrounding their peptidoglycan layers. A **periplasm** (a fluid-filled space) is found between the outer membrane and the **plasma membrane.** The outer membrane also provides resistance to phagocytosis and the action or complement (also part of host defenses). When the cell disintegrates in the host's bloodstream, the lipid portion of the LPS (**lipid A**) is released as an **endotoxin** that can cause illness. Materials may penetrate the outer membrane through channels called **porins.** The **O polysaccharide** extends outward from the core polysaccharide and is composed of sugar molecules, which function as an antigen useful for identifying certain species of gram-negative bacteria.

Cell Walls and the Gram Stain Mechanism

The primary stain, crystal violet, stains both gram-negative and gram-positive bacteria purple because it enters the cytoplasm of both. Iodine, with crystal violet, forms large crystals that cannot be washed through the peptidoglycan wall of gram-positive cells. Alcohol makes the outer membrane of gram-negative cells permeable and allows the crystal violet-iodine crystals to wash out, making them colorless. Safranin counterstain turns gram-negative cells pink. Archaea appear gram-negative because they do not contain peptidoglycans.

Atypical Cell Walls

Mycoplasma bacteria do not have cell walls. They are unique also in having sterols in their plasma membranes. Archaea do not have peptidoglycan in their walls but have a similar substance, **pseudomurein.**

Acid-Fast Cell Walls

These bacteria contain high concentrations of a waxy lipid, **mycolic acid,** that prevents Gram staining. They are stained with red carbolfuchsin dye. The heated dye penetrates the cell wall and resists removal with acid-alcohol.

Damage to the Cell Wall

Lysozyme—an enzyme occurring in tears, mucus, and saliva—damages the cell walls of many gram-positive bacteria. Some bacteria of the genus *Proteus* may spontaneously or, in response to penicillin, lose their cell walls and become **L forms.** Later, they may spontaneously revert to walled bacteria. A bacterium that has lost its cell wall and is surrounded only by the plasma membrane is a **protoplast.** Gram-negative cells treated with lysozyme retain much of the outer membrane layer and are called **spheroplasts.** Both are sensitive to rupture by **osmotic lysis.**

STRUCTURES INTERNAL TO THE CELL WALL

The Plasma (Cytoplasmic) Membrane

The **plasma (cytoplasmic) membrane** is just internal to the cell wall and encloses the cytoplasm. In prokaryotes it consists primarily of phospholipids and proteins. Eukaryotic plasma membranes also contain sterols, making them more rigid, and carbohydrates. Both prokaryotic and eukaryotic membranes have a two-layered structure, molecules in parallel rows, called a **lipid bilayer.** One end (phosphate) is water-soluble, and the other (hydrocarbon) is insoluble. The water-soluble ends are on the outside of the bilayer. Protein molecules are embedded in the membrane; along with phospholipids, they may move freely within the membrane. This arrangement is called the **fluid mosaic model.**

The most important function of the plasma membrane is as a selective barrier. It is **selectively permeable (semipermeable),** and certain molecules and ions pass through, whereas others do not. Several factors affect permeability. Large molecules such as proteins cannot pass; smaller molecules such as amino acids and simple sugars can pass if uncharged. (The phosphate end of the bilayer is charged.) Lipid-soluble substances, because of the phospholipid content, pass more easily. Plasma membranes contain enzymes that help break down nutrients and produce more energy. The **chromatophores** or **thylakoids,** which contain pigments and enzymes for bacterial photosynthesis, are found in the plasma membranes.

Mesosomes are folds in the plasma membrane that may be only an artifact of preparation for electron microscopy.

Movement of Materials across Membranes

Material crosses plasma membranes by *passive processes* such as **simple diffusion** (movement of molecules or ions from an area of higher concentration to an area of lower concentration). At equilibrium, the concentration gradient has been eliminated. **Osmosis** is the net movement of solvent molecules across a selectively permeable membrane. **Osmotic pressure** is the force with which a solvent (such as water) moves from a solution of lower solute concentration (such as dissolved sugar) to a solution of higher solute concentration. **Isotonic (isoosmotic)** solutions have equal solute concentrations on both sides of the membrane. **Hypotonic (hypoosmotic)** solutions have a lower concentration of solutes outside the cell than inside; this is the case with most bacteria. **Hypertonic (hyperosmotic)** solutions have a higher concentration of solutes outside the cell. Bacterial cells placed in such solutions lose water by osmosis and shrink, and the cytoplasm collapses within the cell wall. **Facilitated diffusion,** an *active process,* occurs when a **carrier protein (permease** or **transporter)** combines with and transports a substance across the membrane, but only where a concentration gradient is present. **Active transport** requires cell energy (ATP) and also involves carrier proteins moving substances across the plasma membrane. In **group translocation,** the substance is chemically altered during transport. Once inside, the plasma membrane is impermeable. This is important for low-concentration substances.

Cytoplasm, Nucleoid, Ribosomes, and Inclusions

The term **cytoplasm** refers to the substance of the cell inside the plasma membrane. It has many **inclusions,** such as **metachromatic granules** of stored phosphate (**volutin**), **polysaccharide granules** of glycogen and starch, **lipid inclusions** such as *poly-beta-hydroxybutyric acid,* and sulfur granules. The cytoplasm also contains many **ribosomes,** the sites of protein synthesis. **Carboxysomes** are inclusions found in bacteria that use carbon dioxide as their sole source of carbon. **Gas vacuoles** or **gas vesicles** help some bacteria maintain buoyancy. The **bacterial chromosome,** which contains the genetic information, is a single, long, circular molecule of DNA found in the **nucleoid.** Small circular DNA molecules, **plasmids,** are not connected to the chromosome and replicate independently. Plasmids do not contain normally essential genes but may provide a selective advantage under abnormal conditions—antibiotic resistance, for example. **Magnetosomes** are inclusions of iron oxide formed by a few gram-negative bacteria that aid the microbe in orienting itself environmentally.

Endospores

Endospores are highly resistant bodies formed by a few bacterial species, such as *Bacillus* and *Clostridium*. *Sporulation* or **sporogenesis** is the process of their formation. First, there is an ingrowth of the plasma membrane (**spore septum**). A small portion of the cytoplasm and newly replicated bacterial chromosome is then surrounded by a membrane, the **forespore**. A thick **spore coat** of protein forms around this membrane. The endospore core is dehydrated and contains considerable *dipicolinic acid,* as well as a few essential materials necessary to return it to its vegetative state, which is accomplished through the process of **germination.**

_____ THE EUKARYOTIC CELL _____

FLAGELLA AND CILIA

Eukaryotic **flagella** are relatively long; **cilia** are more numerous and are shorter. Both are involved in locomotion, and both contain small tubules of protein called **microtubules.**

THE CELL WALL AND GLYCOCALYX

Most algae and some fungi have **cell walls** containing *cellulose,* and often fungi have *chitin* as well. Yeast cell walls contain the polysaccharides *glucan* and *mannan.* No eukaryotic cell wall contains peptidoglycans. Protozoa have a flexible outer covering called a *pellicle.* In animal cells, the plasma membrane is covered by sticky carbohydrates called the **glycocalyx.**

THE PLASMA (CYTOPLASMIC) MEMBRANE

In eukaryotic cells, the **plasma membrane,** which contains sterols, may be the external cell covering. Substances cross the membrane by mechanisms similar to those in prokaryotes. In addition, a process of engulfment, **endocytosis,** brings particles, even some viruses, into the cell. Examples are **phagocytosis,** used by white blood cells to engulf and destroy bacteria, and **pinocytosis,** by which liquids and dissolved substances enter cells.

CYTOPLASM

Cytoplasm is the matrix in which various cellular components are found. The complex internal structure of *microfilaments, intermediate filaments,* and *microtubules* is called the **cytoskeleton.** Movement of cytoplasm from one part of a cell to another, **cytoplasmic streaming,** can move a cell over a surface.

RIBOSOMES

Attached to the outer surface of rough endoplasmic reticulum (ER) are **ribosomes** (see Figure 4.25 in the text), which are also found free in the cytoplasm. As in prokaryotes, ribosomes are the sites of protein synthesis in the cell. The ribosomes of eukaryotic ER and cytoplasm are somewhat larger and denser than those of prokaryotic cells. Some ribosomes, called *free ribosomes,* are unattached to any structure in the cytoplasm. Primarily, free ribosomes synthesize proteins used inside the cell. Other ribosomes, called *membrane-bound ribosomes,* attach to the nuclear membrane and the ER. These ribosomes synthesize proteins destined for insertion in the plasma membrane or for export from the cell. Ribosomes located within mitochondria synthesize mitochondrial proteins. Sometimes 10 to 20 ribosomes join together in a string-like arrangement called a *polyribosome*.

ORGANELLES

In eukaryotes, unlike prokaryotes, many important enzymes are found in, and functions carried out by, **organelles.**

Nucleus. The **nucleus** is an oval organelle containing the DNA. It is surrounded by a **nuclear envelope. Nuclear pores** in the nuclear membrane allow the nucleus to communicate with the endoplasmic reticulum of the cytoplasm. The **nucleoplasm** is a gel-like fluid in the nucleus. **Nucleoli,** which may be the center for the synthesis of ribosomal RNA, are present. DNA is combined with protein *histones* and *nonhistones*. The combination is called a **nucleosome.** When the cell is not reproducing, DNA and associated proteins are visible as a mass called **chromatin.** When reproducing, chromatin becomes visible as rodlike bodies called **chromosomes.**

Endoplasmic Reticulum. Within the cytoplasm there is a network of flattened sacs, or **cisterns,** called the **endoplasmic reticulum (ER).** Its function is to synthesize and store lipids and proteins, and to transport them. The ends of the cisterns pinch off into *secretory vesicles,* which transport substances within the cell. **Rough ER** has ribosomes bound to it, and **smooth ER** does not.

Golgi Complex. The **Golgi complex** consists of a network of flattened sacs called cisterns (see previous paragraph), stacked like dishes. Like ER, these function to export substances from the cell and transport substances within it. The Golgi complex receives proteins and lipids from the ER and delivers them to secretory vesicles.

Lysosomes. **Lysosomes** are formed from **Golgi complexes** and look like membrane-enclosed spheres. White blood cells, which destroy bacteria by phagocytosis, contain many **lysosomes.**

Vacuoles. A **vacuole** is a space (cavity) in cytoplasm that serves for storage and other functions.

Mitochondria. **Mitochondria** are organelles with a smooth outer membrane and an inner membrane arranged in a series of folds called **cristae.** The semifluid center of the mitochondrion is called the **matrix.** Enzymes forming ATP are located on the cristae—which is one reason why mitochondria are called "powerhouses" of the cell.

Chloroplasts. Photosynthesizing cells contain membrane-bounded structures called **chloroplasts,** which contain chlorophyll and enzymes involved in photosynthesis. The chlorophyll is found in membranes called **thylakoids.** Stacks of thylakoids are called **grana.**

Peroxisomes. **Peroxisomes** are organelles similar in structure to lysosomes, but smaller. They contain enzymes that oxidize various organic substances. The enzyme catalase that decomposes toxic H_2O_2 (hydrogen peroxide) is also present.

Centrosomes. The **centrosome** consists of a **pericentriolar area,** which is the organizing center for the mitotic spindle that plays a critical role in cell division. Within this area are located **centrioles,** arrays of microtubules that play a role in the formation or regeneration of cilia and flagella.

Table 4.1 outlines the principal differences between prokaryotic and eukaryotic cells.

Table 4.1 Principal Differences Between Prokaryotic and Eukaryotic Cells

Characteristic	Prokaryotic	Eukaryotic
Size of cell	Typically 0.2–2.0 µm in diameter	Typically 10–100 µm in diameter
Nucleus	No nuclear membrane or nucleoli	True nucleus, consisting of nuclear membrane and nucleoli
Membrane-enclosed organelles	Absent	Present; examples include lysosomes, Golgi complex, endoplasmic reticulum, mitochondria, and chloroplasts
Flagella	Consist of two protein building blocks	Complex; consist of multiple microtubules
Glycocalyx	Present as a capsule or slime layer	Present in some cells that lack a cell wall
Cell wall	Usually present; chemically complex (typical bacterial cell wall includes peptidoglycan)	When present, chemically simple (includes cellulose and chitin)
Plasma membrane	No carbohydrates and generally lacks sterols	Sterols and carbohydrates that serve as receptors
Cytoplasm	No cytoskeleton or cytoplasmic streaming	Cytoskeleton; cytoplasmic streaming
Ribosomes	Smaller size (70S)	Larger size (80S); smaller size (70S) in organelles
Chromosome (DNA)	Usually single circular chromosome; typically lacks histones	Multiple linear chromosomes with histones
Cell division	Binary fission	Involves mitosis
Sexual recombination	None; transfer of DNA only	Involves meiosis

THE EVOLUTION OF EUKARYOTES

The theory explaining the origin of eukaryotes from prokaryotes, pioneered by Lynn Margulis, is the **endosymbiotic theory.** Larger bacterial (prokaryotic) cells are presumed to have engulfed smaller bacterial cells (one organism living within another is called *endosymbiosis*) and eventually evolved into eukaryotic cells. Mitochondria and chloroplasts in eukaryotic cells are considered evidence for the theory. They resemble prokaryotic cells and can reproduce independently of their eukaryotic host cell.

SELF-TESTS

In the matching section, there is only one answer to each question; however, the lettered options (a, b, c, etc.) may be used more than once or not at all.

I. Matching

____ 1. Helical; move by flagella, if present.

____ 2. Spherical; in chains.

____ 3. Divide in three regular planes; spheres form cubelike packets.

____ 4. Helical; axial filaments for motility.

____ 5. A simple, commalike curve.

____ 6. Name means "little staff."

____ 7. Ovals.

a. Sarcinae

b. Tetrads

c. Streptococci

d. Spirochetes

e. Vibrios

f. Bacilli

g. Cocci

h. Spirilla

i. Diplococci

j. Coccobacilli

II. Matching

____ 1. Golgi complex.

____ 2. Meiosis occurs in reproduction.

____ 3. Usually single circular chromosome without histones.

____ 4. Sterols generally present in cell membrane.

____ 5. Cell wall almost always contains peptidoglycans.

____ 6. Nucleus bounded by a membrane.

____ 7. DNA contained in a nucleoid.

a. Eukaryotic cell

b. Prokaryotic cell

III. Matching

_____ 1. Contain pigments for photosynthesis by bacteria; found in the plasma membrane.

_____ 2. Gram-negative bacterial cells after their treatment with lysozyme.

_____ 3. Specialized external structures that assist in the transfer of genetic material between cells.

_____ 4. Numerous short, hairlike appendages that help in attachment to mucous membranes.

_____ 5. General term for substances surrounding bacterial cells.

_____ 6. Polysaccharides found in the cell wall of many gram-positive bacteria.

_____ 7. Inclusions of iron oxide.

a. Glycocalyx

b. Flagellin

c. Fimbriae

d. Sex pili

e. Capsules

f. Teichoic acids

g. Spheroplasts

h. Protoplasts

i. Chromatophores

j. Chloroplasts

k. Magnetosomes

IV. Matching

_____ 1. Metachromatic granules of stored phosphate in prokaryotes.

_____ 2. Entrance of fluids and dissolved substances into eukaryotic cells.

_____ 3. Membrane-enclosed spheres in phagocytic cells that contain powerful digestive enzymes.

_____ 4. The "powerhouses" of the cell.

_____ 5. A gel-like fluid found in the eukaryotic nucleus.

_____ 6. A folded inner membrane found in mitochondria.

_____ 7. Sometimes contributes to movement of a cell.

_____ 8. Found in walls of acid-fast bacteria.

a. Volutin

b. Plasmids

c. Cristae

d. Zymogens

e. Ribosomes

f. Nucleoplasm

g. Lysosomes

h. Mitochondria

i. Phagocytosis

j. Pinocytosis

k. Cytoplasmic streaming

l. Mycolic acid

V. Matching

_____ 1. Arrangement of flagella distributed over the entire cell.

_____ 2. Flagella at both poles of the cell.

_____ 3. A widening at the base of the flagellar filament.

_____ 4. An enzyme affecting gram-positive cell walls; found in tears.

_____ 5. A compound found in bacterial endospores.

_____ 6. A compound frequently found in the cell walls of yeasts.

_____ 7. No flagella.

_____ 8. A tuft of flagella at one pole of the cell.

_____ 9. Twitching motility.

a. Exocytosis

b. Dipicolinic acid

c. Chitin

d. Lysozyme

e. Hook

f. Peritrichous

g. Amphitrichous

h. Lophotrichous

i. Monotrichous

j. Atrichous

k. Grappling hook model

l. Flagellin

VI. Matching

_____ 1. Closely involved in protein synthesis.

_____ 2. Structure(s) characteristic of both eukaryotic and prokaryotic plasma membranes.

_____ 3. Found in the flagella and cilia of eukaryotic cells.

a. Phospholipid bilayer

b. Transverse septum

c. Microtubules

d. Ribosomes

VII. Matching

_____ 1. Highly resistant bodies formed by a few bacterial species.

_____ 2. Small circular DNA molecules that are not connected with the main chromosome.

_____ 3. The semifluid center portion of the mitochondrion.

_____ 4. A substance similar to peptidoglycan that is found in the cell wall of archaea.

_____ 5. Bacteria with irregular morphology.

a. Plasmids

b. Endospores

c. Pseudomurein

d. Matrix

e. Pleomorphic

VIII. Matching

_____ 1. Extracellular polymeric substances on some bacterial cells; may help cells adhere to surfaces.

_____ 2. Bacterial cell with thin peptidoglycan layer, outer membrane of lipopolysaccharide.

_____ 3. Protein that forms fimbriae.

_____ 4. Bundles of microtubules that probably play a role in cell division of eukaryotic cells.

_____ 5. Bacteria that have lost their cell walls and may later spontaneously regain them.

_____ 6. Lipid A and O polysaccharide are found on this type of bacteria.

a. Glycocalyx

b. Pilin

c. Gram-positive

d. Gram-negative

e. Centrioles

f. L forms

IX. Matching

_____ 1. ER associated with ribosomes.

_____ 2. Ingrowth of plasma membrane before endospore formation.

_____ 3. Anchors the flagella of bacteria to the cell wall and plasma membrane.

a. Septum

b. Forespore

c. Rough ER

d. Smooth ER

e. Basal body

Fill in the Blanks

1. Chemically, the capsule is a(n) _____, a polypeptide, or both.

2. Capsules protect pathogenic bacteria from _____, a process by which protective host cells engulf and destroy microorganisms.

3. The Golgi complex consists of flattened sacs called _____ that are connected to the endoplasmic reticulum.

4. The _____ complex consists of four to eight flattened sacs connected to the endoplasmic reticulum. The function is largely secretion of proteins, lipids, and carbohydrates.

5. The term _____ means a lower concentration of solutes outside the cell than inside.

6. Three examples of passive diffusion across membranes are _____, _____, and _____.

7. The protein in the flagellar filaments of bacteria is called _____.

8. DNA in eukaryotic cells is combined with protein _____ and nonhistones.

Critical Thinking

1. What is a glycocalyx? How is the presence of a glycocalyx related to bacterial virulence?

2. What substances are able to cross the plasma membrane most easily?

3. Describe how a bacterial cell will respond to the following osmotic pressures: isotonic, hypotonic, hypertonic.

4. How is the presence of peptidoglycan in bacterial cells clinically significant?

ANSWERS

Matching

   I.   1. h   2. c   3. a   4. d   5. e   6 .f   7. j
   II.   1. a   2. a   3. b   4. a   5. b   6. a   7. b
   III.   1. i   2. g   3. d   4. c   5. a   6. f   7. k
   IV.   1. a   2. j   3. g   4. h   5. f   6. c   7. k   8. l
   V.   1. f   2. g   3. e   4. d   5. b   6. c   7. j   8. h   9. k
   VI.   1. d   2. a   3. c
   VII.   1. b   2. a   3. d   4. c   5. e
   VIII.   1. a   2. d   3. b   4. e   5. f   6. d
   IX.   1. c   2. a   3. e

Fill in the Blanks

1. polysaccharide   2. phagocytosis   3. cisterns   4. Golgi 5. hypotonic
6. simple diffusion; osmosis; facilitated diffusion   7. flagellin   8. histones

Critical Thinking

1. The glycocalyx is a sticky, viscous, gelatinous polymer that surrounds some bacterial cells. It may be composed of polysaccharide, polypeptide, or a combination of these two substances. Depending on how the material is arranged and attached to the cell, it may be referred to as a slime layer or a capsule. The glycocalyx is associated with bacterial virulence because it helps protect the bacterium from phagocytosis by white blood cells and helps the bacterium to adhere to and colonize a host.

2. Substances that dissolve easily in lipids can most easily cross the plasma membrane. These include oxygen, carbon dioxide, and nonpolar organic molecules. Also, small molecules such as water are able to cross the plasma membrane easily.

3. There will be no change in a bacterial cell in an isotonic solution; water leaves and enters the cell at the same rate.

   A bacterial cell placed in a hypotonic solution will undergo osmotic lysis because more water will enter the cell than the cell wall can contain.

   A hypertonic solution will cause a bacterial cell to undergo plasmolysis, the osmotic loss of water due to increased solutes outside of the cell.

4. Peptidoglycan is a substance that is found in varying quantities in most prokaryotic cells. Peptidoglycan is unique to prokaryotic cells and is never found in eukaryotic cells. Antibiotics such as the penicillins and the cephalosporins act specifically against peptidoglycan and therefore have low toxicity in humans. These drugs prevent the formation of the peptide cross-bridges of peptidoglycan, preventing synthesis of a functional cell wall.

5 Microbial Metabolism

CATABOLIC AND ANABOLIC REACTIONS

Metabolism is the sum of all chemical reactions within a living organism, including **anabolic** (*biosynthetic*) reactions and **catabolic** (*degradative*) reactions. **Anabolism** is the combination of simpler substances into complex substances and *requires* energy. Catabolism releases energy stored in organic molecules, for example, and *yields* energy. Energy liberated by catabolism is stored in energy-rich bonds of **adenosine triphosphate (ATP).**

ENZYMES

Collision Theory

The **collision theory** explains that in order for chemical reaction to occur, collisions must occur between atoms, ions, or molecules. This requires a specific level of energy in the collision. This collision energy is the **activation energy.** The **reaction rate** is the frequency of collisions of sufficient energy.

Enzymes and Chemical Reactions

Substances that speed up a chemical reaction without being permanently altered themselves are **catalysts.** In living cells, **enzymes** serve as biological catalysts. An enzyme acts on a specific substance; the enzyme's **substrate.** The three-dimensional enzyme molecule has a region that interacts with a specific chemical substance, the *active site*. The enzyme orients the substrate into a position that increases the probability of a reaction. The **enzyme-substrate complex** formed by this temporary binding enables the collisions to be more effective (lowers the activation energy of the reaction).

Enzyme Specificity and Efficiency

Enzymes are generally globular proteins. The rate at which they can catalyze reactions is the **turnover number.**

Naming Enzymes

The names of enzymes usually end in *-ase*. Examples are *dehydrogenases* (remove hydrogen from substrate) and *oxidases* (add molecular oxygen).

Enzyme Components

A nonprotein component of an enzyme is called a **cofactor;** the protein portion is the **apoenzyme.** If the cofactor is an organic molecule, it is a **coenzyme.** Together, the apoenzyme and cofactor are a **holoenzyme,** or a whole, active enzyme. Two of the most important coenzymes are **nicotinamide adenine dinucleotide (NAD^+)** and **nicotinamide adenine dinucleotide phosphate ($NADP^+$).** Both contain derivatives of the B vitamin nicotinic acid (niacin) and both function as electron carriers. The flavin coenzymes, such as **flavin mononucleotide (FMN)** and **flavin adenine dinucleotide (FAD),** contain derivatives of the B vitamin riboflavin and are also electron carriers. **Coenzyme A (CoA)** contains a derivative of pantothenic acid, another B vitamin.

THE MECHANISM OF ENZYMATIC ACTION

Enzymes lower the activation energy of chemical reactions through a general sequence of events (see Figure 5.4a in the text). As a result of these events, an enzyme speeds up a chemical reaction.

Factors Influencing Enzymatic Activity

Temperature.   Most chemical reactions occur more rapidly as the temperature rises, causing an increase in the **reaction rate**—or frequency of—sufficiently energetic collisions. Above a certain point, denaturation of enzyme proteins results in a drastic decline in biological reaction rates. **Denaturation** usually involves breakage of the hydrogen bonds and similar weak bonds that hold the enzyme in its three-dimensional structure.

pH.   Enzymes have a **pH optimum** at which activity is maximal. Extreme pH changes can cause denaturation.

Substrate Concentration.   At high substrate concentrations, the enzyme may have its active site occupied at all times by substrate or product molecules; that is, it may be **saturated.** No further increase in substrate concentrations will have an effect on the reaction.

Inhibitors.   There are two forms of enzyme inhibitors. **Competitive inhibitors** compete with the normal substrate for the active site of the enzyme. These inhibitors have a shape and chemical structure similar to the normal substrate. The action of sulfa drugs depends on competitive inhibition. **Noncompetitive inhibitors** decrease the ability of the normal substrate to combine with the enzyme. The site of a noncompetitive inhibitor's binding is an **allosteric site.** In this process, **allosteric** ("other space") **inhibition,** the inhibitor changes the shape of the active site, making it nonfunctional. Other examples are *enzyme poisons* that tie up metal ions (cofactors) and prevent enzymatic reactions, such as cyanide, which binds iron, and fluoride, which binds calcium or magnesium.

Feedback Inhibition

In some metabolic reactions, several steps are required. In many instances the final product can inhibit enzymatic activity at some step and prevent making of excessive *end-product*. This is called **feedback inhibition** (or **end-product inhibition**).

Ribozymes

Ribozymes that function as enzymelike catalysts act on strands of RNA by removing sections and splicing the remaining pieces.

ENERGY PRODUCTION

Nutrient molecules have energy stored in bonds that can be concentrated into the " high-energy" (or "unstable") bonds of ATP.

Oxidation-Reduction Reactions

Oxidation is the addition of oxygen or, more generally, the removal of electrons (e^-) or hydrogen ions (H^+). Because H^+ are lost, most biological reactions are called **dehydrogenation** reactions. When a compound gains electrons or hydrogen atoms, or loses oxygen, it is **reduced.** Oxidation and reduction in a cell are always coupled—one substance is oxidized and another is reduced; thus the reaction is an **oxidation-reduction** (or *redox*) reaction. NAD^+ and $NADP^+$ commonly carry the hydrogen atoms in these oxidation-reduction reactions. These reactions are usually energy producing. Highly reduced compounds such as glucose, with many hydrogen atoms, contain much potential energy.

The Generation of ATP

The energy from oxidation-reduction is used to form ATP. The addition of a phosphate group is called **phosphorylation.** In **oxidative phosphorylation,** electrons removed from organic compounds are transferred in sequence down an **electron transport chain** to an electron acceptor such as oxygen or another suitable compound, releasing energy in the process. The energy is used to make ATP form ADP by making another phosphate. In **substrate-level phosphorylation,** no oxygen or other inorganic final electron acceptor is required. ATP is generated by the direct transfer of a high-energy phosphate from an intermediate metabolic compound to ADP. Another mechanism is **photophosphorylation,** which occurs in photosynthetic cells. Light liberates an electron from chlorophyll. The electron passes down an electron transport chain, forming ATP.

Metabolic Pathways of Energy Production

A sequence of enzymatically catalyzed chemical reactions in a cell is a **metabolic pathway.** Such pathways are necessary to extract energy from organic compounds; they allow energy to be released in a controlled manner instead of in a damaging burst with a large amount of heat.

CARBOHYDRATE CATABOLISM

Glycolysis

The six-carbon sugar glucose plays a central role in carbohydrate metabolism (Figure 5.1). Glucose is usually broken down to **pyruvic acid** by **glycolysis** (splitting of sugar), which is the first stage of both fermentation and respiration.

Glycolysis, also called the *Embden-Meyerhof pathway,* is a series of ten chemical reactions. The main points are that a six-carbon glucose molecule is split and forms two molecules of pyruvic acid. Two molecules of ATP were needed to start the reaction, and four molecules of ATP are formed by substrate-level phosphorylation; the net yield, therefore, is two ATP molecules.

Alternatives to Glycolysis

An alternative pathway to glycolysis is the **pentose phosphate pathway,** or the **hexose monophosphate shunt** (see Figure A-3 in Appendix A in the main text). It produces important intermediates (pentoses) that act as precursors in the synthesis of nucleic acids, certain amino acids, and glucose from carbon dioxide by photosynthesizing organisms. Another way to oxidize glucose to pyruvic acid is the **Entner-Doudoroff pathway (EDP).** Bacteria (generally gram-negative) that utilize EDP can metabolize without either the glycolysis or the pentose phosphate pathway. This pathway yields NADPH from glucose, which may be used for biosynthetic reactions.

Cellular Respiration

Cellular respiration, or simply *respiration,* is an ATP-generating process in which molecules are oxidized and the final electron acceptor is almost always an inorganic molecule. In **aerobic respiration** the final electron acceptor is oxygen, and in **anaerobic respiration** it is usually an inorganic molecule other than molecular oxygen.

The Krebs Cycle. Note in Figure 5.1 that glycolysis forms pyruvic acid from carbohydrates such as glucose. Pyruvic acid loses a molecule of CO_2 (**decarboxylation**) to form an acetyl group. This then forms a complex, acetyl CoA, that can enter the **Krebs cycle.** The Krebs cycle releases the energy in acetyl CoA in a series of steps. The first step, which requires energy, yields citric acid (other names for the cycle are the *tricarboxylic acid cycle* and the *citric acid cycle*). Subsequent energy-yielding steps result in loss of CO_2 or loss of hydrogen atoms. Four molecules of CO_2 are released into the atmosphere for every two molecules of acetyl CoA that enter the cycle. Most of the energy is contained in six molecules of NADH and two molecules of $FADH_2$.

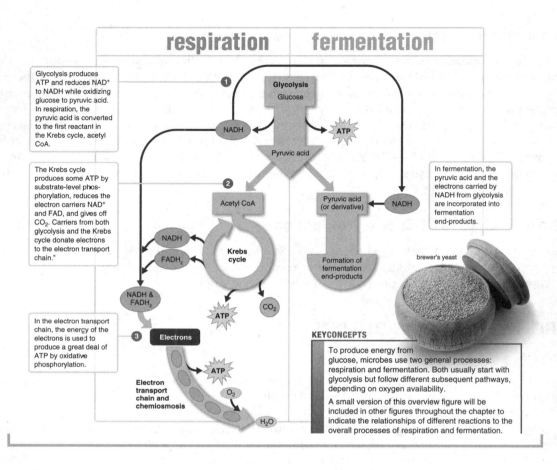

Figure 5.1 An overview of respiration and fermentation.

The Electron Transport Chain (System).   The energy of NADH and FADH$_2$ is generated by the **electron transport chain** (Figure 5.14 in the text). Note that the electrons are accepted by oxygen to form water (for each H$^+$ there is an equivalent electron, 2 H = H$^+$ + 2 e$^-$). Electrons are released from NADH and FADH$_2$, and there is a stepwise release of energy as they pass down the chain. The energy is used to drive the chemiosmotic generation of ATP (described below). In eukaryotic cells, the chain is in mitochondria; in prokaryotic cells, it is in the plasma membrane.

There are three classes of carrier molecules in the chain: **flavoproteins** (a flavin coenzyme is oxidized and reduced), **cytochromes** (iron is oxidized and reduced), and **ubiquinones (coenzyme Q),** which are nonprotein carriers. These carriers accept and release protons (H$^+$) and/or electrons. At places the electrons are pumped from one side of the membrane to another, causing a buildup of protons on one side.

The Chemiosmotic Mechanism of ATP Generation.   The mechanism of ATP synthesis using the electron transport chain is **chemiosmosis**. As electrons from NADH pass down the chain, some protons are pumped across the membrane by carrier molecules called *proton pumps*. The protons accumulate on one side of the membrane, resulting in a *protein motive force* from the positive charges accumulated there. These protons diffuse back across the membrane through special channels, where *ATP synthase* synthesizes ATP.

A Summary of Aerobic Respiration.   Prokaryotes generate 38 molecules of ATP aerobically for each molecule of glucose; eukaryotes produce only 36.

Anaerobic Respiration.   Some bacteria can use oxygen substitutes such as nitrate ion (forming nitrite ion, nitrous oxide, or nitrogen gas), sulfate (forming hydrogen sulfide), or carbonate (forming methane).

Fermentation

After glucose is broken down to pyruvic acid, the pyruvic acid can undergo **fermentation**. Fermentation does not require oxygen or an electron transport chain. It uses an organic molecule as the final electron acceptor. In fermentation, the pyruvic acid accepts electrons (hydrogen) and is turned into various end-products, such as **lactic acid (lactic acid fermentation)** or **ethanol (alcohol fermentation)** (Figure 5.19 in the text). **Homolactic** fermentation produces only lactic acid; **heterolactic** fermentation produces acids and alcohols, as well as lactic acid.

LIPID AND PROTEIN CATABOLISM

Some microorganisms produce extracellular enzymes, *lipases,* that break fats into fatty acids and glycerol. These components are then metabolized separately. Fatty acids are oxidized by **beta oxidation,** in which carbon fragments are removed two at a time to form acetyl coenzyme A, whose molecules then enter the Krebs cycle. Glycerol forms one of the intermediates of glycolysis and then is further oxidized. Proteases and peptidases are extracellular enzymes produced by some microbes to break down proteins, which are much too large for microbes to use unaltered, into component amino acids. Amino acids are first **deaminated** (—NH_2 removed) and **decarboxylated** (—COOH removed). They then enter the Krebs cycle in various ways.

BIOCHEMICAL TESTS AND BACTERIAL IDENTIFICATION

In the **fermentation test,** an inverted *Durham tube* traps gas as a bubble (Figure 5.23a in the text). This reaction is an indication of metabolism of certain proteins or carbohydrates. The fermentation test is often combined with a pH indicator in the medium to indicate acid production. The **oxidase test** identifies organisms that have the enzyme *cytochrome c oxidase,* which transfers electrons to oxygen. The production of **hydrogen sulfide** by the removal of sulfur from amino acids distinguishes *Salmonella* from *E. coli.*

PHOTOSYNTHESIS

We have just discussed organisms that obtain energy by oxidizing organic compounds. This energy source is a result of **photosynthesis,** by which electrons are taken from the hydrogen atoms of water and incorporated into sugar. Formation of ATP in this manner is called **photophosphorylation** and requires light. In **cyclic photophosphorylation,** the electrons return to chlorophyll. In **noncyclic photophosphorylation,** they are incorporated into NADPH.

Photosynthesis also includes **light-independent reactions (dark reactions)** such as the **Calvin-Benson cycle.** In this cycle, the CO_2 used to form sugars is "fixed."

METABOLIC DIVERSITY AMONG ORGANISMS

Phototrophs use light as their primary *energy source*; **chemotrophs** extract their energy from inorganic or organic chemical compounds. The principal *carbon source* of **autotrophs** is carbon dioxide; **heterotrophs** require an organic carbon source. These terms can be combined into terms that reflect the primary energy and carbon sources.

Photoautotrophs

Photoautotrophs include photosynthetic bacteria. **Green sulfur bacteria** are anaerobes that use sulfur compounds or hydrogen gas to reduce carbon dioxide and form organic compounds. Light is the energy source. Most commonly, they produce sulfur from hydrogen sulfide. **Purple sulfur bacteria** also use sulfur compounds or hydrogen gas to reduce carbon dioxide. Neither of these photosynthetic bacteria use water to reduce carbon dioxide, as do plants, and they do *not* produce oxygen gas as a product of photosynthesis (they are **anoxygenic**). Their photosynthetic pigment is *bacteriochlorophyll,* which absorbs longer wavelengths of light than *chlorophyll a.* **Cyanobacteria** use chlorophyll *a* in photosynthesis. They produce oxygen gas (they are **oxygenic**), just as higher plants do.

Photoheterotrophs

Green nonsulfur and **purple nonsulfur bacteria** are **photoheterotrophs.** They use light as an energy source but must use organic compounds instead of carbon dioxide as a carbon source. Otherwise, they are similar to the green and purple sulfur bacteria.

Chemoautotrophs

Chemoautotrophs use *inorganic* compounds such as hydrogen sulfide, elemental sulfur, ammonia, nitrites, hydrogen, and iron as sources of energy. Carbon dioxide is their principal carbon source. These compounds contain energy that may be extracted by oxidative phosphorylation reactions.

Chemoheterotrophs

With **chemoheterotrophs,** the carbon source and energy source are usually the same *organic* compound— glucose, for example. Most microorganisms are chemoheterotrophs. **Saprophytes** live on dead organic matter, and **parasites** derive nutrients from a living host.

METABOLIC PATHWAYS OF ENERGY USE

Energy generated by catabolism may be used in the synthesis of new compounds for the cell (**anabolism**).

Polysaccharides such as glycogen are synthesized from glucose. Glucose is first joined with ATP, forming *adenosine diphosphoglucose (ADPG)*. The energy of the ATP is, in essence, used to fasten together the sequence of glucose molecules to form a polysaccharide.

Lipids such as fats (triglycerides) are formed by combining glycerol with fatty acids. The glycerol is derived from a glycolysis intermediate, and the fatty acids are built up from two-carbon fragments of acetyl coenzyme A.

The **amino acids** required for the biosynthesis of proteins are synthesized by some bacteria; other bacteria require them to be preformed. **Amination** is to add an amine group to a Krebs cycle intermediate. If the amine group comes from a preexisting amino acid it is **transamination.** Intermediates of carbohydrate metabolism are used in the synthesis of amino acids.

DNA and RNA are made up of repeating units called **nucleotides**. These consist of a **purine** or **pyrimidine,** a five-carbon sugar, and a phosphate group. Sugars for nucleotides are derived from the pentose phosphate pathway or the Entner-Doudoroff pathway. Amino acids such as glycine and glutamine furnish the atoms from which are derived the backbone of purines and pyrimidines.

THE INTEGRATION OF METABOLISM

Anabolic and catabolic reactions are integrated through common intermediates. The Krebs cycle, for example, can operate in both anabolic and catabolic reactions. Such pathways are called **amphibolic pathways.**

SELF-TESTS

In the matching section, there is only one answer to each question; however, the lettered options (a, b, c, etc.) may be used more than once or not at all.

I. Matching

_____ 1. Energy-yielding series of reactions.

_____ 2. Means "whole enzyme."

_____ 3. A nonprotein component of an active enzyme.

_____ 4. A measure of the rate of activity of an enzyme.

_____ 5. A protein portion of an enzyme, inactive without a cofactor.

_____ 6. A group of enzymes that function as electron carriers in respiration and photosynthesis.

_____ 7. A mechanism by which fatty acids are degraded.

_____ 8. Fermentation test.

a. Catabolism

b. Anabolism

c. Turnover number

d. Apoenzyme

e. Coenzyme

f. Holoenzyme

g. Beta oxidation

h. Cytochromes

i. Durham tube

II. Matching

_____ 1. Both the carbon source and energy source are usually the same organic compound.

_____ 2. Photosynthetic, but uses organic material rather than carbon dioxide as a carbon source.

_____ 3. The photosynthetic purple nonsulfur bacteria would be classified in this nutritional group.

_____ 4. Photosynthetic bacteria that use carbon dioxide as a carbon source.

_____ 5. Changes the shape of the active site of an enzyme.

_____ 6. Very similar in shape or chemistry to the normal enzyme substrate.

a. Competitive inhibitor

b. Noncompetitive inhibitor

c. Photoautotroph

d. Chemcautotroph

e. Photoheterotroph

f. Chemoheterotroph

III. Matching

_____ 1. Hexose monophosphate shunt.

_____ 2. The final electron acceptor is oxygen.

_____ 3. Produces important intermediates that act as precursors in the synthesis of nucleic acids and so on.

_____ 4. Bacteria use oxygen substitutes such as nitrates.

_____ 5. Pyruvic acid accepts electrons and is turned into various end-products, such as lactic acid or ethanol.

_____ 6. Glucose to pyruvic acid.

a. Fermentation

b. Glycolysis

c. Pentose phosphate pathway

d. Substrate-level phosphorylation

e. Anaerobic respiration

f. Aerobic respiration

IV. Matching

_____ 1. Electrons are removed from an organic compound and are transferred by an electron transport chain to oxygen.

_____ 2. An electron is liberated from chlorophyll and passes down an electron transport chain.

a. Oxidative phosphorylation

b. Substrate-level phosphorylation

c. Photophosphorylation

V. Matching

_____ 1. A dehydrogenase coenzyme derived from nicotinic acid (niacin).

_____ 2. A dehydrogenase coenzyme derived from riboflavin.

_____ 3. In chemiosmosis, protons can diffuse across a membrane only through special channels that contain this enzyme.

_____ 4. Pyruvic acid loses carbon dioxide to form an acetyl group.

a. NAD^+

b. Decarboxylation

c. Coenzyme A

d. FMN

e. ATP synthase

f. Dehydrogenation

VI. Matching

_____ 1. Glycolysis.

_____ 2. A photosynthetic organism that does not produce oxygen.

_____ 3. Removal of electrons.

_____ 4. Uses an inorganic source of energy such as ammonia or elemental sulfur.

a. Embden-Meyerhof

b. Chemoautotrophic

c. Oxidation

d. Reduction

e. Anoxygenic

Fill in the Blanks

1. A chemoheterotroph that lives on dead organic matter is called a(n) _____.

2. When an enzyme's active site is occupied at all times by substrate or product molecules, it is called _____.

3. Cyanide is an example of a general type of inhibitor called _____.

4. Sulfa drugs are an example of a type of inhibitor called _____.

5. In _____ phosphorylation, no oxygen or other inorganic final electron acceptor is required.

6. Cyanobacteria produce _____ gas, just as do higher plants.

7. The amount of ATP yield from aerobic respiration by a prokaryote is _____.

8. The amount of ATP yield from glycolysis is _____.

9. The removal of NH_2 from an amino acid is called _____.

10. The removal of —COOH from an amino acid is called _____.

11. The substance acted upon by an enzyme is called the _____.

12. Coenzyme A is a derivative of the B vitamin _____ acid.

13. A sequence of enzymatically catalyzed chemical reactions in a cell is called a _____ pathway.

14. Glucose is usually broken down to pyruvic acid by _____.

15. In aerobic respiration, pyruvic acid is converted to acetyl _____; this product can then enter the Krebs cycle.

16. DNA and RNA are made up of repeating units called _____.

Label the Art

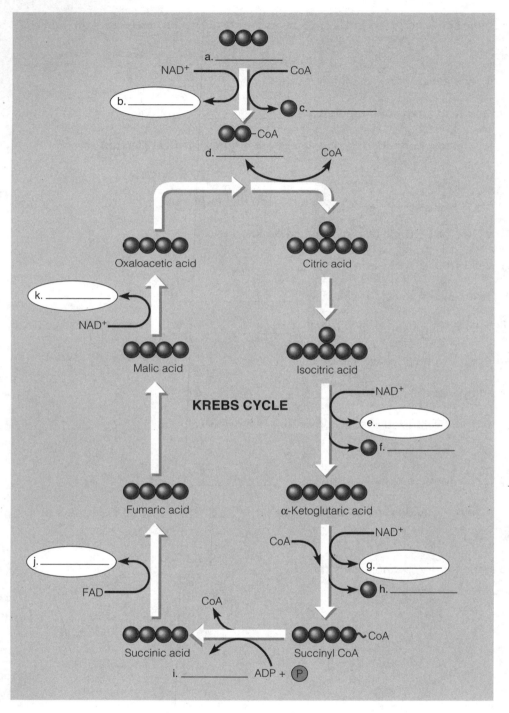

Critical Thinking

1. Why are catabolic and anabolic reactions referred to as coupled reactions?

2. Explain how competitive and noncompetitive enzyme inhibitors work.

3. How does the ultimate fate of electrons liberated differ in cyclic and noncyclic photophosphorylation?

4. What are the key features of the pentose phosphate pathway?

ANSWERS

Matching

I.   1. a   2. f   3. e   4. c   5. d   6. h   7. g   8. i
II. 1. f   2. e   3. e   4. c   5. b   6. a
III. 1. c   2. f   3. c   4. e   5. a   6. b
IV. 1. a   2. c
V.   1. a   2. d   3. e   4. b
VI. 1. a   2. e   3. c   4. b

Fill in the Blanks

1. saprophyte   2. saturated   3. noncompetitive   4. competitive   5. substrate-level   6. oxygen
7. 38   8. 2   9. deamination   10. decarboxylation   11. substrate   12. pantothenic   13. metabolic
14. glycolysis   15. CoA   16. nucleotides

Label the Art

a. Pyruvic acid   b. NADH   c. CO_2 d. Acetyl CoA   e. NADH

f. CO_2 g. NADH   h. CO_2 i. ATP   j. $FADH_2$ k. NADH

Critical Thinking

1. Catabolic and anabolic reactions are referred to as coupled reactions because catabolic reactions furnish the energy necessary to drive anabolic reactions.

2. Competitive enzyme inhibitors bind to and fill the active site of an enzyme. They compete with the substrate for the active site of the enzyme. The inhibitor does not undergo any reaction to form a product. This binding may or may not be reversible.

    Noncompetitive inhibitors interact with some other part of the enzyme, a process that is referred to as allosteric inhibition. The binding of enzyme and inhibitor results in a change in the active site of the enzyme. This prevents binding of the substrate so the reaction cannot occur.

3. In cyclic photophosphorylation, the electrons liberated from chlorophyll pass through the electron transport chain and eventually return to the chlorophyll.

    In noncyclic photophosphorylation, electrons released from chlorophyll pass through the electron transport chain to the electron acceptor, NADP$^+$. Electrons are replaced in chlorophyll from the splitting of water.

4. Provides a means for the breakdown of pentose sugars.

    Produces intermediates that are precursors in the synthesis of nucleic acids, some amino acids, glucose from CO_2 in photosynthesis.

    The process is an important producer of the coenzyme NADPH from NADP$^+$.

6 Microbial Growth

Microbial growth refers to the number of cells, not to the changes in the size of cells. **Colonies** are accumulations of cells large enough to be visible without a microscope.

THE REQUIREMENTS FOR GROWTH

Physical Requirements

Temperature. The **minimum growth temperature** is the lowest temperature at which the species will grow, and the **maximum** is the highest. The **optimum growth temperature** is that at which it grows best.

Psychrophiles are organisms capable of growth at 0°C. Some microbiologists define psychrophiles as having an optimum growth temperature of about 15°C. These usually are found in oceans or arctic environments. The term **psychrotrophs** has been proposed for organisms that grow well at refrigerator temperatures but that have an optimum growth temperature of 20–30°C. It is the term we will use for organisms causing refrigerated-food spoilage. **Mesophiles** are the most common microbes; their optimum temperatures are 25–40°C. **Thermophiles** are capable of growth at high temperatures. Many have an optimum of 50–60°C. Many are not capable of growth below about 45°C. **Hyperthermophiles (extreme thermophiles)** are found among the archaea and have an optimum growth temperature of 80°C or higher.

pH. Most bacteria grow best in a narrow pH range near neutrality, between pH 6.5 and 7.5. Very few grow below pH 4.0. However, many bacteria, such as the **acidophiles** responsible for acid fermentations, are remarkably tolerant of acidity. A **buffer** is sometimes added to media to neutralize acids. Examples of buffers are phosphate salts, peptones, and amino acids.

Osmotic Pressure. If a microbial cell is in a solution in which the concentration of solutes is higher than that found in the cell, water within the cell passes through the cytoplasmic membrane in the direction of the high solute concentration. During the loss of water, the cytoplasmic membrane pulls away from the cell wall, a process called **plasmolysis. Extreme halophiles** (sometimes called **obligate halophiles**) are organisms that have so adapted to high salt concentrations (as high as 30%) that they require them for growth. **Facultative halophiles** do not require high salt concentrations, but they are able to grow at salt concentrations as high as 15%, which tend to inhibit the growth of many other bacteria. Osmotic effects are roughly related to the numbers of molecules in a given volume of solution. For a given weight/volume, therefore, sodium chloride is more effective than sucrose. Under unusually low (*hypotonic*) osmotic pressure, such as distilled water, water tends to enter the cell.

Chemical Requirements

Carbon. Besides water (which is needed because nutrients must be in solution in order for cells to use them), carbon is a primary requirement for cellular growth. It is the structural backbone of living matter. Half the dry weight of a typical bacterial cell is carbon.

Nitrogen, Sulfur, and Phosphorus. Some organisms use proteinaceous material as a nitrogen source; others use ammonium ions (NH_4^+) or nitrate ions (NO_3^-). A few bacteria and *cyanobacteria* are able to use gaseous nitrogen (N_2) directly from the atmosphere. This process is called **nitrogen fixation.** The *Rhizobium* and *Bradyrhizobium* bacteria, in symbiosis with leguminous plants, also fix nitrogen. Sulfur sources are sulfate ion (SO_4^{2-}), hydrogen sulfide (H_2S), and the sulfur-containing amino acids. An important source of phosphorus is the phosphate ion (PO_4^{3-}). Nitrogen and sulfur are required to synthesize proteins. DNA, RNA, and ATP require nitrogen and phosphorus.

Trace Elements.   Mineral elements such as iron, copper, molybdenum, and zinc are referred to as **trace elements.** Although sometimes added to laboratory media, they usually are assumed to be naturally present in water and other media components.

Oxygen.   Microbes that use molecular oxygen are **aerobes;** if oxygen is an absolute requirement, they are **obligate aerobes. Facultative anaerobes** use oxygen when it is present but continue growth by fermentation or anaerobic respiration when it is not available. Facultative anaerobes grow more efficiently aerobically than they do anaerobically. **Obligate anaerobes** are bacteria totally unable to use oxygen for growth and usually find it toxic. Hydrogen atoms in the electron transport chain may be passed to oxygen, forming toxic hydrogen peroxide (H_2O_2). Aerobic organisms usually produce **catalase,** an enzyme that breaks down hydrogen peroxide to water and oxygen; anaerobes usually lack catalase. **Aerotolerant anaerobes** cannot use oxygen for growth but tolerate it fairly well. They will grow on the surface of a solid medium without the special techniques required for cultivation of less oxygen-tolerant anaerobes. Common examples of aerotolerant anaerobes are the bacteria that ferment carbohydrates to lactic acid, a process that occurs in making many fermented foods. A few bacteria are **microaerophilic,** meaning they grow only in oxygen concentrations lower than that found in air. They are aerobic, however, in the sense that they require oxygen. They are probably unusually sensitive to superoxide free radicals and peroxides, which they produce under oxygen-rich conditions.
Oxygen has a number of toxic forms.

1.   **Singlet oxygen** is normal molecular oxygen (O_2) that has been boosted into a higher energy state and is extremely reactive.

2.   **Superoxide radicals** (O_2^-) or **superoxide anions** are formed in small amounts by aerobic organisms; they are so toxic that the bacteria must neutralize them with **superoxide dismutase (SOD).** This enzyme converts superoxide radicals into oxygen and toxic hydrogen peroxide (which contains the **peroxide anion**), which in turn is converted into oxygen and water by the enzyme **catalase:** $2\ H_2O_2 \rightarrow 2\ H_2O + O_2$. Another enzyme that breaks down hydrogen peroxide is **peroxidase.** It does not produce oxygen: $H_2O_2 + 2\ H^+ \rightarrow 2\ H_2O$. Anaerobic bacteria often cannot neutralize the superoxide radicals they produce and do not tolerate atmospheric oxygen.

3.   The **hydroxyl radical** ($OH\cdot$) is formed in cytoplasm by ionizing radiation and as a by-product of aerobic respiration. It is probably the most reactive form.

Organic Growth Factors.   **Organic growth factors** are organic compounds such as vitamins, amino acids, and pyrimidines that are needed for life, but that a given organism is unable to synthesize.

Biofilms

In nature, microorganisms seldom live in the isolated single-species colonies we see on laboratory plates. More typically, they live in communities called **biofilms.** Biofilm microbes reside in a matrix primarily of polysaccharides informally called *slime,* which can be considered a *hydrogel.* Cell-to-cell communication, or *quorum sensing,* allows bacteria to coordinate in a fashion not unlike multicellular organisms. Therefore, biofilms can be thought of as biological systems. Some biofilms may be in the form of the floc that forms in some sewage system processes, or even as filamentous streamers. The advantage of biofilms is that they enable bacteria to share nutrients and shelter them from desiccation and other harmful factors in the environment.
Biofilms form as a free-swimming (*planktonic*) bacterium attaches to a surface. Biofilms usually form pillar-like structures with channels between them that facilitate exchange of nutrients and outgoing wastes. Biofilms play an important role in the digestive system of ruminants. Human infections often involve biofilms; this is especially true of nosocomial infections from indwelling medical devices, such as catheters.

CULTURE MEDIA

Any nutrient material prepared for the growth of bacteria in a laboratory is called a **culture medium.** Microbes growing in a container of culture medium are referred to as a **culture.** When microbes are added to initiate growth, they are an **inoculum.** To ensure that the culture will contain only the microorganisms originally added to the medium (and their offspring), the medium must initially be sterile. When a solid medium is required, a solidifying agent such as **agar** is added. Agar is a polysaccharide derived from a marine seaweed. Few microbes can degrade agar, so it remains a solid. It melts at about the boiling point of water but remains liquid until the temperature drops to about 40°C.

In a **chemically defined medium,** the exact chemical composition is known. Most heterotrophic bacteria and fungi are routinely grown on **complex media,** in which the exact chemical composition varies slightly from batch to batch. Complex media are made up of nutrients such as extracts from yeasts, beef, or plants, or digest of proteins from these and other sources. In many of these media, the energy, carbon, nitrogen, and sulfur requirements of the microorganisms are largely met by partially digested protein products called *peptones*. Vitamins and other organic growth factors are provided by meat extracts or yeast extracts. Such extracts supplement the organic nitrogen and carbon compounds but mainly are sources of soluble vitamins and minerals. This type of medium in liquid form is **nutrient broth;** when agar is added, it is **nutrient agar.**

Anaerobic Growth Media and Methods

Obligately anaerobic bacteria often require **reducing media** for isolation. Because oxygen may be lethal, these media contain ingredients, such as sodium thioglycolate, that chemically combine with dissolved oxygen to deplete the oxygen content of the culture medium. Obligate anaerobes may be grown on the surface of solid media in anaerobic atmospheres produced in special jars in which an oxygen-free atmosphere is generated by a chemical reaction. It is also possible to handle anaerobic organisms in anaerobic glove boxes filled with inert gases and fitted with airtight rubber glove arms and air locks. A new technique in which an enzyme, oxyrase, is added to the growth medium transforms the Petri plate (OxyPlate) into a self-contained anaerobic chamber.

Special Culture Techniques

Many bacteria have never been grown successfully on artificial laboratory media; the leprosy and syphilis organisms are examples. Obligate intracellular parasites such as rickettsias and chlamydias also do not ordinarily grow on artificial media. They, like viruses, require a living host cell. *Carbon dioxide incubators, candle jars,* and plastic bags with self-contained chemical gas generators are used to grow bacteria with special CO_2 concentration requirements. **Capnophiles** are microbes that grow better at high CO_2 concentrations.

Some microorganisms are so dangerous that they can be handled only under *biosafety level 4 (BSL-4)* systems. The personnel are protected by "space suits" connected to an air supply, and the rooms are under negative pressure so that aerosols containing pathogens cannot escape. All intake and exhaust air is filtered through high-efficiency particulate air (HEPA) filters. Less stringent requirements apply to less dangerous organisms and conditions. A basic teaching laboratory is BSL-1. Organisms posing a moderate risk can be handled in BSL-2 labs on open bench tops if personnel wear appropriate gloves, coats, and so on. BSL-3 labs are used for highly infectious airborne pathogens, such as the tuberculosis agent, and require biological safety cabinets similar to an anaerobic chamber. The laboratory is negatively pressurized and has air filters to prevent release of the pathogen.

Selective and Differential Media

Selective media are designed to suppress the growth of unwanted bacteria and encourage the growth of the desired microorganisms. Antibiotics, high concentrations of salt, or high acidity might be used. **Differential media** make it easier to distinguish colonies of the desired organism from other colonies growing on the same plate. The colonies have different colors or cause different changes in the surrounding medium. Sometimes selective and differential functions are combined in the one medium.

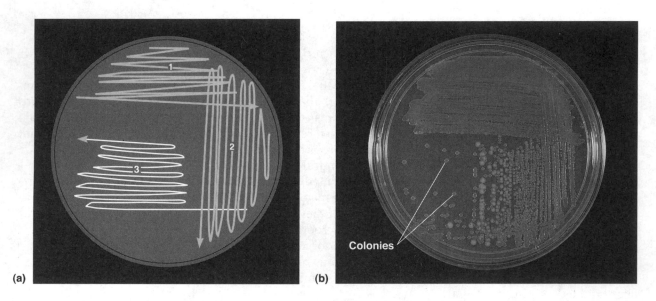

Figure 6.1 Streak plate method for isolating pure bacterial cultures. (a) Arrows indicate the direction of streaking. Streak series 1 is made from the original bacterial mixture. The inoculating loop is sterilized following each streak series. In series 2 and 3, the loop picks up bacteria from the previous series, diluting the number of cells each time. There are numerous variants of such patterns. (b) In series 3 of this example, note that well-isolated colonies of two different types of bacteria have been obtained.

Enrichment Culture

Because bacteria may be present only in small numbers and may be missed, and because the bacterium to be isolated may be of an unusual physiological type, it is sometimes necessary to resort to **enrichment culture,** or **enrichment media.** The conditions or sole nutrient are designed to increase small numbers of a certain organism to a detectable level.

OBTAINING PURE CULTURES

There are several methods for isolating bacteria in **pure cultures,** which contain only one kind of organism.

Streak Plate Method.   Probably the most common method of obtaining pure cultures is the **streak plate** (Figure 6.1). A sterile inoculating needle is dipped into a mixed culture and streaked in a pattern over the surface of the nutrient medium. The last cells rubbed from the needle are wide enough apart that they grow into isolated visible masses called **colonies.**

PRESERVING BACTERIAL CULTURES

In **deep-freezing,** a pure culture of microbes is placed in a suspending liquid and quick-frozen at −50° to −95°C. In **lyophilization (freeze-drying),** a suspension of microbes is quickly frozen at temperatures of −54° to − 72°C and the water removed by a high vacuum. The resulting powder can be stored for many years and the surviving microorganisms cultured by hydrating them with a suitable liquid nutrient medium.

THE GROWTH OF BACTERIA CULTURES

Bacterial Division

Bacteria normally reproduce by **binary fission.** Genetic material becomes evenly distributed; then a transverse wall is formed across the center of the cell, and it separates into two cells. A few bacterial species reproduce by **budding;** that is, an initial outgrowth enlarges to cell size and then separates. Some filamentous species produce **chains of condispores** or simply **fragment** into viable pieces.

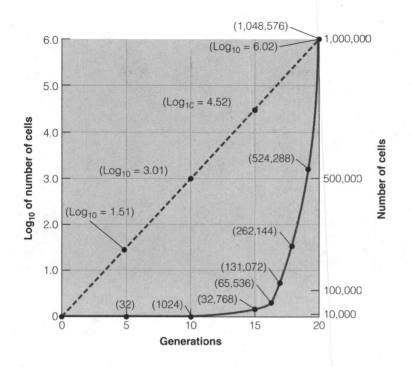

Figure 6.2 **Growth curve for an exponentially increasing population, plotted logarithmically (dashed line) and arithmetically (solid line).**

Generation Time

The time required for a cell to divide or the population to double is called **generation (doubling) time.** Bacterial populations are usually graphed **logarithmically** rather than **arithmetically** to permit the handling of the immense differences in numbers (Figure 6.2).

Phases of Growth

When bacterial population changes are graphed as a **bacterial growth curve** (Figure 6.15 in the text), certain phases become apparent. The **lag phase** shows little or no cell division. However, metabolic activity is intense. In the **log phase,** the cells are reproducing most actively, and their generation time reaches a minimum and remains constant; a logarithmic plot produces an ascending straight line. They are then most active metabolically and most sensitive to adverse conditions. Microbial deaths eventually balance numbers of new cells, and a **stationary phase** is reached. When the number of deaths exceeds numbers of new cells formed, the **death phase,** or **logarithmic decline,** is reached.

Direct Measurement of Microbial Growth

Plate Counts.   Dilutions of a bacterial suspension are distributed into a suitable solid nutrient medium by **serial dilution,** and the colonies appearing on the plates are counted (Figure 6.3). Colonies do not always arise from single cells but from chains or clumps, so counts are often reported as *colony-forming units.* The FDA convention is to count plates with 25 to 250 colonies, but others may specify plates with 30 to 300 colonies.

Filtration.   Bacteria may be sieved out of a liquid suspension onto a thin membrane filter with pores too small for bacteria to pass. This filter can be transferred to a pad soaked in nutrient medium where colonies arise on the surface of the filter.

Most Probable Number (MPN) Method.   In the most probable number method, a sample is diluted out in a series of tubes of liquid medium. The greater the number of bacteria, the more dilutions it takes to dilute them out entirely and leave a tube without growth. Results of such dilutions can be compared to statistical tables, and a cell count can be estimated.

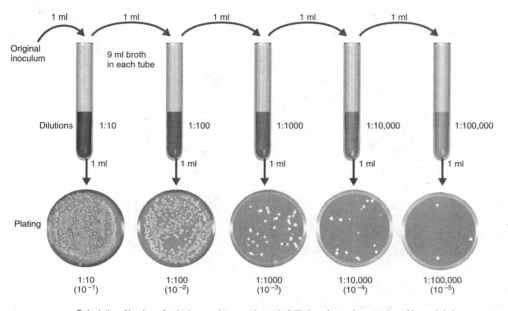

Calculation: Number of colonies on plate × reciprocal of dilution of sample = number of bacteria/ml
(a) (For example, if 54 colonies are on a plate of 1:1000 dilution, then the count is 54 × 1000 = 54,000 bacteria/ml in sample.)

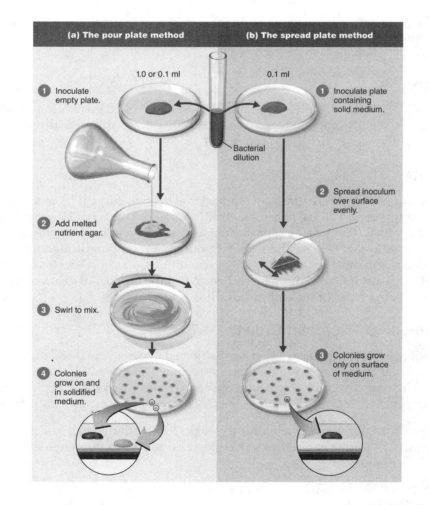

(b)

Figure 6.3 Plate counts and serial dilutions. (a) In serial dilutions, the original inoculum is diluted in a series of dilution tubes. In our example, each succeeding dilution tube will have only one-tenth the number of microbial cells as the preceding tube. Then samples of the dilution are used to inoculate Petri plates, on which colonies grow and can be counted. This count is then used to estimate the number of bacteria in the original sample. (b) Methods for preparation of plates for plate counts.

Direct Microscopic Count.   In a direct microscopic count, a measured volume of a bacterial suspension is placed into a defined area on a special microscope slide, of which there are several designs. A microscope is used to count the cells in microscope fields. The average number per field can be multiplied by a factor that will estimate the total numbers.

Estimating Bacterial Numbers by Indirect Methods

Turbidity.   To estimate turbidity, a beam of light is transmitted through a bacterial suspension to a photoelectric cell. The more bacteria, the less light passes. This is recorded as **absorbance** (sometimes called *optical density*, or *OD*) on the spectrophotometer or colorimeter.

Metabolic Activity.   Microbial numbers can be estimated by the time required to produce acid or other products.

Dry Weight.   Measuring weight is often the most satisfactory method for filamentous organisms such as actinomycete bacteria or fungi.

SELF-TESTS

In the matching section, there is only one answer to each question; however, the lettered options (a, b, c, etc.) may be used more than once or not at all.

I. Matching

_____ 1. Adapted to high salt concentrations, which are required for growth.

_____ 2. The general term used for organisms capable of growth at 0°C.

_____ 3. Capable of growth at high temperatures; optimum 50–60°C.

_____ 4. Used in media to neutralize acids.

_____ 5. A phenomenon that occurs when bacteria are placed in high salt concentration.

_____ 6. Term used in text for organisms that grow well at refrigerator temperatures; optimum growth is at temperatures of 20–30°C.

_____ 7. Microbes that grow better at high CO_2 concentrations.

_____ 8. Members of the archaea with an optimum growth temperature of 80°C or higher.

_____ 9. The matrix that makes up a biofilm.

a. Buffer

b. Mesophile

c. Thermophile

d. Psychrophile

e. Psychrotroph

f. Plasmolysis

g. Extreme halophile

h. Facultative halophile

i. Hyperthermophile

j. Capnophile

k. Hydrogel

II. Matching

____ 1. An enzyme acting upon hydrogen peroxide.

____ 2. *Rhizobium* bacteria do this in symbiosis with leguminous plants.

____ 3. Requires atmospheric oxygen to grow.

____ 4. Requires atmospheric oxygen, but in lower than normal concentrations.

____ 5. Does not use oxygen, but grows readily in its presence.

____ 6. Does not use oxygen and usually finds it toxic.

____ 7. Important source of energy, carbon, nitrogen, and sulfur requirements in complex media.

a. Nitrogen fixation

b. Obligate aerobe

c. Obligate anaerobe

d. Aerotolerant anaerobe

e. Catalase

f. Microaerophile

g. Peptones

h. Agar

III. Matching

____ 1. Breaks down hydrogen peroxide without generation of oxygen.

____ 2. Formed in cytoplasm by ionizing radiation.

____ 3. An enzyme that converts hydrogen peroxide into oxygen and water.

____ 4. The toxic form of oxygen neutralized by superoxide dismutase.

____ 5. A component added to some culture media that makes the Petri plate into a self-contained anaerobic chamber.

____ 6. Synonym for *superoxide anions*.

a. Hydroxyl radical

b. Peroxidase

c. Superoxide dismutase

d. Superoxide radicals

e. Singlet oxygen

f. Catalase

g. Oxyrase

IV. Matching

____ 1. Isolation method for getting pure cultures; uses an inoculating loop to trace a pattern of inoculum on a solid medium.

____ 2. Colonies grow on agar surface for identification.

____ 3. Used to increase the numbers of a small minority of microorganisms in a mixed culture to arrive at a detectable level of microorganisms.

____ 4. Preservation method that uses quick-freezing and a high vacuum.

____ 5. Accumulations of microbes large enough to see without a microscope.

____ 6. Microbes added to initiate growth.

a. Pour plate

b. Streak plate

c. Spread plate

d. Differential medium

e. Reducing medium

f. Enrichment culture

g. Lyophilization

h. Deep-freezing

i. Inoculum

j. Colonies

V. Matching

____ 1. Usual laboratory designation for safe handling of tuberculosis bacteria.

____ 2. Laboratory designation for the most dangerous microorganisms; personnel wear "space suits"

____ 3. A routine microbiology teaching laboratory would be designated as this.

a. BSL-1

b. BSL-2

c. BSL-3

d. BSL-4

VI. Matching

____ 1. New cell numbers balanced by death of cells.

____ 2. No cell division, but intense metabolic activity.

____ 3. A logarithmic plot of the population produces an ascending straight line.

a. Log phase

b. Lag phase

c. Death phase

d. Stationary phase

VII. Matching

____ 1. Used to grow obligate anaerobes.

____ 2. Designed to suppress the growth of unwanted bacteria and to encourage growth of desired microbes.

____ 3. Generally contain ingredients such as sodium thioglycolate that chemically combine with dissolved oxygen.

____ 4. Nutrients are digests or extracts; exact chemical composition varies slightly from batch to batch.

a. Selective media

b. Differential media

c. Complex media

d. Reducing media

e. Chemically defined media

Fill in the Blanks

1. Agar is a(n) _____ derived from a marine alga.

2. A few bacteria and the photosynthesizing _____ are able to use gaseous nitrogen directly from the atmosphere.

3. _____ are the most common microbes; their optimum temperatures are 25–40°C.

4. Osmotic effects are roughly related to the _____ of molecules in a given volume of solution.

5. A complex medium in liquid form is called nutrient _____ .

6. For preservation by _____ , a pure culture of microbes is placed in a suspending liquid and quick-frozen at −50° to − 95°C.

7. Bacteria usually reproduce by _____ fission.

8. Turbidity is recorded in a spectrophotometer as _____ .

9. The growth of filamentous organisms such as fungi is often best recorded by means of _____ .

10. _____ anaerobes grow more efficiently aerobically than they do anaerobically.

11. _____ halophiles do not require high salt concentrations, but they are able to grow at salt concentrations that may inhibit the growth of many other bacteria.

12. Examples of buffers are _____ salts; peptones and _____ found in complex media are also buffers.

13. Any nutrient material prepared for the growth of bacteria in a laboratory is called a _____ .

14. Agar melts at about the boiling point of water but remains liquid until the temperature drops to about _____.

15. Dilutions of a bacterial mixture are poured into a Petri dish and mixed with melted agar. This plate-counting method is called the _____.

16. Partially digested protein products used in complex media are called _____ .

17. To grow obligate intracellular parasites such as rickettsias and chlamydias, it is usually necessary to provide _____ .

18. The general term for tests that estimate microbial growth by the time required for them to deplete oxygen in the medium is _____ tests.

19. The _____ growth temperature is that at which the organism grows best.

20. When a single colony arises from a clump of bacteria, it is recorded as a(n) _____ .

Critical Thinking

1. What conditions that are characteristics of the food tend to retard spoilage in each of the following foods?

    a. Grape jelly

    b. Pickles

    c. Salted fish

    d. Cheddar cheese

2. What kinds of microorganisms (molds, lactic acid bacteria, endospore-forming bacteria, aerobic bacteria, etc.) would be most likely to cause spoilage of each of the foods listed above? (*Hint:* See Chapter 28.)

3. Plate counts are the most common method used to enumerate microbial populations. Discuss the advantages or disadvantages of the use of plate counts for the following:

    a. Milk intended for commercial sale

    b. Molds

4. Draw a bacterial growth curve indicating the four phases of growth. At which phase of growth would exposure to antibiotics cause the most adverse effects on the bacterial population? Why?

5. Complete the following table, indicating where the microorganisms will grow in a tube of a solid medium on the basis of their relationship to oxygen.

Relation to oxygen	Where in the tube does growth occur?	Why?
Obligate aerobe		
Facultative anaerobe		
Obligate anaerobe		
Aerotolerant anaerobe		
Microaerophile		

ANSWERS

Matching

I.	1. g	2. d	3. c	4. a	5. f	6. e	7. j	8. i	9. k
II.	1. e	2. a	3. b	4. f	5. d	6. c	7. g		
III.	1. b	2. a.	3. f	4. d	5. g	6. d			
IV.	1. b	2. c	3. f	4. g	5. j	6. i			
V.	1. a	2. d	3. c						
VI.	1. d	2. b	3. a						
VII.	1. d	2. a	3. d	4. c					

Fill in the Blanks

1. polysaccharide 2. cyanobacteria 3. mesophiles 4. number 5. broth 6. deep-freezing 7. binary 8. absorbance (also optical density) 9. dry weight measurement 10. Facultative 11. Facultative 12. phosphate; amino acids 13. culture medium 14. 40°C 15. pour plate method 16. peptones 17. living host cells 18. reduction 19. optimum 20. colony-forming units

Critical Thinking

1. a. Fruit jelly is acidic and also has a relatively high osmotic pressure from added sugars.

   b. Pickles are acidic.

   c. Salted fish have high osmotic pressures.

   d. Hard cheeses are acidic and have relatively low moisture.

2. a. Molds; they are relatively tolerant of acidity, high osmotic pressure, and low moisture. (Yeasts have similar characteristics but are much less common in the environment. Acidophilic bacteria will grow in acidic foods but are usually not considered spoilage organisms—in fact, they very likely were used to make the food acidic, for example, cheese and pickles.)

   b. Molds

   c. Molds

   d. Molds

3. a. Milk is highly perishable, and the delay required for incubation of plates would often be too lengthy for practical use.

   b. Molds are filamentous, and plate counts would often arise from mold spores or fragments of mold filaments, which would not indicate the mold growth very well.

4. See figure below.

   Lag phase—The period immediately following inoculation to fresh media in which little or no growth occurs. A time of intense metabolic activity as the cells gear up for reproduction.

   Log or exponential phase—The period of growth in which cellular reproduction is most active. Generation time is at a minimum.

   Stationary phase—The number of new cells being produced equals the number of cell deaths; the period of equilibrium.

   Death phase—The number of dead cells exceeds the number of living cells until only a small portion of the population exists or the population dies out completely.

   Exposure to antibiotics during log or exponential phase would cause the most adverse effects on the bacterial population. This is because antibiotics are most effective against growing cells.

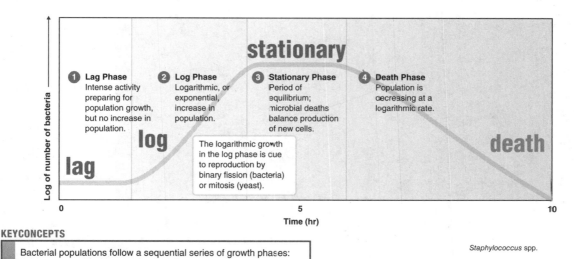

Staphylococcus spp.

KEYCONCEPTS

Bacterial populations follow a sequential series of growth phases: the lag, log, stationary, and death phases.

Knowledge of the bacterial growth curve is critical to understanding population dynamics and population control in the course of infectious diseases, in food preservation and spoilage, as well as in industrial microbiology processes, such as ethanol production.

5.

Relation to oxygen	Where in the tube does growth occur?	Why?
Obligate aerobe	Near the top of the medium	Dissolved oxygen diffuses only a short distance in the medium.
Facultative anaerobe	Best near surface but throughout tube	They can survive without oxygen but grow better in the presence of oxygen.
Obligate anaerobe	At the bottom	Oxygen is poisonous to these organisms.
Aerotolerant anaerobe	Evenly throughout tube	These organisms don't use oxygen.
Microaerophile	In a narrow band of the medium	They will grow at the depth having the optimum oxygen concentration.

7 The Control of Microbial Growth

THE TERMINOLOGY OF MICROBIAL CONTROL

Sterilization is the removal or destruction of all living microorganisms. **Commercial sterilization** subjects canned food to only enough heat to destroy the endospores of *Clostridium botulinum.* **Disinfection** is the destruction of vegetative pathogens on a surface, usually with chemicals. Spores and viruses are not necessarily destroyed. **Antisepsis** is the chemical disinfection of living tissue, such as skin or mucous membrane. **Asepsis** is the absence of pathogens on an object or area, as in **aseptic surgery. Degerming** (degermation) is the removal of transient microbes from skin by mechanical cleansing or by an antiseptic. **Sanitization** is the reduction of microbial populations on objects to safe public health levels. In general, the suffix *-cide* indicates the killer of a specified organism. A **biocide** or **germicide** kills microorganisms. **Fungicides** kill fungi, **virucides** kill viruses, and so on. The suffix *-stat* or *-stasis* used in this way indicates only that the substance inhibits—for example, **bacteriostasis.**

THE RATE OF MICROBIAL DEATH

Bacterial populations killed by heat or chemicals tend to die at constant rates—for example, 90% every 10 minutes. Plotted logarithmically, these figures form straight descending lines.

ACTIONS OF MICROBIAL CONTROL AGENTS

Alteration of Membrane Permeability

The plasma membrane controls the passage of nutrients and wastes into and out of the cell. Damage to the plasma membrane causes leakage of cellular contents and interferes with cell growth.

Damage to Proteins and Nucleic Acids

Chemicals may denature proteins by reacting, for example, with disulfide bonds (or disulfide bridges), which give proteins their three-dimensional active shape. Chemicals and radiation may prevent proper replication or functioning of DNA or RNA.

PHYSICAL METHODS OF MICROBIAL CONTROL

Heat

Thermal death point is the lowest temperature required to kill a liquid culture of a certain species of bacteria in 10 minutes. **Thermal death time** is the length of time required to kill all bacteria in a liquid culture at a given temperature. **Decimal reduction time** (*D value*) is the length of time, in minutes, required to kill 90% of the population of bacteria at a given temperature.

Moist Heat Sterilization and Pasteurization.   Boiling (100°C) kills vegetative forms of bacterial pathogens, many viruses, and fungi within 10 minutes. Endospores and some viruses survive boiling for longer times. **Steam under pressure** allows temperatures above boiling to be reached. **Autoclaves,** retorts, and pressure cookers are vessels in which high steam pressures can be contained. A typical operating

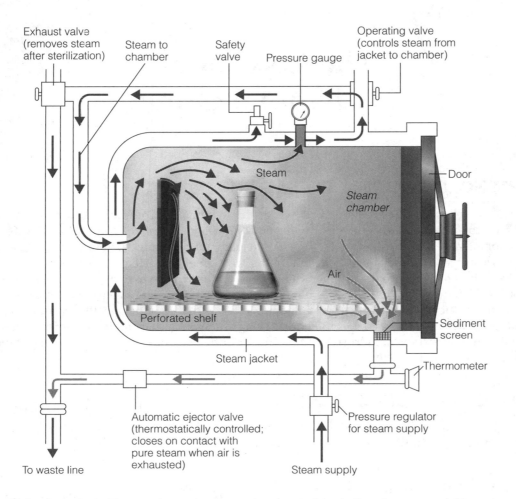

Figure 7.1 Autoclave. The entering steam forces the air out of the bottom (gray arrows). The automatic ejector valve remains open as long as an air-steam mixture is passing out of the waste line. When all the air has been ejected, the higher temperature of the pure steam closes the valve, and the pressure in the chamber increases.

condition for sterilization is 15 psi (pounds per square inch) at 121°C for 15 minutes. Moisture must touch all surfaces in order to bring about sterilization. Air must be completely exhausted from the container. An autoclave is shown in Figure 7.1.

Pasteurization is mild heating that is sufficient to kill particular spoilage or disease organisms without seriously damaging the taste of the product. Dairies use the *phosphatase test* to determine whether products have been properly pasteurized. **High-temperature, short-time pasteurization** uses temperatures of at least 72°C for about 15 seconds to pasteurize milk. **Equivalent treatments** are illustrated by the following example: the heat of 115°C acting on an organism for 70 minutes is equivalent to heat of 125°C acting on an organism for only 7 minutes; that is, applying a higher temperature for a shorter time may kill the same number of microbes as a lower temperature for a longer time. Milk can be sterilized and stored without refrigeration when given **ultra-high-treatment (UHT).**

Dry Heat Sterilization.   **Incineration,** as in direct **flaming,** is efficient for limited purposes. **Hot-air sterilization,** as in an oven, requires higher temperatures (such as 170°C) and longer times (such as 2 hours) to ensure sterilization. Moist heat is generally more efficient.

Filtration

Liquids sensitive to heat can be passed through a thin **membrane filter** that has carefully controlled pore sizes to retain microorganisms. Operating theaters and special clean rooms receive air passed through **high-efficiency particulate air (HEPA) filters.**

Low Temperatures

Refrigerator temperatures (0–7°C) slow the metabolic rate of microbes; however, psychrotrophic species still grow slowly. Some organisms grow at temperatures slightly below freezing, but microbes at the usual temperatures of freezer compartments are completely dormant.

High Pressure

High pressure applied to liquid suspensions such as fruit juices can kill vegetative bacterial cells while preserving flavors, colors, and nutrient values. Endospores are relatively resistant.

Desiccation

Microbes require water for growth, and adequately dried (**desiccated**) foods will not support their growth.

Osmotic Pressure

High salt or sugar concentrations cause water to leave the cell; this is an example of **osmosis** (see *plasmolysis* in Chapter 6). Generally, molds and yeasts resist osmotic pressures better than bacteria.

Radiation

Ionizing radiation such as *X rays, gamma rays,* and *high-energy electron beams* carry high energy and break DNA strands. Ionizing radiation forms reactive hydroxyl radicals. Such radiation is used to sterilize pharmaceuticals. **Nonionizing radiation** such as ultraviolet (UV) light has a longer wavelength and less energy. UV light causes bonds to form between adjacent thymines (*thymine dimers*) in DNA chains. Penetration is low. Sunlight has some biocidal activity, mainly due to formation of singlet oxygen in the cytoplasm.

CHEMICAL METHODS OF MICROBIAL CONTROL

Principles of Effective Disinfection

By reading the label, we can learn a great deal about a disinfectant's properties. Usually the label indicates what groups of organisms the disinfectant is effective against. Remember that the concentration of a disinfectant affects its action, so it should always be diluted exactly as specified by the manufacturer. To be effective, a disinfectant might need to be left on a surface for several hours.

Evaluating a Disinfectant

Bacteria are tested under standard conditions against concentrations of phenol and the test disinfectant.

Use-Dilution Tests.   In the **American Official Analytic Chemist's use-dilution test,** a series of tubes containing increasing concentrations of the test disinfectants is inoculated and incubated. The more the chemical can be diluted and still be effective, the higher its rating.

The Disk-Diffusion Method.   A disk of filter paper is soaked in a chemical agent, which is placed on an inoculated surface of an agar plate. A clear zone around the disk indicates inhibition.

Types of Disinfectants

Phenol and Phenolics.   Phenol (carbolic acid) is seldom used today. Derivatives of the phenol molecule, however, are widely used. **Phenolics** injure plasma membranes, inactivate enzymes, or denature proteins.

They are stable, persistent, and are not sensitive to organic matter. **O-phenylphenol,** a *cresol*, is the main ingredient in most formulations of Lysol.

Bisphenols.   Phenol derivatives called **bisphenols** contain two phenolic groups connected by a bridge. **Hexachlorophene** is the main ingredient in pHisoHex and is used in nurseries to control gram-positive skin bacteria such as staphylococci and streptococci. Excessive use can cause neurological damage. **Triclosan** is a widely used bisphenol found in many household products. It has a broad spectrum of activity, especially against gram-positive bacteria. It is also effective against gram-negative bacteria and fungi.

Biguanides.   **Chlorhexidine,** a member of the biguanide group, is not a phenol, but its structure and applications resemble those of hexachlorophene. It is frequently used for surgical skin preparation and surgical hand scrubs. *Alexidine* is similar but more rapid in action.

Halogens.   The **halogens,** especially iodine and chlorine, are effective antimicrobial agents. **Iodine** impairs protein synthesis. A **tincture** is a solution of iodine in water, and an **iodophore** is a combination of iodine and an organic molecule from which the iodine is slowly released. An example of *povidone-iodine* is Betadine. **Chlorine** is a widely used disinfectant, either as a gas or in a chemical combination. When added to water, it forms *hypochlorous acid*, which is the germicidal form. It is a strong oxidizing agent that inhibits enzymatic function. *Calcium hypochlorite* (chloride of lime) is used to disinfect utensils. *Sodium hypochlorite* (household bleach) is a widely used disinfectant. *Sodium dichloroisocyanurate* (ChlorFloc) is a water disinfectant issued by the U.S. military. *Chlorine dioxide* is a gas used for area disinfection, most notably to kill endospores of anthrax bacteria. It is also used in solution as a surface disinfectant. *Chloramines* are a combination of chlorine and ammonia. They are more stable than other forms of chlorine and are used as a sanitizer and for disinfection in municipal water systems.

Alcohols.   Both **ethanol** and **isopropanol** (rubbing alcohol) are widely used, normally at a concentration of about 70%. Concentrations of 60–95% are effective. They are bactericidal and fungicidal but are not effective against endospores or nonenveloped viruses. Alcohols enhance the effectiveness of other chemical agents. Purell, a widely used hand cleaner, contains 62–65% ethanol.

Heavy Metals and Their Compounds.   The fact that tiny amounts of heavy metals are effective antimicrobials can be illustrated by **oligodynamic action.** A silver coin on an inoculated nutrient medium will inhibit growth for some distance. A *1% silver nitrate* solution has been used to prevent gonorrheal eye infections in newborns. *Silver-sulfadiazine* is used in wound dressings. Silver combines with sulfhydryl groups on proteins, denaturing them. *Surfacine* is a water-insoluble silver iodide in a carrier that is a persistent disinfectant for surfaces. *Mercuric chloride* is highly bactericidal, but it is toxic and corrosive and is inactivated by organic matter. *Copper sulfate* is often used to destroy green algae in reservoirs or other waters. *Zinc chloride* is used in mouthwashes. *Zinc pyrithione* is an ingredient in antidandruff shampoo.

Surface-Active Agents.   **Surface-active agents,** or **surfactants,** decrease the surface tension of a liquid. Soaps and detergents are examples. They emulsify oils and are good degerming agents. *Acid-anionic* sanitizers are important for cleaning dairy equipment.

Quaternary Ammonium Compounds (Quats).   The **quaternary ammonium compounds** (Figure 7.2) are most effective against gram-positive bacteria, less so against gram-negative bacteria. These *cationic* detergents have good fungicidal, amoebicidal, and virucidal (enveloped virus) activity, but they are not sporicidal. They are colorless, odorless, tasteless, nontoxic, and stable, but they are inactivated by organic matter, soaps, detergents, and surfaces such as gauze. They may even support the growth of *Pseudomonas* bacteria. They act by disrupting the plasma membranes and by denaturing enzymes. Widely used examples of quats are *benzalkonium chloride* (Zephiran) and *cetylpyridinium chloride* (Cepacol).

Chemical Food Preservatives.   *Sorbic acid (potassium sorbate)* inhibits mold spoilage in foods such as cheese. *Benzoic acid (sodium benzoate)* is an antifungal used in soft drinks and other acidic foods. Methylparaben and propylparaben, which are derivatives of benzoic acid, work at a neutral pH.

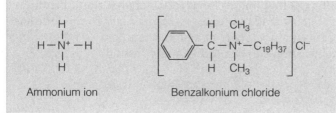

Figure 7.2 The ammonium ion and a quaternary ammonium compound, benzalkonium chloride (Zephiran). Note how other groups replace the hydrogens of the ammonium ion.

They inhibit molds in liquid cosmetics and shampoos. *Calcium propionate* prevents mold growth in bread. All of these organic acids inhibit enzymatic or metabolic activity; their activity is not related to their acidity. *Sodium nitrate* and *sodium nitrite* are added to meats to produce a red color and to inhibit outgrowth of botulism endospores. (The active principle is the nitrite ion; bacteria in the meat reduce the nitrate to nitrite.) **Nitrosamines,** formed by a reaction between certain amino acids and nitrite, are possibly carcinogenic (cancer causing).

Antibiotics. A few antibiotics, not medically useful, are used as food preservatives. These include *nisin*, a bacteriocin (a protein produced by a bacterium that inhibits another), which is added to cheese to inhibit endospore formers. Another is *Natamycin (pimaricin)*, an antifungal mostly used in cheese.

Aldehydes. Among the most effective antimicrobials are the aldehydes such as *formaldehyde*. In the form of an aqueous solution, this gas is called *formalin* and is used to preserve biological specimens. *Glutaraldehyde* is a less irritating form; in a 2% solution (Cidex), it is bactericidal, tuberculocidal, and virucidal in 10 minutes. It is sporicidal after about 3 to 10 hours of contact. Both glutaraldehyde and formaldehyde are used for embalming. A possible replacement for glutaraldehyde is *ortho-phthalaldehyde* (OPA).

Chemical Sterilization. *Ethylene oxide* gas, which requires a closed chamber similar to an autoclave, is the most familiar example of a chemical sterilizer. Its activity depends on *alkylation*, which leads to a cross-linking of nucleic acids and proteins. Heated *hydrogen peroxide* can also be used as a gaseous sterilant. *Chlorine dioxide*, usually manufactured at the site, is a gas used to fumigate enclosed building areas.

Plasmas. Plasma represents a state of matter in which a gas is excited to make a mixture of nuclei with assorted electrical charges and free electrons. *Plasma sterilization* is a method available for the difficult task of sterilizing surgical devices with small interior diameters. Such plasmas have many radicals that destroy even endospores at relatively low temperatures.

Supercritical Fluids. When carbon dioxide is compressed into a "supercritical" state, it has properties of both a liquid and a gas. *Supercritical carbon dioxide* is used for decontamination of foods and medical implants. It inactivates even endospores at temperatures of only about 45°C.

Peroxygens and Other Forms of Oxygen. *Hydrogen peroxide* effectively disinfects inanimate objects; at high concentrations it is sporicidal. *Peracetic acid (peroxyacetic acid, or PAA)* is one of the most effective liquid chemical sporicides available. The U.S. Food and Drug Administration (FDA) has approved PAA for washing fruits and vegetables, and it is used to disinfect medical equipment. Other oxidizing agents include *benzoyl peroxide* (used to treat acne) and *ozone*, a reactive form of oxygen used to supplement chlorine for water treatment. It is produced on site by electrical discharges.

MICROBIAL CHARACTERISTICS AND MICROBIAL CONTROL

Chemical antimicrobials are not uniformly effective against all microorganisms. *Gram-positive bacteria* are relatively more resistant than *gram-negative*. *Viruses* with an envelope (lipophilic) are more susceptible than those with only a protein coat. *Endospores*, and *cysts* and *oocysts* of protozoa, are affected by few liquid chemical agents. *Mycobacteria* have a waxy cell wall that makes them relatively resistant. **Prions,** infectious proteins that cause neurological diseases such as mad cow disease, are difficult to render non-infectious. Normal autoclaving is inadequate, and at least autoclaving at a higher temperature, 134°C in a solution of sodium hydroxide, is recommended. Incineration can be used in some cases.

SELF-TESTS

In the matching section, there is only one answer to each question; however, the lettered options (a, b, c, etc.) may be used more than once or not at all.

I. Matching

____ 1. A suffix meaning "to kill."

____ 2. Destroying or removing *all* forms of microbial life.

____ 3. The absence of pathogens on an object or area.

____ 4. The reduction of microbial populations to safe public health levels.

____ 5. The chemical disinfection of living tissue, such as skin or a mucous membrane.

____ 6. The removal of transient microbes from skin by mechanical cleansing or by an antiseptic.

____ 7. Heat sufficient only to kill endospores of the botulism bacterium.

a. Disinfection

b. Sterilization

c. Antisepsis

d. Asepsis

e. Sanitization

f. Degerming

g. *-cide*

h. *-stat*

i. Commercial sterilization

II. Matching

____ 1. The lowest temperature required to kill a liquid culture of a certain species of bacteria in 10 minutes.

____ 2. The time in minutes required to kill 90% of a bacterial population.

____ 3. Mild heating to destroy particular spoilage organisms or disease organisms in milk or similar products.

____ 4. A test for the effectiveness of a chemical disinfectant.

____ 5. The absence of water, resulting in a condition of dryness.

a. Thermal death time

b. Decimal reduction time

c. Thermal death point

d. Phenol coefficient

e. Pasteurization

f. Desiccation

g. Incineration

III. Matching

____ 1. Ethylene oxide.

____ 2. Sodium hypochlorite.

____ 3. Copper sulfate.

____ 4. Silver nitrate.

____ 5. Benzalkonium chloride.

____ 6. Acid-anionic detergents.

____ 7. Sorbic acid.

____ 8. Benzoyl peroxide.

____ 9. Hexachlorophene.

____ 10. Isopropanol.

a. Bisphenol

b. Halogen

c. Alcohol

d. Heavy metal

e. Quaternary ammonium compound

f. Surface-active agents

g. Organic acid

h. Aldehydes

i. Gaseous chemosterilizer

j. Oxidizing agent

IV. Matching

____ 1. An effective liquid sporicide.

____ 2. A bacteriocin classified as an antibiotic.

____ 3. Pimaricin.

____ 4. A biguanide.

____ 5. A bisphenol found in many household products.

____ 6. An antibiotic antifungal.

a. Peracetic acid

b. Chlorhexidine

c. Triclosan

d. Natamycin

e. Nisin

V. Matching

____ 1. Added to chlorine to form chloramines.

____ 2. An antibacterial effect of ultraviolet radiation on DNA.

____ 3. Formaldehyde in an aqueous solution.

____ 4. An example would be iodine in an aqueous-alcohol solution.

____ 5. For example, povidone-iodine solution.

a. Iodophore

b. Formalin

c. Thymine dimers

d. Ammonia

e. Tincture

VI. Matching

___ 1. Chlorine in tablet form issued as a water purifier by the U.S. military.

___ 2. Name of a test that determines if milk has been properly pasteurized.

___ 3. Used as an antiseptic in certain mouthwashes.

___ 4. Used in many water treatment plants as a disinfectant; produced by electrical discharges at the site.

___ 4. Antifungal organic compound used in food.

___ 5. Ingredient in antidandruff shampoo.

a. Sodium dichloroisocyanurate

b. Phosphatase

c. Zinc chloride

d. Sodium benzoate

e. Ozone

f. Zinc pyrithione

Fill in the Blanks

1. Ultraviolet light is an example of _____ radiation.

2. Sunlight owes its biocidal activity mainly to the formation of _____ oxygen.

3. A good example of ionizing radiation is _____ .

4. Ethanol is usually used in a concentration of about _____ .

5. A less irritating form of formaldehyde is _____ .

6. A compound that would only inhibit the growth of a fungus would be a fungi _____ (supply the suffix).

7. Steam _____ allows temperatures above boiling to be reached.

8. Steam under pressure is obtained in retorts, pressure cookers, and _____ .

9. Supercritical _____ is used for decontaminating foods and medical implants.

10. Generally speaking, the group of organisms that is more resistant to osmotic pressure than bacteria is _____ .

Critical Thinking

1. What physical method of control would be most effective in each of the following situations?

   a. To eliminate endospore-forming pathogens.

   b. To sterilize milk for storage at room temperature.

   c. To sterilize vaccines.

   d. To sterilize microbiological media.

2. What chemical agent would be most effective in each of the following situations?

    a. A puncture wound acquired while gardening.

    b. For presurgical scrubbing.

    c. To sterilize packaged bandages.

    d. To prevent the growth of molds in liquid cosmetics.

3. Compare and contrast sterilization and sanitation.

4. Discuss the advantages and disadvantages associated with each of the following physical methods of control.

    a. Osmotic pressure

    b. Desiccation

    c. Refrigeration

    d. Filtration

5. Discuss the advantages and disadvantages of UV light as a method to control microbial growth.

ANSWERS

Matching

    I.   1. g 2. b 3. d 4. e 5. c 6. f 7. i
   II.   1. c 2. b 3. e 4. d 5. f
  III.   1. i 2. b 3. d 4. d 5. e 6. f 7. g 8. j 9. a 10. c
  IV.   1. a 2. e 3. d 4. b 5. c 6. d
   V.   1. d 2. c 3. b 4. e 5. a
  VI.   1. a 2. b 3. c 4. e 5. d 6. f

Fill in the Blanks

1. nonionizing 2. singlet 3. X rays, gamma rays, high-energy electrons 4. 70% 5. glutaraldehyde 6. -stat 7. under pressure (as in an autoclave) 8. autoclaves 9. carbon dioxide 10. fungi, such as molds and yeasts

Critical Thinking

1. a. Autoclaving at 121°C, 15 psi for 15 minutes will kill all organisms and their endospores.

   b. The milk should be sterilized by ultra-high-temperature (UHT) treatment.

   c. Vaccines are heat-sensitive and must be filter-sterilized.

   d. Most media can be safely autoclaved. Heat-sensitive media can be filter-sterilized.

2. a. An oxidizing agent such as hydrogen peroxide would be a good choice. Oxidizing agents are especially effective against anaerobic bacteria.

   b. Chlorhexidine is useful for surgical scrubbing because it is bactericidal against both gram-positive and gram-negative organisms.

   c. Ethylene oxide would be appropriate because it is 100% effective and can penetrate the wrapping material covering the bandage.

   d. The addition of a compound such as methylparaben would inhibit mold growth.

3. *Sterilization* refers to the destruction or the removal of *all* microbial life, including endospores. There are many ways to achieve sterilization, including the use of heat, chemical agents, or filtration. *Sanitation* is the *reduction* of pathogens on inanimate objects (such as eating utensils) to "safe" levels. This may be achieved by mechanical cleaning or with chemical agents.

4. a. The use of high concentrations of salt or sugar creates a hypertonic environment that results in the osmotic loss of water from microbial cells. The advantage is that this is a simple way to preserve meat and fruit. Applications include jams and jellies. Disadvantages are that molds may grow on foods prepared this way and that it isn't a practical way to preserve many foods.

   b. Desiccation involves drying food (for example, meat and fruit). The lack of water retards the growth and reproduction of microbes. The advantage is that it is an easy way to preserve some foods. The disadvantages are that many microorganisms are able to survive desiccation for long periods of time and are revived upon the addition of moisture. Applications are beef jerky and sun-dried tomatoes.

   c. Refrigeration is a simple and relatively effective way to retard the spoilage of food. Although many bacteria can survive and even reproduce at refrigerator temperatures, the rate of chemical reactions is slowed.

   d. Filtration is the passage of gas or liquid through a screenlike material with pores small enough to retain microbes. There are many applications of filtration, such as sterilizing heat-sensitive materials. It is difficult to filter-sterilize viscous materials such as some media.

5. Nonionizing radiation (for example, UV light) damages the DNA of exposed cells and is used to control microbes in air and to sterilize vaccines, serums, and toxins. A serious disadvantage of nonionizing radiation is that because of its relatively low energy content, it penetrates poorly. Organisms protected by practically anything are not affected.

8 Microbial Genetics

STRUCTURE AND FUNCTION OF THE GENETIC MATERIAL

Genetics is the science of heredity; it includes the study of what genes are, how they carry information, how they are replicated and passed to subsequent generations of cells or passed between organisms, and how the expression of their information within an organism determines the particular characteristics of that organism. The genetic information in a cell is called the **genome**. A cell's genome includes its chromosomes and plasmids. **Chromosomes** are structures containing DNA that physically carry hereditary information; the chromosomes contain the genes. **Genes** are segments of DNA (except in some viruses, in which they are made of RNA) that code for functional products.

In Chapter 2 we saw that DNA is composed of repeating **nucleotides** containing the bases adenine (A), thymine (T), cytosine (C), or guanine (G); a deoxyribose sugar; and a phosphate group. Bases occur in specific *complementary* **base pairs,** the hydrogen bonds from which connect strands of DNA: adenine with thymine, and cytosine with guanine. The information in DNA can be transcribed into RNA (*transcription*) and this information can be, in turn, translated into protein (*translation*).

Genotype and Phenotype

The **genotype** is an organism's genetic makeup, the information that codes for all the characteristics and *potential* properties of the organism. The genotype is its gene collection—its DNA. The **phenotype** refers to an organism's *actual expressed* properties, such as its ability to perform a chemical reaction. The phenotype is the collection of enzymatic or structural proteins.

DNA and Chromosomes

DNA in chromosomes is in the form of one long double helix. In prokaryotes, DNA is not found within a nuclear membrane. The chromosome takes up only about 10% of the cell's volume because the DNA is *supercoiled*. There are noncoding regions called *short tandem repeats (STRs)*. They are repeating sequences of two- to five-base sequences and are used in DNA fingerprinting.

The Flow of Genetic Information

DNA replication (Figure 8.2 in the text) makes possible the flow of genetic information from one generation to the next. The DNA of a cell replicates before cell division so that each offspring cell receives a chromosome identical to the parent's. Within each metabolizing cell, the genetic information contained in DNA also flows in another way: it is transcribed into mRNA and then translated into protein.

DNA Replication

In DNA *replication* (Figure 8.3 in the text), the two helical strands unravel and separate from each other at a **replication fork**, where the synthesis of new strands begins. The complementary pairing of bases—for example, adenine with thymine—yields a complementary copy of the original DNA. Segments of new nucleotides are joined to form short strands of DNA by **DNA polymerase** enzymes. Short strands of DNA are then joined into continuous DNA by action of **DNA ligase** enzymes. When DNA replication begins, the supercoiling is relaxed by **topoisomerase** or **DNA gyrase,** and the two strands of parental

DNA are unwound by **helicase** and separated from each other in one small DNA segment after another. Because each new double-stranded DNA molecule has one original strand and one new strand, the process is called **semiconservative replication.**

In bacteria, replication begins at an *origin of replication,* and in some cases two replication forks move in opposite directions. DNA replication makes few mistakes, largely because of the proofreading capability of DNA polymerase.

RNA and Protein Synthesis

Transcription.   In **transcription** (Figure 8.7 in the text), a strand of **messenger RNA (mRNA)** is synthesized from the genetic information in DNA. (Adenine in the DNA dictates the location of uracil, which replaces thymine in mRNA.) If DNA has the base sequence ATGCAT, the mRNA will have UACGUA. The region where *RNA polymerase* (needed for synthesis) binds to DNA and transcription begins is known as the **promoter site.** The **terminator site** is where the RNA polymerase and newly formed mRNA are released from the DNA, signaling the endpoint for transcription of the gene.

Translation.   Protein synthesis is called translation because it translates the language of nucleic acids into the language of proteins. The language of mRNA is in **codons,** groups of three nucleotides such as AUG. Each codon "codes" for a particular amino acid. There are 64 possible codons, but only 20 amino acids. Therefore, an amino acid has more than one codon (see Figure 8.8 in the text); this is referred to as **degeneracy** of the code.

Sense codons code for amino acids; **nonsense codons** (or *stop codons*) signal the end of synthesis of a protein.

The sites of translation are **ribosomes** that move along mRNA. The amino acids are transported to the ribosome by **transfer RNA (tRNA).** Each tRNA molecule is made specific for an amino acid by an **anticodon** that is complementary to a codon. That is, the codon AUG would be complementary to the anticodon UAC. Figure 8.9 in the text shows how the amino acids are positioned in the growing protein chain.

In eukaryotic cells, transcription takes place in the nucleus. The regions that code for proteins are often interrupted by noncoding DNA. In eukaryotic genes, **exons** are regions of DNA that are expressed, and **introns** are the intervening regions of DNA that do not encode protein. In the nucleus, particles called *small nuclear ribonucleoproteins,* or **snRNPs** (pronounced "snurps"), remove introns and splice exons together.

THE REGULATION OF BACTERIAL GENE EXPRESSION

The cell conserves energy by making only those proteins needed at the time. If a gene produces a product at a fixed rate, it is *constitutive.*

Pre-Transcriptional Control

Repression and Induction.   An **inducer** is a substance (substrate) whose presence results in the formation, or increase in the amount, of an enzyme. Such enzymes are called **inducible enzymes**; this genetically controlled response is termed **enzyme induction.** (Lactase production in response to lactose is an example.) Genetic regulation that decreases enzyme synthesis is **enzyme repression.** Proteins that mediate this are **repressors.** Repression occurs if cells are exposed to an overabundance of a particular end-product of a metabolic pathway.

The Operon Model of Gene Expression.   Protein synthesis in bacteria is controlled by a system called the **operon model.** For example, three enzymes are involved in uptake and utilization of the sugar lactose in *E. coli.* The genes for these enzymes, **structural genes,** are close together on the bacterial chromosome. There is also an **operator site** next to these structural genes, and more remotely located is a gene that codes for a **repressor protein.** The operator and promoter sites plus the structural genes are the **operon.**

When lactose is absent, the repressor protein prevents the operator from making lactose-utilizing enzymes. A small molecule, the **corepressor,** is required to bind repressor to operator genes. When lactose is present, some diffuses into the cells and binds with the repressor protein so it cannot bind to the operator site. The operator then induces the structural genes to produce enzymes to utilize lactose, an example of an **inducible enzyme** (Figure 8.12 in the text). Many genes are not regulated in this manner but are constitutive and usually represent functions needed for major life processes. An example of constitutive enzymes are those for utilization of glucose.

Positive Regulation.   Cells prefer glucose to lactose, but if the level of glucose is too low, the cell responds with a cellular alarm signal. These are called **alarmones; cyclic AMP (cAMP)** is an example. It is produced when glucose is no longer available and is required for the beginning of changes in the operon system for the use of lactose. Other sugars use similar mechanisms. Inhibition of metabolism of alternative carbon sources when glucose is present is called **catabolic repression** (or the *glucose effect*).

Epigenetic Control.   Eukaryotic and bacterial cells can turn genes off by methylating (adding a methyl group) certain nucleotides. Unlike mutations, this is not permanent and the genes can be turned on in a later generation, which is called *epigenetic inheritance*.

Post-Transcriptional Control

Some regulatory mechanisms stop protein synthesis after transcription has occurred. For example some single-stranded RNA molecules called *microRNAs (miRNAs)* inhibit protein production in eukaryotic cells. Bacterial miRNAs allow the cell to cope with environmental stress such as low temperatures or oxidative damage.

MUTATION: CHANGE IN THE GENETIC MATERIAL

A **mutation** is a change in the base sequence of DNA.

Types of Mutations

The most common mutation is a **base substitution,** or **point mutation,** in which a single base in DNA is replaced with a different one. Such a substitution is likely to result in the incorporation of an incorrect amino acid in the synthesized protein, a result known as a **missense mutation.** Such an error may create a stop codon, which stops protein synthesis before completion, resulting in a **nonsense mutation.** Deletion or addition of base pairs results in a **frameshift mutation.** In this mutation, there is a shift in the "translational reading frame" (the three-by-three grouping of nucleotides), and a long stretch of missense and an inactive protein product result. **Spontaneous mutations** occur without the known intervention of mutation-causing agents. Many chemicals and radiation bring about mutations; these are called **mutagens.**

Mutagens

Chemical Mutagens.   Nitrous acid is a **base-pair mutagen.** It causes adenine to pair with cytosine instead of thymine. Other mutagens are **nucleoside analogs,** which are structurally similar to bases and are incorporated into DNA by error. Examples are 2-aminopurine and 5-bromouracil, which are analogs of adenine and thymine. Some antiviral drugs are nucleoside analogs. Examples of **frameshift mutagens,** many of which are carcinogens, are benzopyrene (found in smoke and soot), aflatoxin (a mold toxin), and acridine dyes.

Radiation.   **Ionizing radiation**—such as X rays and gamma rays, which are mutagens—damages DNA. Ultraviolet light (an example of **nonionizing radiation**) is another mutagen that affects DNA. Certain

light-repair enzymes can repair ultraviolet damage in a process stimulated by visible light. **Nucleotide excision repair** (Figure 8.21 in the text) uses enzymes to cut out distorted DNA and synthesize replacements.

The Frequency of Mutation

The **mutation rate** is the probability that a gene will mutate when a cell divides. DNA replication is very faithful, and only about once in 1 billion base pair replications does an error occur. Mutagens increase the rate of such errors 10 to 1000 times.

Identifying Mutants

Mutants, which occur at low rates, can be identified more easily with bacteria because bacteria produce very large populations very quickly. **Positive (direct) selection** is illustrated by plating out bacteria on a medium containing penicillin. Survivors, which are penicillin-resistant mutants, can be isolated. Nutritional mutants called **auxotrophs** are unable to synthesize a nutritional requirement such as an amino acid—an activity the parent type is capable of doing. To isolate auxotrophs, colonies growing on a master plate containing a complete medium can be transferred by **replica plating** (Figure 8.22 in the text). This is an example of **negative (indirect) selection.** A sterile velvet pad is pressed onto the master plate, and the colonies are transferred simultaneously to a *minimal medium,* which lacks essential nutrients such as the required amino acid. An auxotrophic mutant will fail to appear on the minimal medium.

Identifying Chemical Carcinogens

Carcinogens cause cancer in animals, including humans. The **Ames test** (Figure 8.23 in the text) is based on the ability of a mutated cell to mutate again and to revert to its original form. An auxotroph of *Salmonella,* which has lost the ability to synthesize the amino acid histidine, is plated out on a minimal medium without histidine. The test chemical (together with a rich source of activation enzymes found in rat liver extract) is placed on this plate also. Mutations of the *Salmonella* to the normal histidine-synthesizing form are indicated by colonies growing near the test chemical. High mutation rates are characteristic of the effect of carcinogens.

GENETIC TRANSFER AND RECOMBINATION

Genetic recombination is the rearrangement of genes to form new combinations. If two chromosomes break and are rejoined in such a way that some of the genes are reshuffled between the two chromosomes, the process is called **crossing over.** The original chromosomes each have been recombined so that they carry a portion of genes from the other chromosome. Recombination is more likely to be beneficial than mutation and can occur in several ways. In all cases, however, the **donor cell** gives a portion of its total DNA to a different **recipient cell.** The recipient, called the **recombinant,** has DNA from the donor added to its own DNA. **Vertical gene transfer** occurs when genes are passed from an organism to its offspring. Bacteria can also pass genes laterally to other microbes of the same generation—known as **horizontal gene transfer.**

Transformation in Bacteria

In **transformation,** "naked" DNA in solution is transferred from one bacterial cell to another. The process occurs naturally among very few genera of bacteria, and it usually occurs when donor and recipient are closely related and in the log phase of growth. To take up the DNA fragment, the recipient cell must be **competent;** that is, its cell wall must be permeable to large DNA molecules.

Conjugation in Bacteria

Conjugation requires contact between living cells of opposite mating types. The donor cell, called an F$^+$ **cell** (corresponding to maleness), has extra DNA pieces called F (or **fertility**) **factors.** These are a type of plasmid (discussed below), a free genetic element in the cell. When F$^+$ and F$^-$ (corresponding to femaleness) cells are mixed, the F$^+$ cells attach by sex pili to F$^-$ cells. F factors are duplicated by the donor, and the new copy is transferred to the F$^-$ cell, which becomes an F$^+$ cell. The bacterial chromosome of the F$^+$ is not passed, and no recombinants are produced. However, some F$^+$ cells have the F factor integrated into their chromosome and are called **Hfr** cells—(*high frequency of recombination*). The chromosome of these cells may be transferred along with the integrated F factor if conjugation proceeds without interruption (a copy is retained by the donor). In this case, the recipient cell becomes an Hfr donor cell. Considerable amounts of the donor chromosome can be transferred even if interruption occurs. In this case, the recipient cell may acquire a sizable portion of the donor's genome: but unless conjugation is completed, the F factor will not be transferred.

Transduction in Bacteria

In **generalized transduction,** the **phage** (short for **bacteriophage,** a bacterial virus) attaches to the bacterial cell wall and injects DNA into the bacterium. Normally this directs the synthesis of new viruses. Sometimes, however, bits of bacterial chromosome are accidentally incorporated into the viral DNA. When these viruses infect a new host cell, they can incorporate this genetic information from the previous bacterial host into the new bacterial host. In **specialized transduction,** only certain bacterial genes are transferred.

Plasmids and Transposons

Plasmids.   **Plasmids,** circular pieces of DNA that replicate independently from the cell's chromosome, usually carry only genes that are not essential for growth of the cell. **Dissimilation plasmids** code for enzymes to utilize unusual sugars and hydrocarbons. Plasmids cause the synthesis of **bacteriocins,** toxic proteins that kill other bacteria. The F factor is a **conjugative plasmid. Resistance factors (R factors)** carry genes for antibiotic resistance and other antimicrobial factors. Sometimes R factors contain two groups of genes: **resistance transfer factor,** coding for plasmid replication, and **r-determinant,** coding for enzymes that inactivate antimicrobials.

Transposons.   **Transposons** are small segments of DNA that can move (are *transposed*) from one region of the chromosome to another (jumping genes). Such movement is uncommon, about the same as spontaneous mutation rates in bacteria. The simplest transposons are called **insertion sequences;** they code only for an enzyme (*transposase*) that cuts and ligates DNA in recognition sites. **Recognition sites** are short regions of DNA that the enzyme recognizes as recombination sites between transposon and chromosome. **Complex transposons** may carry other genes, such as those conferring antibiotic resistance. Transposons are a powerful mechanism for moving genes from one chromosome to another, even between species.

SELF-TESTS

In the matching section, there is only one answer to each question; however, the lettered options (a, b, c, etc.) may be used more than once or not at all.

I. Matching

____ 1. Where the RNA polymerase and the newly formed mRNA are released.

____ 2. Enzymes that assemble the nucleotides of DNA into chains.

____ 3. Formation of protein from the genetic information contained in mRNA.

____ 4. Formation of mRNA from the genetic information contained in DNA.

____ 5. Enzymes that bind short strands of DNA together into longer strands.

____ 6. Where transcription begins on mRNA.

a. DNA ligases

b. DNA polymerases

c. Transcription

d. Translation

e. Promoter site

f. Terminator site

II. Matching

____ 1. A sequence of three bases coding for the position of an amino acid in the assembly of a protein chain.

____ 2. A cluster of related genes together with the operator and promoter sites on mRNA.

____ 3. A sequence of three bases on tRNA that locates the codon on the mRNA at the ribosome.

____ 4. A sequence of bases that does not code for an amino acid, but that terminates the protein or polypeptide chain.

____ 5. DNA region of eukaryotic cell that is expressed.

a. Anticodon

b. Stop codon

c. Codon

d. Operon

e. Exon

f. Intron

g. Short tandem repeats

III. Matching

___ 1. The actual template upon which the protein or polypeptide chain is assembled.

___ 2. The product of transcription.

___ 3. One of these is specific for each of the 20 amino acids.

___ 4. The original genetic information in a bacterial cell.

___ 5. Relaxes supercoiling ahead of the replication fork.

___ 6. Unwinds double-stranded DNA.

___ 7. Stops protein synthesis after transcription has occurred.

a. mRNA

b. tRNA

c. rRNA

d. DNA

e. DNA gyrase

f. Helicase

g. miRNA

IV. Matching

___ 1. The probability of a gene mutation each time a cell divides.

___ 2. Usually a result of the deletion or addition of a base pair.

___ 3. A mutation caused by a chemical that is structurally similar to nucleotide components such as adenine or thymine.

___ 4. A mutagen that would, for example, make the base adenine pair with cytosine instead of thymine.

a. Missense mutation

b. Frameshift mutation

c. Spontaneous mutation

d. Base pair type of mutagen

e. Nucleotide-analog type of mutagen

f. Mutation rate

V. Matching

___ 1. DNA transferred between cells in solution in the suspending medium.

___ 2. Requires contact between living cells of opposite bacteria mating types.

___ 3. Requires a sex pilus.

___ 4. Hfr cells.

___ 5. The method by which plasmids such as F factors are transferred between cells.

___ 6. Turns genes off by methylating certain nucleotides.

a. Conjugation in bacteria

b. Transformation in bacteria

c. Transduction in bacteria

d. Epigenic control

VI. Matching

____ 1. Contain genes coding for enzymes that catabolize unusual sugars or hydrocarbons, for example.

____ 2. Contain genes for synthesis of toxic proteins lethal for other bacteria.

a. Dissimilation plasmids

b. Bacteriocinogenic plasmids

c. Conjugative plasmids

VII. Matching

____ 1. Ultraviolet light.

____ 2. In the operon model, the place on the mRNA at which the repressor binds to prevent transcription of structural genes into a protein.

____ 3. In the operon model, the regulator gene codes for a protein of this name.

____ 4. X ray.

a. Ionizing radiation

b. Nonionizing radiation

c. Repressor

d. Operator

e. Transposons

VIII. Matching

____ 1. In replicating a strand of DNA, where adenine is on the original strand, there will be this on the new strand.

____ 2. When a strand of mRNA is made from DNA, this is found where adenine is located on the original DNA.

____ 3. A nutritional mutant.

____ 4. In the Ames test, the *Salmonella* bacterium has lost the ability to synthesize this.

____ 5. Small segments of DNA that can move from one region of the chromosome to another.

a. Thymine

b. Uracil

c. Histidine

d. Transposons

e. Auxotroph

Fill in the Blanks

1. Colonies growing on a master plate containing a complete medium can be transferred simultaneously to minimal medium by the _____ technique.

2. A segment of DNA that codes for a functional product is a(n) _____.

3. The site at which the replicating DNA strands separate is called the _____.

4. The organism's entire genetic potential is the _____.

5. A cell with a cell wall permeable to soluble DNA is _____.

6. Bacteria that have the F factor integrated into their chromosome and that tend to transfer F factor and chromosome together are called _____ cells.

7. Enzymes that are always present in the cytoplasm are called _____ enzymes.

8. Some R factors have a set of genes called the r-determinant that codes for resistance, and another set of genes called the _____ that codes for replication and conjugation.

9. A bacterial virus is known, for short, as a(n) _____.

Critical Thinking

1. How do you account for the fact that *E. coli* can replicate faster than two replication forks can separate the strands of DNA in the circular chromosome?

2. Distinguish base mutations (point mutations) from frameshift mutations. Which type of mutation is more likely to result in termination of translation?

3. Discuss the mutations caused by ionizing and nonionizing radiation. By what mechanism can bacteria repair the damage caused by radiation?

4. What are "jumping genes," and what role do they play in the evolution of bacteria?

5. Explain the clinical significance of using antibiotics as a food supplement in animal feeds.

ANSWERS

Matching

I.   1. f   2. b   3. d   4. c   5. a   6. e
II. 1. c   2. d   3. a   4. b   5. e
III. 1. a   2. a   3. b   4. d   5. e   6. f   7. g
IV. 1. f   2. b   3. e   4. d
V.   1. b   2. a   3. a   4. a   5. a   6. d
VI. 1. a   2. b
VII. 1. b   2. d   3. c   4. a
VIII. 1. a   2. b   3. e   4. c   5. d

Fill in the Blanks

1. replica plating   2. gene   3. replication fork   4. genotype   5. competent   6. Hfr   7. constitutive
8. resistance transfer factor   9. phage

Critical Thinking

1. When conditions are optimal—for example, in log phase—bacterial cells can initiate multiple replication forks on the origin of the chromosome. A new pair begins before the original pair finishes.

2. Base substitutions, also called point mutations, are common mutations in which a single base of a DNA molecule is replaced with a different base. Because of the redundancy in the genetic code, this may not cause a problem, or it may result in a missense or a nonsense mutation.

   Frameshift mutations involve the insertion or deletion of one or more nucleotide pairs and can shift the reading frame (the three-by-three groupings) of nucleotides. This usually results in a long stretch of missense and production of a nonfunctional protein. Translation is likely to be terminated when a nonsense codon is encountered.

3. Ionizing radiation, such as X rays and gamma rays, penetrate well and have high energy contents. They ionize molecules, making them very reactive. Some of the affected ions combine with bases in DNA, resulting in replication errors. Ionizing radiation may even cause breakage of the covalent bonds in the sugar-phosphate backbone of DNA, resulting in physical breaks in the chromosome. Bacteria do not possess a mechanism to repair this very serious damage.

   Nonionizing radiation such as UV light does not carry as much energy or penetrate as effectively. UV light will, however, cause thymine dimers to form in the DNA strand. If the dimers are not repaired, then proper transcription and replication of DNA cannot occur.

Bacteria (and other organisms) have photoreactivating enzymes that can repair damage caused by UV light by splitting the dimers. Occasionally this repair process will result in errors and is yet another source of mutation.

4. Jumping genes, or transposons, are small segments of DNA that can move from one region of a DNA molecule to another, to another DNA molecule, or to a plasmid. This is not a common occurrence. They may contain genes that allow bacteria to resist toxins or antibiotics. Because transposons can be carried between cells and even species via viruses or plasmids, they can play a significant role in evolution.

5. Some bacterial cells have a type of plasmid referred to as an R factor. R factors carry genes that provide resistance to one or more antibiotics and can be passed from cell to cell and even to other species through the process of conjugation. When antibiotics are used as a supplement in animal feeds, bacteria able to resist these drugs are selected by preferentially diluting the effectiveness of antibiotics.

9 Biotechnology and DNA Technology

INTRODUCTION TO BIOTECHNOLOGY

Recombinant DNA Technology

Recall from Chapter 8 that recombination, the reshuffling of genes between two DNA molecules, forming **recombinant DNA,** occurs naturally in microorganisms. It is also possible to manipulate DNA artificially to combine genes from two different sources, even from vertebrates to bacteria. Artificial gene manipulation is called **recombinant DNA (rDNA) technology,** or sometimes **genetic engineering,** and the term **biotechnology** usually means the industrial use of genetically modified microorganisms.

An Overview of Recombinant DNA Procedures

The gene of interest is first inserted into **vector DNA** (or *cloning vector*). This DNA molecule used as a carrier must be self-replicating, such as a plasmid or viral genome. This recombinant vector DNA must enter a cell where it can multiply, forming a **clone** of genetically identical cells. The desired product may be the gene itself, or it may be a protein product expressed by the gene (Figure 9.1 in the text).

TOOLS OF BIOTECHNOLOGY

Selection

In nature, organisms with characteristics that enhance their survival are more likely to reproduce, which is called *natural selection.* Humans use **artificial selection** to select for desirable breeds of animals or strains of plants.

Mutation

A bacterium with a mutation that confers resistance to an antibiotic will survive and reproduce when exposed to that antibiotic. This procedure can be enhanced by exposing microbes to mutagens. **Site-directed mutagenesis** can be used to make a specific change in a gene, for example, one amino acid in an enzyme.

Restriction Enzymes

A special class of DNA-cutting enzymes, **restriction enzymes,** are the technical basis of genetic engineering. Restriction enzymes have the natural function of protecting the bacteria from attack by phages by hydrolyzing their DNA. The DNA of the bacteria is protected from the enzyme by addition of a methyl group to some cytosines, **methylates.** The enzyme recognizes and cuts only one particular sequence of nucleotide bases. Many restriction enzymes make *staggered* cuts in the two DNA strands—cuts that are not directly opposite each other (Figure 9.2 in the text). The stretches of single-stranded DNA at the ends of the DNA fragments are called **sticky ends.** They stick to complementary stretches of single-stranded DNA by base pairing. If two fragments of DNA from different sources have been produced by the same restriction enzymes, the sets of sticky ends can be spliced (recombined) easily. **DNA ligase** then links the DNA pieces.

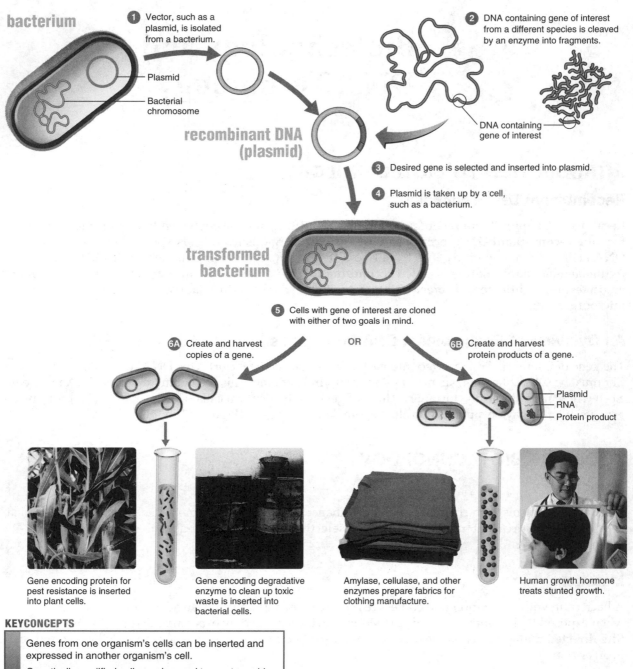

Figure 9.1 A typical genetic modification procedure, with examples of applications.

Vectors

Figure 9.1 diagrams the cloning of a DNA fragment using a **plasmid** for a vector. The host cell can be induced to take up the plasmid vector by chemical treatment. Plasmids that can exist in several species are **shuttle vectors**.

Polymerase Chain Reaction

A development in DNA analysis is the **polymerase chain reaction (PCR)** (Figure 9.4 in the text). Starting with just one gene-sized piece of DNA, PCR can make billions of copies in only a few hours. The target piece of DNA is heated to separate the DNA strands, which serve as templates for DNA synthesis. **DNA polymerase** enzyme, which forms DNA by linking the nucleotides, is supplied with DNA's four nucleotides and short pieces of primer nucleic acid. Each newly synthesized piece of DNA serves in turn as a template for more new DNA.

TECHNIQUES OF GENETIC MODIFICATION

Inserting Foreign DNA into Cells

Plasmid vectors can be inserted into many cells by chemical treatments that make **transformation** possible. Such cells are then called *competent,* or able to take up external DNA.

The walls of cells can be enzymatically removed (forming a **protoplast**) and exchange DNA by **protoplast fusion.** Polyethylene glycol may be used to improve efficiency. Transfer of DNA in this manner can also be enhanced by using an electric field to form minute pores in the protoplast membranes—**electroporation.** Foreign DNA can be introduced into plant cells by coating microscopic particles of tungsten with DNA and firing it through the plant cell wall using a "gene gun." DNA can be introduced into a cell through a minute glass micropipette by **microinjection.**

Obtaining DNA

Genes from a particular organism are isolated by cutting up the entire genome with restriction enzymes, splicing as many as possible into vectors, and then introducing the genes into bacterial cells. A collection of bacterial clones containing different DNA fragments is called a **genomic library.**

Cloning genes from a eukaryotic organism generally requires removal of **introns,** stretches of DNA that do not code for protein. This can be done by **splicing;** introns are removed when an RNA transcript of a gene is converted to mRNA. What remains are **exons,** stretches of DNA that code for protein. An artificial gene that contains only exons can be made with the enzyme **reverse transcriptase.** This will synthesize **complementary DNA (cDNA)** from an mRNA template. This is the most common method of obtaining eukaryotic genes.

Genes of **synthetic DNA** can also be made with the help of DNA-synthesizing machines. The smaller chains of about 40 nucleotides synthesized in this way can be linked together to make an entire gene.

Selecting a Clone

One of the most common methods of selecting a desired gene is **blue/white screening**—from the color of the bacterial colonies formed at the end of the process. Briefly, only white colonies are of interest, because they contain foreign DNA incorporated into a plasmid. Further work is then needed to identify this foreign DNA.

For example, **colony hybridization** is a common method. Short segments of single-stranded DNA, consisting of a sequence of nucleotides unique to the gene sought, can be synthesized. These molecules, called **DNA probes,** are radioactively labeled so they can be located later. The clone-carrying bacteria from the library are grown into colonies on a plate of nutrient medium treated to break open the cells and separate the DNA into single strands. The labeled probe, added to the plate, will react with DNA in any bacterial colony that base-pairs with the probe (*colony hybridization*). The radioactive tag allows the colonies containing the desired gene to be identified. A similar probe based on labeled antibodies against protein products of the cells is also used.

Making a Gene Product

Escherichia coli is often used as the genetically modified organism to produce a desired gene product. It has disadvantages; it produces endotoxins that cause fever and shock in animals. Also, it does not secrete the product; the cells must be harvested and ruptured, and the product recovered.

Organisms such as the gram-positive bacterium *Bacillus subtilis* and yeasts are more likely to secrete their products. Animal viruses, such as the vaccinia virus, have been genetically modified to produce vaccines. Mammalian cells and plant cells are often modified to produce useful products.

APPLICATIONS OF DNA TECHNOLOGY

Medically important products made by genetic modification include insulin, somatotropin growth hormone, **DNA vaccines,** and **subunit vaccines. Gene therapy** is the mostly theoretical process of removing cells, replacing any defective genes, and returning the cells to the body.

Therapeutic, Scientific, and Medical Applications

Recombinant DNA technology is also the basis of DNA analysis of genetic abnormalities responsible for various diseases, and it contributes to advances in gene therapy, in which abnormal genes might be replaced with normal genes in a living individual.

To isolate a fragment of DNA containing a gene, **Southern blotting** is commonly used. DNA fragments from each candidate clone are treated with the same restriction enzyme, and the resulting fragments separated by **gel electrophoresis.** The fragments are called **restriction fragment length polymorphisms (RFLPs).** The bands representing the fragments are then blotted onto a special filter, which is bathed with radioactively tagged probes of DNA complementary to the gene desired. The desired bands can be cut out of the gel and recovered by soaking in solvent. The genes may be studied by **DNA sequencing** to determine the sequence of nucleotides. **Proteomics** is the science of determining all the proteins expressed in a cell. **Reverse genetics** can be used to identify the function of a gene from a genetic sequence by blocking the gene and looking for the lost characteristic.

In **random shotgun sequencing,** small pieces of the genome are sequenced and then assembled by computer analysis. These techniques were vital to the Human Genome Project. DNA sequencing, which is often highly automated, can be combined with PCR. This permits recovery of detectable, identifiable DNA from extremely small samples. These procedures are used for so-called **DNA fingerprinting,** which is used for analysis of crime-scene (forensic) samples, tissue identification, and paternity testing.

Gene silencing is a natural process that is a defense against viruses and transposons. It is similar to miRNA in that a gene encoding a small piece of RNA is transcribed. An enzyme called *Dicer* processes RNA (following transcription) into RNAs called **small interfering RNAs (siRNAs).** The siRNAs bind to mRNA and cause its destruction by proteins called the **RNA-induced silencing complex (RISC).** This silencing of a gene (shown in Figure 9.14 in the text) is called **RNA interference (RNAi),** which holds promise for gene therapy in treating cancer and for viral infections.

Genome Projects

Completed in 2003, the **Human Genome Project** sequences the entire human genome. The next goal is the **Human Proteome Project.** This will map all the proteins expressed in human cells.

Agricultural Applications

The most elegant method of introducing recombinant DNA into a plant cell is by the **Ti plasmid.** A bacterium that infects plants, *Agrobacterium tumefaciens,* normally carries this plasmid. The infection causes a tumorlike growth called a crown gall (Ti means tumor-inducing). The plasmid also serves as a vehicle for inserting genetically modified DNA into a plant. Other applications are to make crop plants resistant to herbicides that then selectively kill weeds and to improve the ability to fix nitrogen in certain symbiotic bacteria. A bacterium, *Bacillus thuringiensis (Bt),* has been engineered into plants to produce a toxin that kills certain plant pathogens that feed on the plant. A genetically modified product, bovine growth hormone, increases milk production in dairy herds.

SAFETY ISSUES AND THE ETHICS OF USING DNA TECHNOLOGY

Laboratories engaged in recombinant DNA research must meet rigorous safety standards to avoid accidentally releasing genetically engineered microbes. The microbes may also be engineered to contain suicide genes that prevent them from surviving outside the laboratory environment. Genetic screening for hereditary diseases and birth defects in the fetus introduces ethical questions not yet resolved.

SELF-TESTS

In the matching section, there is only one answer to each question; however, the lettered options (a, b, c, etc.) may be used more than once or not at all.

I. Matching

___ 1. DNA-cutting enzymes that often form sticky ends.

___ 2. A self-replicating DNA molecule used as a carrier to transmit a gene from one organism to another.

___ 3. An enzyme that links short pieces of DNA into longer pieces.

___ 4. An enzyme that links nucleotides to form DNA.

a. Restriction enzymes

b. DNA ligase

c. DNA polymerase

d. Vector

II. Matching

___ 1. Probably the most common cloning vector used in genetic engineering.

___ 2. The reshuffling of genes between two DNA molecules forms DNA of this type.

___ 3. The kind of DNA synthesized by using mRNA as a template.

___ 4. DNA exchange between cells in this process uses polyethylene glycol to improve efficiency.

___ 5. Short segments of single-stranded DNA used to recognize a DNA sequence in a gene.

a. Recombinant

b. Complementary

c. Probe

d. Protoplast fusion

e. Plasmid

III. Matching

____ 1. In the blue/white screening procedure, the foreign DNA is in these colonies.

____ 2. These stretches of DNA of eukaryotes do not code for proteins.

a. Blue

b. White

c. Exon

d. Intron

Fill in the Blanks

1. Sticky ends stick to each other by complementary stretches of single-stranded DNA by _____ pairing.

2. A collection of bacterial clones each containing a different DNA fragment is called a(n) _____.

3. To isolate a fragment of DNA containing a gene, DNA fragments of clones are separated by gel electrophoresis. This is an early step in the _____ technique for DNA analysis.

4. The procedure by which billions of copies of a sequence of DNA can be made in a few hours is called the _____ reaction.

5. The most elegant way of introducing recombinant DNA into a plant cell is by means of the _____ plasmid, carried naturally by *Agrobacterium tumefaciens*.

6. The goal of the Human Proteome Project is to identify all the _____ produced by human cells.

7. The full term is *restriction fragment length* _____.

8. _____ is the science of determining all the proteins expressed in a cell.

9. In Southern blotting, the gene fragments are called _____ *fragment length* _____ .

10. In the acronym RISC, the letters stand for *RNA-induced* _____.

Label the Art (Supply the terms needed to complete the sentences correctly.)

1 a. _____ cuts (arrows) double-stranded DNA at its particular b. _____, shown in dark gray.

2 These cuts produce a DNA fragment with two c. _____.

DNA from another source, perhaps a plasmid, cut with the same restriction enzyme

3 When two such fragments of DNA cut by the same restriction enzyme come together, they can join by d. _____.

4 The joined fragments will usually form either a linear molecule or a circular one, as shown here for a plasmid. Other combinations of fragments can also occur.

5 The enzyme e. _____ is used to unite the backbones of the two DNA fragments, producing a molecule of f. _____.

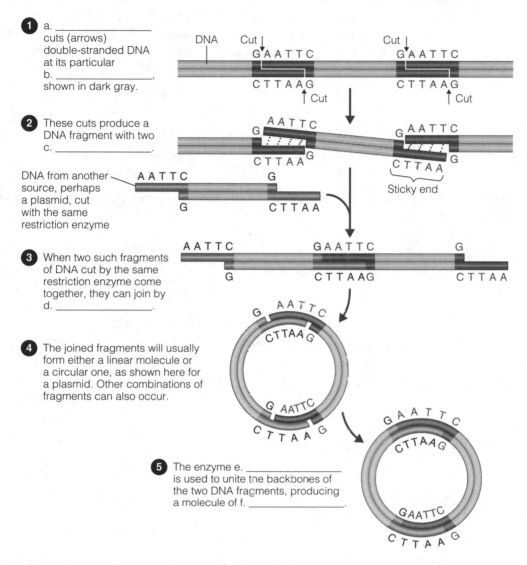

Critical Thinking

1. During the investigation of a robbery, police discover a small quantity of blood and skin on a piece of broken glass from the window through which the perpetrator gained entry. There was insufficient blood on the glass to type conventionally. No usable fingerprints were obtained. How might the police tie their prime suspect to the crime?

2. What are some of the advantages of using genetic modification to produce human hormones such as insulin and somatotropin?

3. What method would be most appropriate for inserting foreign DNA in each of the following examples?

   a. Inserting DNA into an animal cell.

   b. Inserting DNA into a plant cell.

   c. Inserting DNA into a yeast.

4. You are asked to develop a protocol for the industrial production of a protein from a genetically modified bacterium. A bacterium of what Gram designation would most likely be the best choice? Why?

5. What are subunit vaccines? Are they safer than avirulent vaccines? Why?

ANSWERS

Matching

I.   1. a   2. d   3. b   4. c

II. 1. e   2. a   3. b   4. d   5. c

III. 1. b   2. d

Fill in the Blanks

1. base   2. gene library   3. Southern blot   4. polymerase chain   5. Ti   6. proteins
7. polymorphisms   8. Proteomics   9. restriction, polymorphisms   10. silencing, complex

Label the Art

a. Restriction enzyme   b. recognition sites   c. sticky ends

d. base pairing e. DNA ligase f. recombinant DNA

Critical Thinking

1. Police could use recombinant DNA techniques to amplify the DNA from the skin sample or from the white cells in the blood sample. This material would be compared to the suspect's DNA using the Southern blot method to do a DNA fingerprint. If the samples match, then the police have positive identification tying the suspect to the crime—that is, unless the suspect has an identical twin!

2. Genetic modification produces human hormones that are less expensive, less allergenic, available in large quantities, and pose no risk of transmitting disease.

3. a. Animal cells may be made "competent" to take up external DNA by soaking them in a solution of calcium chloride. After this treatment, many of the animal cells will take up the DNA; recombination will occur in some of the cells. Direct introduction of foreign DNA into animal cells can be achieved by microinjection using a glass micropipette.

   b. Plant cells may have foreign DNA introduced by first enzymatically removing the cell wall to create protoplasts. Protoplasts are then fused by adding polyethylene glycol to form a hybrid cell. The DNA in the hybrid cell may then undergo recombination naturally. Another method utilizes a "gene gun" that shoots DNA-coated tungsten or gold particles through the cell wall. Some cells will incorporate and express the new DNA.

   c. Yeast cells may also be made competent by soaking them in a calcium chloride solution or through the process of electroporation. This involves using an electrical current to form microscopic pores in the cell membrane. DNA is able to enter through these pores.

4. A gram-positive bacterium would be the best choice for two reasons. First, gram-positive bacteria lack a cell-wall component called endotoxin that is found in gram-negative bacteria. Endotoxin causes fever and shock in animals and would pose a serious problem if present in gene products. Second, gram-negative cells such as *E. coli* don't usually secrete protein products, and the process used to harvest them is too expensive for industrial applications.

5. Subunit vaccines consist of a protein portion of a pathogen produced by a genetically modified organism—for example, a yeast. Because the protein is made by a yeast rather than a treated pathogen or a portion of the actual pathogen, there is no chance that the disease will be transmitted.

10 Classification of Microorganisms

The science of classification, especially of living forms, is called **taxonomy.**

THE STUDY OF PHYLOGENETIC RELATIONSHIPS

Organisms are arranged into taxonomic categories called **taxa,** which reflect degrees of relatedness among organisms. The evolutionary history of a group is **systematics** or **phylogeny.**

The Three Domains

New techniques in molecular biology have revealed that there are two types of prokaryotic cells and one type of eukaryotic cell (Figure 10.1 in the text). In particular, this classification scheme depends on comparing sequences of nucleotides in ribosomal RNA (rRNA). These cell types are now generally called **domains.** The domain **Bacteria** includes many of the nonpathogenic prokaryotes found in soil and water as well as all of the pathogenic and photoautotrophic prokaryotes. The domain **Archaea** includes the prokaryotes that never have peptidoglycan in their cell walls. They often live in extreme environments. The three major groups within the domain Archaea are the methanogens, the extreme halophiles, and hyperthermophiles. The domain **Eukarya** contains the kingdoms of animals, plants, fungi, and protists (single-celled animals).

A Phylogenetic Hierarchy

In a phylogenetic hierarchy, grouping organisms according to common properties implies that a group of organisms evolved from a common ancestor; each species retains some of the characteristics of the ancestor.

CLASSIFICATION OF ORGANISMS

Scientific Nomenclature

The system of scientific nomenclature used today was developed in the eighteenth century by Carolus Linnaeus. All organisms have two names: the **genus name** and the **specific epithet (species).** Both names are underlined or italicized. This system of two names for an organism is called **binomial nomenclature.**

The Taxonomic Hierarchy

For **eukaryotic** organisms, a **species** is a group of closely related organisms that breed among themselves and not with other species. A **genus** (plural: *genera*) consists of different species related by descent. Related genera make up a **family.** A group of similar families constitutes an **order,** and a group of similar orders makes up a **class.** Related classes in turn constitute a **phylum.** (In zoology, the term *division* is used.) All phyla or divisions that are related to each other make up a **kingdom.** Related kingdoms are grouped into a **domain.**

Classification of Prokaryotes

The term *species*, as applied to higher organisms, does not apply well to prokaryotes, usually referred to as bacteria. Bacterial morphological traits are limited. A **prokaryotic species** is defined as a population of cells with similar characteristics. Sometimes the same species contains distinguishable groups within the species called **strains.** A **clone** is a population of cells derived from a single parent cell.

Bergey's Manual of Systematic Bacteriology is the standard reference for bacterial taxonomic classification.

Classification of Eukaryotes

Simple eukaryotic organisms, mostly unicellular, are grouped as the Kingdom **Protista.** The Kingdom **Fungi** includes the unicellular yeasts, multicellular molds, and the mushrooms. These organisms obtain nutrients through absorption. The Kingdom **Plantae** includes some algae and all mosses, ferns, conifers, and flowering plants. All are multicellular. These organisms obtain energy from photosynthesis and convert carbon dioxide and water into the organic molecules used by the cell. The Kingdom **Animalia** includes sponges, worms, insects, and vertebrate animals. They obtain nutrients and energy by ingesting organic matter. (All eukaryotes are discussed in Chapter 12.)

Classification of Viruses

Viruses (Chapter 13) are not classified under any of the kingdoms in which prokaryotes are grouped. However, a **viral species** is a population of viruses with similar characteristics (morphology, genes, and enzymes) occupying a particular ecological niche.

METHODS OF CLASSIFYING AND IDENTIFYING MICROORGANISMS

Morphological Characteristics and Differential Staining

Literally hundreds of bacterial species appear in the shape of small rods or small cocci. Nonetheless, these characteristics are still of some use in classifying bacteria. Sometimes the presence of endospores or flagella can be helpful. One of the first steps in identification is **differential staining,** such as the Gram stain or acid-fast stain.

Biochemical Tests

Enzymatic activity is widely used to differentiate bacteria. Tests that determine the bacteria's ability to ferment an assortment of selected carbohydrates are particularly common. Bacteria also may be subjected to other biochemical tests. The use of selective or differential media has been discussed for use in bacterial isolation or identification. Some medically important bacteria have rapid identification tests that perform several biochemical tests simultaneously. Results of tests are assigned numbers: for example, a positive test is 1, and a negative test is 0. Test results are assigned numbers ranging from 1 to 4 based on the relative reliability and importance of each test. This system is sometimes called **numerical identification.**

Serology

Serology is the science that studies serum (fluid from clotted blood) and the immune responses in serum. If an animal is exposed to a microbe, its body forms specific *antibodies* against the microbe. These antibodies in the animal's serum can be used to identify the microbe. Solutions of such antibodies are commercially available and are called **antiserum.** In a **slide agglutination test,** an unknown bacterium is placed in a drop of known antiserum under a microscope. A positive identification is made if the antiserum causes bacterial agglutination. Serological testing can differentiate between microbial strains with different antigens; called **serotypes, serovars,** or **biovars.** A serological test called **enzyme-linked immunosorbent assay (ELISA)** is fast and can be read by computer scanners. In **Western blotting** tests, proteins from a microbe are separated by electrophoresis. If a patient's serum has antibodies to any of the proteins it will become visible as a colored band.

Phage Typing

Phages, or bacterial viruses (Chapter 8), like antibodies, are highly specific for species or strains within a species. By means of stocks of different phages, this characteristic can be used to identify bacteria or even to trace epidemics. **Phage typing** is commonly used in outbreaks of illness caused by staphylococci.

Fatty Acid Profiles

Commercial systems that can identify bacteria by their cellular fatty acids are available. Fatty acid profiles (**FAME,** for Fatty Acid Methyl Ester) are used in clinical microbiology laboratories.

Flow Cytometry

Some bacteria can be identified without culturing. For example, in **flow cytometry** a moving fluid containing the bacteria is forced through a small opening (Figure 18.12 in the text). A laser beam scattered by the bacteria gives information about the bacteria, which may also be fluorescently labeled.

DNA Base Composition and DNA Fingerprinting

A technique that may be able to suggest evolutionary relationships among bacteria is the determination of the **DNA base composition.** The number of guanine (G) and cytosine (C) base pairs (the **G + C ratio**) can be determined as a percentage. By inference, this will give the percentage of adenine (A) and thymine (T) as well. On the one hand, if there is a difference of more than 10% in the percentage of G + C pairs, the organisms are probably unrelated. On the other hand, two organisms with the same percentage of G + C need not be closely related; other supporting data are needed.

Restriction enzymes (Chapter 9) can cut bacterial DNA into certain base sequences. The restriction fragments produced can be separated by electrophoresis and the DNA patterns of different organisms compared—**DNA fingerprints.**

Nucleic Acid Amplification Tests (NAATs)

When a microorganism cannot be cultured by conventional methods, the causative agent of an infectious disease might not be recognized. However, **nucleic acid amplification tests (NAATs)** can be used to increase the amount of microbial DNA to levels that can be tested by gel electrophoresis. NAATs use polymerase chain reaction (PCR), reverse-transcription PCR, and real-time PCR (see Chapter 9). If a primer for a specific microorganism is used, the presence of amplified DNA indicates that microorganism is present.

Nucleic Acid Hybridization

If a double-stranded molecule of DNA is heated, the complementary strands will separate as the hydrogen bonds connecting the bases break.

However, because the strands are complementary, they will reunite if allowed to incubate together. If the separated strands of two different organisms are incubated together, the closer the relationship, the greater the amount of DNA that will join together by complementary base pairing. This technique is known as **nucleic acid hybridization** and assumes that if two species are similar or related, a major portion of their nucleic acid sequences also will be similar. **Southern blotting** (see Figure 9.16 in the text) makes use of nucleic acid hybridization to identify microbes. **DNA probes,** based on these techniques, are being developed for rapid identification of bacteria. **DNA chips** are composed of DNA probes and will fluoresce if there is hybridization between the probe DNA and the DNA sample. Microbes in the environment that cannot be recovered by cultivation can sometimes be discovered or identified by **fluorescent in situ hybridization (FISH).** Probes of rRNA labeled with fluorescent dye can react with the ribosome of a target cell. Compared to DNA, there are advantages to using **ribotyping.** All cells contain rRNA; it has undergone fewer changes over time, and the cells do not have to be cultured.

Putting Classification Methods Together

Information concerning points of microbial identification can be placed into a **dichotomous key.** The worker answers successive questions, each of which has two possible answers, which then directs the worker to the next question. **Cladograms** are maps that show evolutionary relationships. Each branch point is defined by some feature. In bacteria, recently, rRNA sequences have been used to make such cladograms. All species beyond a branch point have similar rRNA sequences, suggesting that they arose from an ancestor at that node.

SELF-TESTS

In the matching section, there is only one answer to each question; however, the lettered options (a, b, c, etc.) may be used more than once or not at all.

I. Matching

___ 1. Bacteria.

___ 2. Elephants.

___ 3. Multicellular algae.

___ 4. Mushrooms.

___ 5. Yeasts.

___ 6. Cyanobacteria.

a. Prokaryotes

b. Protista

c. Animalia

d. Plantae

e. Fungi

II. Matching

___ 1. Related kingdoms.

___ 2. Made up of all phyla or divisions that are related to each other.

___ 3. In bacteria, all cells with similar characteristics in common.

___ 4. In eukaryotes, closely related organisms that are capable of breeding among themselves.

___ 5. A group of similar orders in eukaryotic cells.

a. Domain

b. Species

c. Genus

d. Kingdom

e. Class

f. Order

III. Matching

____ 1. A serological test for bacterial identification.

____ 2. Makes use of bacterial viruses to identify bacteria.

____ 3. A sequence of questions with only two possible answers.

____ 4. Determines the taxonomic relationship of bacteria by allowing complementary strands of DNA from different organisms to reassemble as a complementary pair.

____ 5. Taxonomic relationship of bacteria found by base-pair percentage in DNA.

____ 6. Maps of branching lines that show evolutionary relationships.

____ 7. Used to discover or identify microbes that cannot be cultivated.

a. Dichotomous key

b. Phage typing

c. Nucleic acid hybridization

d. Amino acid sequencing

e. G + C ratio

f. Slide agglutination test

g. Cladogram

h. Fluorescent in situ hybridization

IV. Matching

____ 1. A technique that may be able to suggest evolutionary relationships among bacteria based on the determination of the base composition of the DNA.

____ 2. A system we use for naming biological organisms with two names.

____ 3. The taxonomic categories into which organisms are arranged that reflect degrees of relatedness among them.

____ 4. The science of classification.

____ 5. Plural of *genus*.

a. Taxa

b. G + C ratio

c. Genera

d. Binomial nomenclature

e. Taxonomy

V. Matching

____ 1. Giraffes.

____ 2. Microbe that causes anthrax.

____ 3. Mushrooms.

____ 4. Prokaryotes without peptidoglycan in cell walls.

a. Domain Bacteria

b. Domain Archaea

c. Domain Eukarya

VI. Matching

____ 1. Identifies by fatty-acid profile.

____ 2. A blotting test that identifies nucleic acids.

____ 3. A serological test that is fast and can be read by computer scanners.

____ 4. Can be used to discover microbes in the environment that cannot be cultivated; hybridizes rRNA.

a. ELISA

b. FAME

c. FISH

d. Western blotting

e. Southern blotting

f. Slide agglutination test

Fill in the Blanks

1. The taxonomic classification scheme for bacteria may be found in the book called _____ *Manual*.

2. _____ typing is a technique commonly used to trace the epidemiology of outbreaks of illness caused by staphylococci.

3. The evolutionary history of a taxonomic group is called _____.

4. Solutions of antibodies against specific bacteria are called _____.

5. The cell walls of the archaea differ from those of bacteria by lack of _____.

Critical Thinking

1. What is the endosymbiont theory? What evidence exists to support the endosymbiont theory?

2. How do archaea differ from bacteria?

3. Distinguish a eukaryotic species, a bacterial species, and a strain.

4. How are DNA probes used to identify bacteria?

ANSWERS

Matching

I.   1. a   2. c   3. d   4. e   5. e   6. a
II. 1. a   2. d   3. b   4. b   5. e
III. 1. f   2. b   3. a   4. c   5. e   6. g   7. h
IV. 1. b   2. d   3. a   4. e   5. c
V.   1. c   2. a   3. c   4. b
VI. 1. b   2. e   3. a   4. c

Fill in the Blanks

1. *Bergey's*   2. Phage   3. phylogeny   4. antisera   5. peptidoglycan

Critical Thinking

1. The endosymbiont theory attempts to explain the evolution of eukaryotic cells. The theory suggests that eukaryotic cells evolved from prokaryotic cells living inside one another as endosymbionts. Evidence to support the theory includes the following:

   a. Similarities between prokaryotes and eukaryotic organelles such as mitochondria and chloroplasts. All have 70S ribosomes.

   b. Prokaryotic cells, mitochondria, and chloroplasts all multiply by binary fission.

   c. Fossil evidence shows that prokaryotes evolved first, approximately 3.5 billion years ago, whereas eukaryotic cells evolved only 1.4 billion years ago.

   d. Modern-day examples of endosymbionts exist—for example, the protozoan *Myxotrichia paradoxa.*

2. a. Archaea lack peptidoglycan in their cell walls, a substance that is found in the cell walls of most bacteria.

   b. Archaea can live in more extreme environments than most bacteria can tolerate.

   c. Even more metabolic diversity is seen within the archaea than in the bacteria.

3. Eukaryotic species refers to a group of closely related organisms that are capable of interbreeding. A bacterial species is a population of cells with similar characteristics indistinguishable from each other. Strains are distinguishable groups within a species.

4. First, DNA fragments of the bacteria that you wish to identify are produced with restriction enzymes. These fragments should react with the bacteria of interest but not with closely related organisms. After cloning sufficient numbers of the fragments, they are radioactively or fluorescently tagged. These probes are mixed with single-stranded DNA from the sample to be tested. If the organism in question is present, the DNA probes will hybridize with its DNA and can be detected by the radioactivity or fluorescence of the probes.

11

The Prokaryotes: Domains Bacteria and Archaea

Bergey's Manual of Systematic Bacteriology is the standard reference for bacterial taxonomy. At this time a second edition, based on a phylogenetic system, is replacing the phenotypic system based on morphologic characteristics. The new phylogenetic system is based mostly on differences in **ribosomal RNA (rRNA)**. Ribosomal RNA is slow to change and performs the same functions in all microbes. Comparisons of DNA are sometimes useful at the species and genus level.

THE PROKARYOTIC GROUPS

In the second edition of *Bergey's Manual* the prokaryotes are grouped into two domains, the **Archaea** and the **Bacteria** (formerly Eubacteria). Both domains are bacteria in the sense that they are prokaryotic cells. Eukaryotic multicellular and unicellular organisms are in the domain **Eukarya.**

_____ DOMAIN BACTERIA _____

THE PROTEOBACTERIA

The proteobacteria, which include most of the gram-negative, chemoheterotrophic bacteria, are presumed to have arisen from a common photosynthetic ancestor. Few are now photosynthetic.

The Alphaproteobacteria

Pelagibacter.   *Pelagibacter ubique,* discovered by the FISH technique in the ocean, is one of the most abundant microorganisms on Earth. It is exceptionally small and has a minimal genome.

Azospirillum.   *Azospirillum* is a soil bacterium that grows in close association with roots, especially of tropical grasses. They use excreted nutrients and aid plant growth by fixation of atmospheric nitrogen.

Acetobacter* and *Gluconobacter.   The genera *Acetobacter* and *Gluconobacter* are industrially important aerobic bacteria that convert ethanol into acetic acid (vinegar).

Rickettsia.   Rickettsias are gram-negative, obligately intracellular parasites, frequently pathogenic; they are responsible for diseases known as the spotted fever group. They are often transmitted to humans by bites of insects or ticks. Among rickettsial diseases are epidemic typhus caused by *Rickettsia prowazekii* (louseborne), endemic murine typhus caused by *Rickettsia typhi* (rat-flea-borne), and Rocky Mountain spotted fever caused by *Rickettsia rickettsii* (tickborne).

Ehrlichia.   The ehrlichiae are gram-negative, rickettsialike bacteria that live obligately within white blood cells. They cause the tickborne disease ehrlichiosis.

Caulobacter and Hyphomicrobium.   Both *Caulobacter* and *Hyphomicrobium* produce prominent **prosthecae,** a term applied to protrusions such as stalks or buds. Caulobacteria are found in aquatic environments such as lakes. They feature stalks that anchor them to surfaces, which increases their nutrient uptake in these low-nutrient environments. *Caulobacter* reproduction results in one stalked cell and one flagellated swarmer cell, which eventually becomes another stalked cell. *Hyphomicrobium* bacteria, which divide by budding, increase in size until separating into new cells, and are found in low-nutrient waters.

Rhizobium, Bradyrhizobium, and Agrobacterium.   The genera *Rhizobium* and *Bradyrhizobium* infect roots of leguminous plants such as peas, beans, or clover. This causes formation of nodules, within which these rhizobia symbiotically fix nitrogen for the plant. *Agrobacterium tumefaciens* also infects plants. It introduces bacterial DNA in a plasmid that results in a disease called crown gall. This bacterium's plasmid is also used in genetic engineering of plants to introduce foreign DNA.

Bartonella.   The genus *Bartonella* contains several members that are human pathogens. This includes *Bartonella henselae,* the cause of cat-scratch disease.

Brucella.   The *Brucella* species are the cause of the disease brucellosis in mammals. They are able to survive phagocytosis by the host's defenses.

Nitrobacter and Nitrosomonas.   *Nitrobacter* and *Nitrosomonas* are agriculturally important chemoautotrophs. *Nitrobacter* species oxidize ammonium (NH_4^+) to nitrite (NO_2^-), which is in turn oxidized by *Nitrosomonas* species to nitrates (NO_3^-). The process is called **nitrification.** (*Nitrosomonas* is discussed here for convenience, but is actually a member of the betaproteobacteria.)

Wolbachia.   *Wolbachia* may be the most common infectious bacterial genus. They live only inside the cells of their hosts, usually insects, a relationship called *endosymbiosis.*

The Betaproteobacteria

Thiobacillus.   *Thiobacillus* species and other sulfur-oxidizing bacteria are parts of the important sulfur cycle (Figure 27.7 in the text). These chemoautotrophic bacteria oxidize reduced forms of sulfur, such as hydrogen sulfide (H_2S) or elemental sulfur (S^0) into sulfates (SO_4^{2-}).

Spirillum.   These are helical bacteria that are motile by conventional polar flagella. Species such as *Spirillum volutans* are relatively large, gram-negative, aerobic bacteria whose habitat is fresh water.

Sphaerotilus.   Sheathed bacteria, including *Sphaerotilus natans,* are found in sewage and fresh water. These gram-negative bacteria with polar flagella form a hollow, filamentous sheath in which to live.

Burkholderia.   The genus *Burkholderia* was formerly classified with the genus *Pseudomonas*. These bacteria are aerobic, gram-negative rods that are motile with a single polar flagellum or tuft of flagella. Nutritionally, they are capable of degrading a wide spectrum of organic molecules. *Burkholderia pseudomallei* causes the disease meliodosis.

Bordetella.   The cause of pertussis, or whooping cough, is *Bordetella pertussis.*

Neisseria.   Members of the genus *Neisseria* are aerobic, gram-negative cocci that include *Neisseria gonorrhoeae,* which causes gonorrhea, and *Neisseria meningitidis,* which causes meningococcal meningitis.

Zoogloea.   The genus *Zoogloea* is important in the operation of activated-sludge sewage systems.

The Gammaproteobacteria

Beggiatoa.   The only species in the genus *Beggiatoa* is *Beggiatoa alba,* which morphologically resembles cyanobacteria but is not photosynthetic. It uses hydrogen sulfide as an energy source.

Francisella.   *Francisella* is a genus of small, pleomorphic bacteria that includes the cause of tularemia, *Francisella tularensis.*

Pseudomonadales

Pseudomonas.   The genus *Pseudomonas* is characterized by one or several polar flagella. Many species excrete extracellular, water-soluble pigments. *Pseudomonas aeruginosa* produces a blue-green pigment and can infect the urinary tract, burns, and wounds. Pseudomonads are common in soil and can often grow at refrigerator temperatures. They are able to decompose chemicals such as pesticides in soil, and their ability to grow in antiseptic solutions and whirlpool baths, as well as their resistance to antibiotics, make them troublesome. Many pseudomonads substitute nitrate for oxygen during anaerobic respiration, depleting nitrate fertilizers, which escape as nitrogen gas. *Pseudomonas syringae* is a plant pathogen.

Azotobacter* and *Azomonas.   The genera *Azotobacter* and *Azomonas* are heavily capsulated aerobes that fix nitrogen from the atmosphere.

Moraxella.   *Moraxella lacunata* is considered a cause of conjunctivitis, an eye infection.

Acinetobacter.   *Acinetobacter* is of increasing concern in the medical community, where a member, *Acinetobacter baumanii,* develops antibiotic resistance very rapidly. It is environmentally hardy for a gram-negative bacterium.

Legionellales

Legionella.   *Legionella* species are the cause of the respiratory disease legionellosis.

Coxiella.   *Coxiella burnetii* resembles the rickettsias, requiring a mammalian host cell to reproduce. It is the cause of Q fever, which is transmitted to cattle by ticks and to humans by contact or aerosols.

Vibrionales

Vibrio.   Morphologically, the bacteria of the genus *Vibrio* are usually slightly curved. Important pathogens are *Vibrio cholerae,* the cause of cholera, and *Vibrio parahaemolyticus,* the cause of a gastroenteritis transmitted mostly by ingestion of shellfish.

Enterobacteriales

Escherichia.   *Escherichia coli* is one of the most common inhabitants of the human intestinal tract and a familiar laboratory bacterium. It is used as an indicator organism for fecal pollution. It can cause urinary tract infections. Occasionally, enterotoxins produced by *E. coli* cause traveler's diarrhea and even, in the case of *E. coli* O157:H7, very serious foodborne disease.

Salmonella.   Almost all members of the genus *Salmonella* are pathogenic. Typhoid fever is caused by *Salmonella typhi;* most salmonellae cause a less serious gastrointestinal disease called salmonellosis. Taxonomically, the salmonellae are divided into about 2300 **serovars** (or **serotypes**) by serological means; serovars are further differentiated by biochemical, physiological properties into biovars (or biotypes). For most purposes all are considered a single species, *Salmonella enterica.* Many are known by antigenic formulas according to the Kauffman-White scheme. For example, *Salmonella typhimurium* (more properly written as *Salmonella enterica* serovar Typhimurium) is represented as O-antigens 1, 4, (5), 12 H-antigens i, 1, 2. *S. bongori* is a second species but is not a human pathogen.

Shigella.   Species of *Shigella* cause bacillary dysentery, or shigellosis, and are responsible for many cases of traveler's diarrhea. They infect only humans.

Klebsiella.   Members of the genus *Klebsiella* are common in soil and water. Many fix nitrogen from the air. *Klebsiella pneumoniae* causes pneumonia in humans.

Serratia. *Serratia marcescens* produces a distinctive red pigment and occasionally causes infections of the urinary and respiratory tracts in hospital patients.

Proteus. Colonies of *Proteus* growing on agar show a swarming type of growth with the appearance of concentric rings. They are responsible for many urinary tract and wound infections.

Yersinia. *Yersinia pestis*, transmitted by fleas, is the cause of plague.

Erwinia. *Erwinia* species are primarily plant pathogens, causing soft-rot diseases by hydrolyzing the pectin between plant cells.

Enterobacter. Urinary tract infections and hospital-acquired infections are often caused by *Enterobacter cloacae* and *Enterobacter aerogenes.*

Pasteurellales

Pasteurella. Members of the genus *Pasteurella* are nonmotile bacteria primarily causing diseases in domestic animals. *Pasteurella multocida* is transmitted to humans by animal bites.

Haemophilus. *Haemophilus influenzae* is a common cause of meningitis, earaches, and a number of other important diseases. *Haemophilus* bacteria are cultured on media enriched by hemoglobin or culture media containing X factors and V factors. *Haemophilus ducreyi* is the cause of the sexually transmitted disease chancroid.

The Deltaproteobacteria

Bdellovibrio. *Bdellovibrio* is a predator on other bacteria, reproducing within gram-negative bacteria, which it kills.

Desulfovibrionales

Desulfovibrio. *Desulfovibrio* is the best-studied genus of the sulfur-reducing deltaproteobacteria. It is found in anaerobic sediments and in the intestinal tracts of mammals. It uses hydrogen sulfide (H_2S) as part of photosynthesis or as an energy source. Because the H_2S is not assimilated as a nutrient, this type of metabolism is termed *dissimilatory.*

Myxococcales

Myxococcus. Members of the myxobacteria move by gliding and leave a slime trail. For nutrition they ingest and lyse bacteria. In their life cycle, large numbers of vegetative cells converge on a single point, where they aggregate and differentiate into a stalked body that carries resting cells called myxospores. These myxospores eventually germinate and renew the cycle.

The Epsilonproteobacteria

Campylobacter. The campylobacteria are microaerophiles. Morphologically they are vibrioids—a term applied to bacteria that are helical or curved. *Campylobacter fetus* causes spontaneous abortion in domestic animals; *Campylobacter jejuni* causes foodborne intestinal disease.

Helicobacter. The species *Helicobacter pylori* has been linked to gastric ulcers in humans.

THE GRAM-POSITIVE BACTERIA

Firmicutes (Low G + C Gram-Positive Bacteria)

A low G + C ratio means that the G + C ratio is on the order of, as an example, that of the genus *Streptococcus,* whose ratio is 33–44%.

Clostridiales

Clostridium.   Members of the genus *Clostridium* are obligately anaerobic rod-shaped cells containing endospores that usually distend the cells. *Clostridium botulinum* causes botulism, *Clostridium tetani* causes tetanus, and *Clostridium perfringens* causes gas gangrene. *Clostridium difficile* is an inhabitant of the intestinal tract that may cause serious diarrhea when antibiotic therapy alters the normal intestinal flora.

Epulopiscium.   The bacterium *Epulopiscium fishelsoni* is extraordinarily large, over half a millimeter in length. It does not reproduce by binary fission. Daughter cells formed within the cell are released through a slit opening in the parent cell—a process possibly related to the evolutionary development of sporulation.

Bacillales

Bacillus.   Members of the genus *Bacillus* are gram-positive rod-shaped bacteria, aerobic or facultatively anaerobic, and are distinguished by possession of endospores. *Bacillus anthracis* is the cause of anthrax in humans and animals; *Bacillus thuringiensis* is an insect pathogen; *Bacillus cereus* is an occasional cause of food poisoning.

Staphylococcus.   Staphylococci are gram–positive cocci that occur in grapelike clusters. The most important species is *Staphylococcus aureus*. Staphylococci are able to survive and grow at high osmotic pressures. *S. aureus* produces a yellow pigment and many toxins, including an enterotoxin that causes food poisoning.

Lactobacillales

Lactobacillus.   These gram-positive bacteria lack a cytochrome system and are unable to use oxygen as an electron acceptor. They are aerotolerant, however, and grow in the presence of oxygen. The genus *Lactobacillus* is found in the vagina, intestinal tract, and oral cavity. Industrially, these bacteria are used in producing sauerkraut, pickles, buttermilk, and yogurt.

Streptococcus.   The streptococci, gram-positive cocci that typically appear in chains, are metabolically similar to the lactobacilli. *Streptococcus* bacteria cause a great variety of diseases. A useful basis for classification of some streptococci is the appearance of their colonies on blood agar. **Beta-hemolytic streptococci** form a clear zone of hemolysis. This hemolytic group includes *Streptococcus pyogenes* (also known as the antigenic *group A streptococcus*), which is the principal streptococcal pathogen. It causes scarlet fever, pharyngitis (sore throat), erysipelas, impetigo, and rheumatic fever. *Streptococcus agalactae* is the only antigenic *group B streptococcus* and causes neonatal sepsis of the newborn. **Non-beta hemolytic streptococci** (also called **alpha-hemolytic**) produce colonies surrounded by greening. *Streptococcus pneumoniae* causes pneumococcal pneumonia. Also included in this grouping are the *viridans streptococci,* although not all species produce the greening on blood agar. *Streptococcus mutans* is the cause of dental caries (cavities).

Enterococcus.   The enterococci (*Enterococcus faecalis* and *E. faecium* especially) are adapted to areas rich in nutrients but low in oxygen, such as the gastrointestinal tract, vagina, and oral cavity. They develop high resistance to antibiotics and are a cause of nosocomial infections, entering the bloodstream through indwelling catheters, surgical wounds, etc.

Listeria.   *Listeria monocytogenes* survives within phagocytic cells, can grow at refrigerator temperatures, and can cause serious damage to a fetus.

Mycoplasmatales

The mycoplasmas are highly pleomorphic because they lack a cell wall. They are exceptionally small, having one of the smallest genomes of any bacteria. Related by DNA analysis to certain gram-positive bacteria, they appear to have lost genetic material (*degenerative evolution*). Among the mycoplasmas is

Mycoplasma pneumoniae, which causes a mild pneumonia; *Spiroplasma,* which is a plant pathogen and insect parasite; and *Ureaplasma,* which enzymatically splits urea and is occasionally associated with urinary tract infections.

Actinobacteria (High G + C Gram-Positive Bacteria)

Bacteria of the Phylum Actinobacteria have a high G + C ratio; for example, the genus *Corynebacterium* has a G + C content of 51–63%.

Mycobacterium.   The mycobacteria are aerobic rods that stain acid-fast. This staining characteristic is related to their cell wall, in which the outer lipopolysaccharide layer of most gram-negative bacteria is replaced by mycolic acids. This makes it difficult for stains, nutrients, and antimicrobials to enter the cell but makes them relatively resistant to environmental stresses. Important pathogens are *Mycobacterium tuberculosis,* the cause of tuberculosis, and *Mycobacterium leprae,* the cause of leprosy.

Corynebacterium.   The corynebacteria tend to be pleomorphic. The best-known species is *Corynebacterium diphtheriae,* the cause of diphtheria.

Propionibacterium.   Some species of the genus *Propionibacterium,* which produces propionic acid, are used in the fermentation of Swiss cheese. *Propionibacterium acnes* is a skin bacterium implicated as the primary bacterial cause of acne.

Gardnerella.   *Gardnerella vaginalis* is one of the most common forms of vaginitis. Highly pleomorphic and gram-variable, its taxonomic position has always been uncertain.

Frankia, Streptomyces, Actinomyces, Nocardia.   Bacteria of the genera *Streptomyces, Actinomyces, Frankia,* and *Nocardia* are filamentous and often resemble molds by using externally carried asexual spores for reproduction. One genus, *Frankia,* causes formation of nitrogen-fixing nodules on alder tree roots. *Streptomyces* form spores at the ends of filaments. Strict aerobes, these bacteria are important degraders of proteins, starch, and cellulose in soil. They produce **geosmin,** a gas that gives soil its odor. Most commercial antibiotics are produced by *Streptomyces.* The genus *Actinomyces* consists of facultative anaerobes found in the mouth and throat of humans and animals. These bacteria occasionally form filaments that can fragment for reproduction. *Actinomyces israelii* causes actinomycosis. *Nocardia* morphologically resembles *Actinomyces;* however, its members are aerobic. Their cell wall resembles the mycobacteria and they are often acid-fast. *Nocardia asteroides* is an occasional cause of pulmonary infections or mycetoma (tissue destruction) of the hands or feet.

THE NONPROTEOBACTERIA GRAM-NEGATIVE BACTERIA

Cyanobacteria (the Oxygenic Photosynthetic Bacteria)

The cyanobacteria are photosynthetic aerobes that carry out oxygen-producing photosynthesis, much as do higher plants. Many species of cyanobacteria are capable of fixing nitrogen from the atmosphere. For this, they use enzymes carried in structures called **heterocysts.** They are morphologically varied; unicellular and filamentous forms are common.

Chlamydiae

Members of the Phylum Chlamydiae are grouped with other genetically similar bacteria that do not contain peptidoglycan in the cell walls.

Chlamydia* and *Chlamydophilia.   The chlamydia, which are gram-negative bacteria, have a unique developmental cycle that is illustrated in Figure 11.23 in the text. It requires a host mammalian cell. The

elementary body attaches to a host cell, enters, and is housed in a cell vacuole. There the elementary body becomes a **reticulate body** that divides successively. Eventually, these condense into infectious elementary bodies that are released to infect surrounding host cells. *Chlamydia trachomatis* causes trachoma, as well as the sexually transmitted diseases nongonococcal urethritis and lymphogranuloma venereum. *Chlamydophilia psittaci* causes psittacosis (ornithosis). *Chlamydophilia pneumoniae* causes a mild form of pneumonia. Chlamydia are cultivated in laboratory animals, in cell cultures, or in the yolk sac of chicken embryos.

Planctomycetes

The **planctomycetes** are a group of gram-negative, budding bacteria said to "blur the definition of what bacteria are." Their cell walls resemble those of archaea, and some (*Gemmata obscuriglobus*) have organelles that resemble the nucleus of a eukaryotic cell.

Bacterioides

Bacterioides.   The genus *Bacteroides* is a member of a group of anaerobic, gram-negative bacteria that live in the human intestinal tract in huge numbers. Infections by *Bacteroides* bacteria often result from wounds or surgery.

Cytophaga.   The motility of *Cytophaga* is due to their ability to glide over surfaces, and they are important cellulose degraders in soil.

Fusobacteria

Fusobacterium.   Another genus of gram-negative anaerobic bacteria is *Fusobacterium*. These slender microbes have pointed rather than blunt ends. They are found in the gingival crevices of the gums and may cause some abscesses.

Purple and Green Photosynthetic Bacteria (the Anoxygenic Photosynthetic Bacteria)

The purple and green photosynthetic bacteria are taxonomically confusing: There are purple sulfur bacteria and purple nonsulfur bacteria, and green sulfur bacteria and green nonsulfur bacteria. The important **purple sulfur** bacteria are gammaproteobacteria. The **purple nonsulfur** bacteria are alphaproteobacteria. The **green sulfur** and **green nonsulfur** bacteria are nonproteobacteria. For simplicity we group them all here. These are photosynthetic bacteria, which, unlike plants, do not produce oxygen (are anoxygenic) and are generally anaerobic. Their habitat is deep aquatic sediments. *Chromatium* is an important genus.

Spirochaetes

***Treponema, Borrelia, Leptospira* (commonly called spirochetes).**   The spirochetes have a coiled morphology and are motile by means of **axial filaments** (endoflagella). Axial filaments, attached at the end of the cell, are wound around the body of the cell in the space between an outer sheath and the body of the cell. The cell moves by stretching and relaxing the axial filaments. *Treponema pallidum*, the cause of syphilis; the genus *Borrelia*, which causes relapsing fever and Lyme disease; and *Leptospira* species, which cause leptospirosis, are all spirochetes.

Deinococci

The deinococci include two species of bacteria that are distinctive because of their resistance to environmental extremes. *Deinococcus radiodurans* is exceptionally resistant to radiation. *Thermus aquaticus* is unusually heat stable. It is the source of the heat-resistant enzyme *taq polymerase* which is essential to the polymerase chain reaction (PCR).

DOMAIN ARCHAEA

The Domain Archaea includes the extreme halophiles, such as *Halobacterium* or *Halococcus,* which require high concentrations of sodium chloride to grow. *Sulfolobus* also thrives in extreme environments such as acidic, sulfur-rich hot springs. The methane-producing bacteria, *Methanobacterium,* which derive their energy by combining hydrogen (H_2) with CO_2 to form methane (CH_4), are important in anaerobic sewage treatment. The archaea are grouped by related ribosomal RNA sequences; their cell walls lack peptidoglycan.

MICROBIAL DIVERSITY

A single gram of soil may contain 10,000 or so bacterial types—about twice as many as have ever been described. Illustrating such diversity, recently in the coastal waters of Africa another example of a giant bacterium, *Thiomargarita namibiensis,* has been discovered. It is a fluid-filled sphere that in this way minimizes nutrient absorption problems. It obtains its energy mainly from oxidation of hydrogen sulfide.

New methods of identifying bacteria have greatly expanded our knowledge of prokaryote diversity (see the discussion of the FISH technique on page 292 in Chapter 10 in the text). For example, an extremely small bacterium, *Pelagibacter ubique,* was only recently discovered, although it may constitute 20% of the prokaryote population of the Earth's seas. The size of bacteria can vary enormously; they can be very large, such as *E. fishelsoni* or *Thiomargarita namibiensis,* for example. At the other extreme of size, *Carsonella ruddii* has only 182 genes (it has fewer genetic requirements because it is not free-living like the mycoplasmas, but an endosymbiont of a host insect). There are reports of **nanobacteria** (as small as 0.02–0.03 nm) in deep rock formations, but most microbiologists now think these are artifacts and have proposed calling them **nanomes.**

SELF-TESTS

In the matching section, there is only one answer to each question; however, the lettered options (a, b, c, etc.) may be used more than once or not at all.

I. Matching

____ 1. Found in some nitrogen-fixing *Cyanobacteria.*

____ 2. Found in cell wall of *Mycobacterium.*

____ 3. Provide motility to spirochetes.

____ 4. Produced by many *Streptomyces.*

a. Heterocysts

b. Endoflagella

c. Flagella

d. Mycolic acids

e. Geosmin

II. Matching

_____ 1. Serovars, typhoid fever.

_____ 2. Cause of Q fever.

_____ 3. Several *Pseudomonas* species have been reclassified into this genus.

_____ 4. Grow obligately in white blood cells; cause a tickborne disease.

_____ 5. Endosymbionts of insects.

a. *Burkholderia*

b. *Ehrlichia*

c. *Coxiella*

d. *Wolbachia*

e. *Salmonella*

III. Matching

_____ 1. Endospores.

_____ 2. Anaerobic, gram-negative, slender rods with pointed ends.

_____ 3. Filamentous bacteria that produce most of our commercial antibiotics.

_____ 4. Gram-positive cocci that form grapelike clusters.

_____ 5. Gram-positive cocci that are aerotolerant anaerobes.

_____ 6. Cause of cat-scratch disease.

_____ 7. Cause of melioidosis.

a. *Clostridium*

b. *Streptococcus*

c. *Staphylococcus*

d. *Streptomyces*

e. *Fusobacterium*

f. *Bartonella*

g. *Burkholderia*

IV. Matching

_____ 1. Many of these are plant pathogens, causing plant soft-rot diseases.

_____ 2. Infectious by elementary bodies.

_____ 3. Filamentous bacteria, aerobes; cell wall resembles mycobacteria; often stain acid-fast.

_____ 4. Spirochetes.

_____ 5. Some of these are stalked and attach themselves to aquatic surfaces.

_____ 6. Many are capable of fixing nitrogen from air.

a. *Erwinia*

b. *Caulobacter*

c. *Leptospira*

d. *Nocardia*

e. *Klebsiella*

f. *Chlamydia*

V. Matching

___ 1. Genus *Homo*.

___ 2. Genus *Sulfolobus*.

___ 3. Genus *Staphylococcus*.

___ 4. Genus *Chlamydia*.

a. Domain Bacteria

b. Domain Archaea

c. Domain Eukarya

VI. Matching

___ 1. A genus of gliding bacteria that is an important cellulose degrader.

___ 2. A sheathed bacterium.

___ 3. A chemoautotrophic bacterium that participates in nitrification in soil.

___ 4. Photosynthetic bacteria that may fix nitrogen.

___ 5. Photosynthetic, anoxygenic bacteria. Often use reduced-sulfur compounds for energy and sulfur granules accumulate in the cells.

___ 6. Has organelle that resembles the nucleus of a eukaryotic cell.

a. *Nitrosomonas*

b. *Cyanobacteria*

c. *Sphaerotilus natans*

d. Purple sulfur or green sulfur bacteria

e. *Cytophaga*

f. *Beggiatoa*

g. *Gemmata obscuriglobus*

VII. Matching

___ 1. Cause of whooping cough (pertussis).

___ 2. Produces a food-poisoning enterotoxin.

___ 3. Endospores.

___ 4. Plague.

___ 5. Important for operation of an activated-sludge sewage system.

___ 6. Observed to fix nitrogen while living in close association with certain tropical grasses.

___ 7. A filamentous bacterial pathogen.

a. *Pseudomonas aeruginosa*

b. *Bordetella pertussis*

c. *Escherichia coli*

d. *Yersinia pestis*

e. *Staphylococcus aureus*

f. *Clostridium tetani*

g. *Zoogloea* spp.

h. *Nocardia asteroides*

i. *Azospirillum*

VIII. Matching

____ 1. *Bdellovibrio.*   a. Alphaproteobacteria

____ 2. *Helicobacter.*   b. Betaproteobacteria

____ 3. *Pseudomonas.*   c. Gammaproteobacteria

____ 4. *Escherichia.*   d. Deltaproteobacteria

____ 5. *Rhizobium.*   e. Epsilonproteobacteria

____ 6. *Neisseria.*   f. Nonproteobacteria

Fill in the Blanks

1. *Serratia marcescens* colonies produce a _____-colored pigment.

2. The term _____ is applied to bacteria that are helical or curved.

3. The cell walls of the archaea do not contain _____.

4. _____-hemolytic types of bacteria form a narrow, greenish zone of hemolysis on blood agar plates (supply prefix).

5. _____-hemolytic types of bacteria form a clear zone of hemolysis on blood agar plates (supply prefix).

6. *Streptococcus pyogenes* is an example of _____-hemolytic bacteria (supply prefix).

7. Appendages such as stalks or buds on bacteria are called _____.

8. Usually, nutrients are assimilated during metabolism; when they are not assimilated and external products such as hydrogen sulfide gas are formed, this is termed _____ metabolism.

9. A gram-positive bacterium with a G + C content of 35% would be considered a member of the _____ G + C gram-positive bacteria.

10. The genus *Rhizobium* is important for agriculture because it allows _____ plants such as peas to fix nitrogen.

Critical Thinking

1. Which morphological type of bacterium, a spherical cell or a filamentous cell, would be the most efficient in taking up nutrients, given that they have the same volume? Briefly explain your answer.

2. Assuming that both types of bacteria were present on the Earth before other life arose, which photosynthesizing bacterium would have been the most valuable to support other life when it appeared, cyanobacteria or purple sulfur bacteria?

3. Why do many gastric ulcer patients respond favorably to treatment with antibiotics?

4. What characteristics do you think would favor the proliferation of a bacterial species
   a. In the human intestinal tract?

   b. On a rock of newly cooled liquid lava that is just protruding above the sea?

ANSWERS

Matching

I.	1. a	2. d	3. b	4. e		
II.	1. e	2. c	3. a	4. b	5. d	
III.	1. a	2. e	3. d	4. c	5. d	6. f 7. g
IV.	1. a	2. f	3. d	4. c	5. b	6. e
V.	1. c	2. b	3. a	4. a		
VI.	1. e	2. c	3. a	4. b	5. d	6. g
VII.	1. b	2. e	3. f	4. d	5. g	6. i 7. h
VIII.	1. d	2. e	3. c	4. c	5. a	6. b

Fill in the Blanks

1. red   2. vibrioid   3. peptidoglycan   4. Alpha   5. Beta   6. beta   7. prosthecae   8. dissimilatory   9. low   10. leguminous

Critical Thinking

1. All things being equal, the ratio of surface area to volume is critical in determining the uptake of nutrients. A sphere has the smallest surface area for a given volume. Filamentous shapes generally have a much larger surface area for a given volume and take up nutrients more efficiently. However, the spherical shape is stronger for resistance to environmental stresses.

2. First, the cyanobacteria produce oxygen as they photosynthesize—a factor essential to mammalian life as we know it. Second, many cyanobacteria fix nitrogen, which would be valuable.

3. If antibacterial drugs lead to improvement, the assumption is that some gastric ulcers must be caused by stomach bacteria sensitive to the antibiotic. Do you recall whether there is such a bacterium?

4. a. Conditions in the human intestinal tract would be rich in preformed organic nutrients, so little synthesizing from simple compounds such as carbon dioxide would be required. Environmental stresses such as low moisture or radiation would be low, so cellular resistance to such stresses would be minimal. Oxygen would be limited, so strictly aerobic bacteria would not thrive.

   b. On newly cooled rock there would be minimal amounts of nutrients. The ability to obtain energy from sunlight would be very advantageous. Among the nutrients that would be especially limited are carbon and nitrogen. A bacterium that used carbon dioxide from the atmosphere and could extract nitrogen from the atmosphere would outgrow bacteria that had to obtain carbon and nitrogen from organic material. Nitrogen-fixing strains of cyanobacteria should come to mind.

12 The Eukaryotes: Fungi, Algae, Protozoa, and Helminths

FUNGI

The study of fungi is called **mycology. Molds** are multicellular, filamentous organisms that include mildews, rusts, and smuts. **Fleshy fungi** include the mushrooms and puffballs. **Yeasts** are unicellular fungi. Almost all plants depend on a symbiotic relationship with a group of fungi, the **mycorrhizae,** to efficiently absorb minerals and water.

Characteristics of Fungi

Vegetative Structures. Fungal colonies are described as **vegetative** structures. Fungal filaments are called **hyphae** and form the body (or **thallus**) of molds or fleshy fungi. Crosswalls in the **septate hyphae** are **septa; coenocytic hyphae** have no septa. A mass of hyphae is a **mycelium;** the **vegetative hyphae** are concerned with obtaining nutrients, and the **reproductive** or **aerial hyphae** are involved in reproduction. Fragments of hyphae can elongate to form new hyphae.

Yeasts. Yeasts usually reproduce by **budding,** in which a protuberance on the cell enlarges and eventually separates as another cell. If daughter cells do not detach immediately, they form a short chain, a **pseudohypha.** Some yeasts, **fission yeasts,** undergo fission. Many yeasts are facultative anaerobes and ferment carbohydrates to produce CO_2 and ethanol. Fungi that, under different conditions, grow in either yeastlike or moldlike forms (**dimorphism**) are known as **dimorphic fungi.** Features differentiating fungi from bacteria are summarized in Table 12.1 in the text.

Life Cycle. **Asexual spores** arise from one organism only. **Sexual spores** result from the fusion of nuclei from two opposite mating strains.

Asexual Spores. Fungi produce a number of different asexual spore types. A **condiospore,** or **conidium** (plural: *conidia*), is a spore not enclosed in a sac. Conidia are produced in a chain at the end of a **conidiophore.** Conidia formed by fragmentation of a septate hyphae are called **arthroconidia.** Conidia formed by buds from a parent cell are **blastoconidia.** A **chlamydoconidium** is a thick-walled spore formed by rounding and enlargement of a hyphal segment. A **sporangiospore** is formed within a **sporangium** (sac) at the end of a **sporangiophore.**

Sexual Spores. A fungal sexual spore results from the three phases of sexual reproduction: a haploid nucleus of a donor (+) cell penetrates the cytoplasm of a recipient (−) cell (**plasmogamy**). These fuse to form a diploid zygote nucleus (**karyogamy**). These, by **meiosis,** give rise to haploid nuclei in sexual spores. Most fungi exhibit only asexual spores. Sexual spores are the criteria for classifying fungi into divisions.

Nutritional Adaptations.

1. Fungi usually prefer an acidic pH, too acid for most bacteria.
2. Molds are almost all aerobic. Most yeasts are facultative anaerobes.
3. Fungi are relatively resistant to osmotic pressures; for example, they grow well at high sugar and salt concentrations.
4. Fungi are capable of growing on substances with a low moisture content, where bacteria are unable to grow.

5. Fungi require somewhat less nitrogen for equivalent weight of growth than do bacteria.
6. Fungi are capable of using many complex carbohydrates such as lignin.

Medically Important Fungi

Zygomycota.   The **Zygomycota,** or conjugation fungi, form sexual **zygospores** (resulting from fusion of nuclei of two cells), and asexual sporangiospores. Their hyphae are coenocytic.

Microsporidia.   **Microsporidia** are unusual eukaryotes because they lack mitochondria. Microsporidia do not have microtubules (see Chapter 4), and they are obligate intracellular parasites. In 1857, when they were discovered, microsporidians were classified as fungi. They were reclassified as protists in 1983 because they lack mitochondria. Recent genome sequencing, however, reveals that the microsporidians are fungi.

Ascomycota.   Members of the **Ascomycota,** or sac fungi, form septate hyphae. Their sexual spores are **ascospores** produced in an **ascus** (sac).

Basidiomycota.   The **Basidiomycota,** or club fungi, have septate hyphae. This group includes the mushrooms. They form sexual basidiospores, and some form asexual conidiospores. **Basidiospores** are formed externally on a base pedestal called a **basidium.**

The fungi we have discussed thus far are **teleomorphs;** that is, they produce both sexual and asexual spores. Some ascomycetes have lost the ability to reproduce asexually and are called **anamorphs.** Historically, fungi whose sexual cycle has not yet been observed have been placed in the **Deuteromycota.** Most of these have been found, by modern rRNA sequencing, to be **Ascomycota,** but a few are basidiomycetes.

Fungal Diseases

A fungal infection is a **mycosis.** A **systemic mycosis** occurs deep within the patient in various tissues and organs. Infections usually are transmitted by inhalation and often begin in the lungs. Fungal infections just beneath the skin are **subcutaneous mycoses.** They usually result from puncture wounds and often form disfiguring subcutaneous abscesses. An **opportunistic pathogen** is normally harmless.

Dermatophytes are fungi that infect the epidermis, hair, and nails, causing **cutaneous mycoses.** They secrete *keratinase,* which degrades a protein (**keratin**) found in hair, skin, and nails.

Superficial mycoses are localized along hair shafts and the superficial epidermal cells. **Mucormycosis** is an opportunistic mycosis caused by *Rhizopus* and *Mucor,* primarily in patients with ketoacidosis resulting from diabetes, leukemia, or treatment with immunosuppressive drugs. **Aspergillosis** is caused by *Aspergillus* and usually occurs in individuals with lung disease or cancer. **Candidiasis** is usually caused by an overgrowth of *Candida albicans.* Candidiasis, a **yeast infection,** usually occurs in the mouths and throats of newborns; vulvovaginal candidiasis may occur.

Economic Effects of Fungi

All the nutritional adaptations described in this chapter are important in understanding the reasons for fungal spoilage of dry cereals; acidic foods with high sugar content, such as jelly; growth on painted walls or on shower curtains; and the importance of fungi as plant pathogens. Examples of plant fungal diseases are Dutch elm disease and the chestnut blight; the latter essentially eliminated chestnut trees in the United States.

LICHENS

A **lichen** is a combination of a green alga (or cyanobacterium) and a fungus. This is a *mutualistic* relationship, from which each member benefits. **Crustose** lichens grow flush on a surface; **foliose** lichens are leaflike in shape; **fruticose** lichens have fingerlike projections. The photosynthetic alga provides carbohydrates, and the fungus provides protection from desiccation and a means of attachment.

In a lichen the fungal hyphae grow around algal cells to form the **medulla.** Fungal hyphae projecting below the lichen body form **rhizines,** or *holdfasts.* They also form a protective covering, the **cortex.**

ALGAE

Characteristics of Algae

Algae are photosynthetic autotrophs; that is, they use light to convert atmospheric carbon dioxide into carbohydrates for energy. Oxygen is a by-product of photosynthesis. The body of a multicellular alga is a **thallus,** which may collectively function as **holdfasts** to anchor them. They often have stemlike **stipes** and leaflike **blades** (Figure 12.12b in the text). Algae are buoyed by gas-filled **bladders.**

Asexual reproduction in multicellular or filamentous algae is accomplished by fragmentation of the thallus. Following mitosis, unicellular algae divide by simple fission. Some algae reproduce sexually.

Selected Phyla of Algae

Brown algae, or **kelp,** may be 50 meters in length. **Algin,** a product used as a food thickener and in the production of other goods, is extracted from their cell walls. **Red algae** are branched and multicellular and live at greater depths than do other algae. **Agar** and **carrageenan** are gelatinous thickeners derived from red algae. **Green algae** are usually classified as microscopic plants. **Diatoms** are unicellular or filamentous algae with walls of pectin or silica that fit together as do the two halves of a Petri dish.

Dinoflagellates are unicellular algae, referred to as **plankton** because they are free-floating. Some produce neurotoxins (called **saxitoxins**), which may be ingested by shellfish and eventually poison humans (**paralytic shellfish poisoning**). **Domoic acid intoxication** is another example. **Euglenoids** are unicellular, flagellated algae that are facultative chemoheterotrophs. They resemble protozoa, lacking a cell wall and moving with a flagellum. In the dark they may ingest organic matter.

Roles of Algae in Nature

Algae are important to the aquatic food chain because they convert carbon dioxide into consumable organic molecules. A large increase in numbers of planktonic algae is called a **bloom.** Blooms of dinoflagellates cause **red tides** in the ocean. Fish that ingest toxin-containing dinoflagellates can, when eaten by humans, cause a disease called **ciguatera.** When large numbers of algae die and decompose, oxygen dissolved in water is depleted. Some algae are symbionts of animals, providing carbohydrates for a large clam, for example. **Water molds** or **oomycota** are decomposers that form cottony masses, usually in fresh water. They resemble the zygomycete fungi and have even been classified as fungi. Many are plant parasites, such as *Phytophthora infestans,* which caused failure of the potato crop and famine in Ireland in the 1800s.

PROTOZOA

Characteristics of Protozoa

Protozoa are one-celled eukaryotic organisms that mostly feed on bacteria and small particulate nutrients. The feeding stage is called a **trophozoite.** They are classified largely on the basis of their means of motility.

Life Cycle.   Protozoa reproduce asexually by fission, budding, or **schizogony.** The last involves multiple fission of the nucleus prior to cell division. Daughter cells eventually form around each nucleus. Sexual reproduction such as **conjugation** has been observed in some ciliates. Two cells fuse, and a haploid nucleus from each migrates to the other cell. Both are now fertilized and produce daughter cells. Other protozoa produce **gametes** or **gametocytes**—haploid sex cells that fuse to form a zygote.

Encystment.   Some protozoa produce a protective, hardened capsule called a **cyst.** The members of the Apicomplexa (see below) form **oocysts.**

Nutrition.   Most protozoa are aerobic heterotrophs, but most intestinal protozoa are capable of anaerobic metabolism. A few, like *Euglena,* contain chlorophyll and are photoautotrophs. Some protozoa absorb food through the cell membrane, but many have a coating, the **pellicle,** that does not allow them to do so. Ciliates wave cilia (other than those used for motility) to bring food to an opening called the **cytostome.** Amebae engulf food and phagocytize it. Digestion in all protozoa takes place in **vacuoles,** and waste is eliminated through an **anal pore.**

Medically Important Protozoa

Archaezoa are eukaryotes that lack mitochondria. Typically they are spindle shaped with flagella on the front, pulling the cell. Examples are *Trichomonas vaginalis* and *Giardia lamblia* (also called *G. intestinalis* or *G. duodenalis*).

   Microsporidia also lack mitochondria; they also do not have microtubules.

   Amebae move by extending projections of the cytoplasm called **pseudopods.** Examples are *Entamoeba histolytica* and the genus *Acanthamoeba.*

   Apicomplexa are not motile in their mature forms and are obligate intracellular parasites. Examples are the malaria-causing members of the genus *Plasmodium.* This genus has a complex life cycle involving an animal host and a mosquito (see Figure 12.20 in the text). The sexual cycle takes place in the mosquito, resulting in **sporozoites** that enter a host by means of mosquito bites. There they enter the liver cells, undergo **schizogony,** and leave the liver cells as **merozoites.** These merozoites infect red blood cells, again undergo schizogony, and are released as more merozoites when the red blood cells rupture. This release is periodic and causes the episodes of fever and chills of malaria. Some merozoites develop into sexual forms called **gametocytes.** These are picked up by feeding *Anopheles* mosquitoes, in which a new sexual cycle begins. In the transmission of malaria, the mosquito is the **definitive host** (it harbors the sexually reproducing stage) and the human is the **intermediate host** (where the parasite undergoes reproduction).

   Other important apicomplexa are *Babesia microti* and *Toxoplasma gondii.* This latter organism reproduces in domestic cats. The trophozoite stage is called a **tachyzoite. Oocysts** are excreted with cat feces and are infective when ingested.

   Cryptosporidium causes intestinal infections when oocysts are ingested—usually from contaminated water. *Cyclospora cayetanensis* is a recently identified cause of diarrhea.

   Ciliates have cilia, shorter than flagella, for motility. The only human parasite is *Balantidium coli.*

   The ciliates, apicomplexa, and dinoflagellates may be placed in their own phylum, or kingdom, called **Alveolata.** All have membrane-bound cavities (alveoli) and rRNA sequences in common.

   The **Euglenozoa** all have common rRNA sequences, disk-shaped mitochondria, and no sexual reproduction. **Euglenoids** are photoautotrophs. Most have a red eyespot that is light sensitive. **Hemoflagellates** such as the genus *Trypanosoma,* the cause of Chagas' disease and African sleeping sickness, have long, slender bodies and an undulating membrane. They are transmitted by blood-feeding insects.

SLIME MOLDS

Slime molds have both fungal and animal characteristics and are classified as protists. **Cellular slime molds** resemble amebae at one stage. When conditions are unfavorable for growth, large numbers of ameboid cells aggregate to form a single structure, a **slug.** Cyclic AMP produced by some amebae is the attractant toward which they migrate to form the slug. The slug moves toward light and eventually forms a stalked structure with a spore cap at the top. Under favorable conditions, this spore cap differentiates into single ameba-like spores, repeating the cycle.

   Plasmodial slime molds are a mass of protoplasm called a **plasmodium.** The entire plasmodium moves like a giant ameba and engulfs organic debris and bacteria. **Cytoplasmic streaming,** which apparently distributes oxygen and nutrients, can be observed in these slime molds. When conditions are unfavorable for growth, the plasmodium separates, and stalked sporangia with spores are formed. When conditions improve, the spores germinate and the plasmodium is again formed.

HELMINTHS

Helminths are multicellular eukaryotic animals. Many are parasitic.

Characteristics of Helminths

Adult helminths may be **dioecious;** that is, there are male and female individuals. Some, however, are **hermaphroditic;** one animal has both male and female reproductive organs. The **definitive host** harbors the adult, sexually mature helminth. **Intermediate hosts** may be necessary for **larval** or developmental stages.

Platyhelminths

The **platyhelminths,** or **flatworms,** are flattened from front to back (dorsoventrally). The digestive system is incomplete; there is only one opening for entry of food and exit of waste. Trematodes and cestodes are members of this group.

Trematodes.   **Trematodes,** or **flukes,** have flat, leaf-shaped bodies with a ventral sucker and an oral sucker to hold them in place and suck fluids from the host. They also may obtain food by absorption through the outer covering, the **cuticle.** A typical life cycle is that of the lung fluke, *Paragonimus westermani.* It begins with the excretion of an egg into water. A miracidial larva (**miracidium**) develops from the egg and enters a snail. There it produces **rediae,** each of which develops into **cercariae,** which, in turn, penetrate a crayfish. There they encyst as **metacercariae.** Humans are infected by ingesting undercooked crayfish. Cercariae of diseases such as schistosomiasis are not ingested but burrow through the skin of the human host.

Cestodes.   **Cestodes,** or **tapeworms,** are intestinal parasites. The head, or **scolex,** has suckers and usually attachment hooks. The worm consists of segments called **proglottids,** which contain both male and female organs. Mature proglottids contain fertilized eggs that are shed in the feces. The egg-laden proglottids of the beef tapeworm, for example, are ingested by grazing animals. The eggs hatch in the intestine, releasing larvae that migrate to muscles, where the larval form is encysted as **cysticerci.** These may be ingested by humans, thus initiating an infestation.

In some dog and cat tapeworms, humans are only intermediate hosts. The eggs are picked up from contamination of the hands by dog feces or a dog's tongue. The cyst forming in these infections of the host tissue, usually the lung or liver, is called a **hydatid cyst** in the case of the tapeworm *Echinococcus granulosus.*

Nematodes

The **nematodes,** or **roundworms,** are cylindrical and tapered at the ends. They have a complete digestive system of mouth, intestine, and anus. Males are smaller than females and have one or two **spicules** (guides for sperm at the posterior end). Many human parasites are found in the nematode group. Parasitic nematodes (many are free-living in soil and water) do not have the succession of larval stages found in flatworms.

In some cases, the eggs are infective for humans. For example, the pinworm *Enterobius vermicularis* lives near the human anus. *Ascaris lumbricoides* lives in the intestines on semidigested food and excretes eggs in the feces. After the ingested eggs hatch in the intestine, the larvae mature in the lungs before migrating to the intestines. It is dioecious with **sexual dimorphism;** that is, males and females look distinctly different.

In hookworms the larvae are the infective form. Hookworm larvae in the soil penetrate the human skin on contact, eventually reaching the intestines. Eggs shed in feces become larvae in the soil and continue the cycle. In trichinosis, the encysted larvae are ingested in meats such as pork. In the human host, the cysts mature into adults in the intestine, and eventually eggs from the adults become larvae and encyst in human muscles.

ARTHROPODS AS VECTORS

Arthropods are joint-legged animals and include the **Arachnida,** which have eight legs (spiders, mites, ticks); **Crustacea** (crabs, crayfish); and **Insecta,** which have six legs (bees, flies). Ticks and mites (Arachnida) and lice, fleas, and mosquitoes (Insecta) are particularly important in medicine as **vectors** of disease; that is, they carry and transmit disease-causing microorganisms.

SELF-TESTS

In the matching section, there is only one answer to each question; however, the lettered options (a, b, c, etc.) may be used more than once or not at all.

I. Matching

_____ 1. The sac produced at the end of an aerial hypha called a sporangiophore.

_____ 2. Formed by the fragmentation of a hypha.

_____ 3. Buds.

_____ 4. Produced in a chain; not enclosed; asexual.

_____ 5. Formed on a base pedestal called a basidium.

_____ 6. A sexual spore resulting from the fusion of nuclei of two cells, morphologically either similar or dissimilar; produced in a sac.

_____ 7. A result of fusion of the nuclei of two cells morphologically similar to each other.

_____ 8. Fungi that produce both sexual and asexual spores.

a. Arthroconidia

b. Blastoconidia

c. Conidiophores

d. Conidiospores

e. Chlamydospores

f. Sporangiospores

g. Sporangium

h. Basidiospores

i. Zygospores

j. Ascospores

k. Teleomorphs

l. Anamorphs

II. Matching

___ 1. Tapeworm.

___ 2. Roundworm.

___ 3. Scolex.

___ 4. Proglottids.

___ 5. Hydatid cysts.

___ 6. Fluke.

a. Trematode

b. Nematode

c. Cestode

III. Matching

___ 1. A resistant form of a protozoan.

___ 2. Organelles of movement by amebae.

___ 3. An organism (e.g., a mosquito) that transmits a disease-causing organism.

___ 4. A term describing helminths with both male and female individuals.

___ 5. An outer covering on some helminths.

___ 6. An outer covering on some protozoa.

___ 7. Male and female helminths distinctly different in appearance.

a. Dioecious

b. Cuticle

c. Hermaphroditic

d. Pellicle

e. Vector

f. Pseudopods

g. Ergot

h. Sclerotia

i. Cyst

j. Schizogony

k. Sexual dimorphism

IV. Matching

___ 1. Nonmotile in their mature forms.

___ 2. *Toxoplasma gondii.*

___ 3. *Balantidium coli.*

___ 4. The cause of malaria.

___ 5. *Entamoeba histolytica.*

a. Ciliophora

b. Archaezoa

c. Amebae

d. Apicomplexa

V. Matching

_____ 1. Walls of pectin or silica fit together much like the two halves of a Petri dish.

_____ 2. May be 50 meters in length; algin is extracted from them.

_____ 3. Agar is a product of this type of organism.

_____ 4. Crustose, foliose, fruticose.

_____ 5. Cyclic AMP is involved in aggregation of individual cells.

_____ 6. Composed of a mass called a plasmodium.

_____ 7. Some species are responsible for red tides in the ocean.

_____ 8. A flagellated alga sometimes classified as a form of protozoan.

a. Dinoflagellates

b. Cellular slime molds

c. Kelp

d. Red algae

e. Plasmodial slime molds

f. Diatoms

g. Euglenoids

h. Lichens

VI. Matching

_____ 1. The vegetative, feeding form of protozoa.

_____ 2. Haploid sex cells that fuse to form a zygote.

_____ 3. In the malarial life cycle, the red blood cell eventually ruptures and releases these.

_____ 4. In the malarial life cycle, these are transmitted from the mosquito to the human host.

_____ 5. The larval form, in schistosomiasis, that burrows through the skin, transmitting the disease.

a. Merozoites

b. Gametocytes

c. Sporozoites

d. Cercaria

e. Trophozoite

VII. Matching

_____ 1. Mushrooms.

_____ 2. Conjugation fungi.

_____ 3. Club fungi.

_____ 4. Sac fungi.

a. Basidiomycota

b. Zygomycota

c. Ascomycota

VIII. Matching

___ 1. The body of a multicellular alga.

___ 2. Hyphae with no septa.

___ 3. Site of digestion in protozoa.

___ 4. Food enters protozoa through this.

___ 5. Mass of fungal hyphae.

a. Cytostome

b. Mycelium

c. Coenocytic

d. Vacuoles

e. Thallus

Fill in the Blanks

1. A fungal infection is called a(n) _____.

2. _____ are fungi that infect the epidermis, hair, and nails.

3. Fungi that are sometimes yeastlike and other times filamentous are called

   _____.

4. The common name for candidiasis of the mouth and throat is _____.

5. Insects have _____ legs.

6. Mites have _____ legs.

7. In helminths the adult, sexually mature stage is found in the _____ host.

8. In helminths the larval or developmental stages are found in the _____ host.

9. An animal with both male and female reproductive organs is called _____.

10. Fungal infections just beneath the skin, usually resulting from a puncture wound, are called

    _____ mycoses.

11. A form of sexual reproduction called _____ is found in some ciliated protozoa
    and involves two cells fusing together and exchanging haploid nuclei.

12. One form of division by protozoa involves repeated fission of nuclei prior to cell division. This is
    called _____.

13. The stemlike structures of a multicellular alga are called _____.

14. A lichen represents a(n) _____ type of association between an alga and a
    fungus.

15. The structures that buoy an alga in water are called _____.

Label the Art

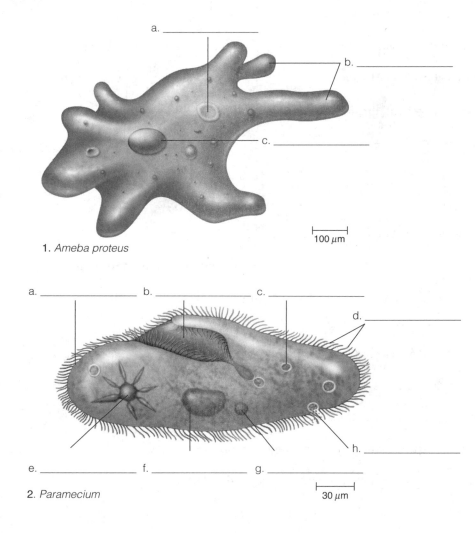

a. _____

b. _____

c. _____

1. *Ameba proteus*

100 μm

a. _____ b. _____ c. _____

d. _____

e. _____ f. _____ g. _____

h. _____

2. *Paramecium*

30 μm

Critical Thinking

1. Why has it been difficult to determine the taxonomic position of *Pneumocystis jirovecii*? Where does current research suggest it should be classified? Why?

2. List and discuss at least three ways in which algae are economically important.

3. What type of symbiotic relationship is displayed by lichens? What two organisms make up a lichen? What does each partner contribute to the relationship?

4. A patient is admitted to the hospital with dysentery and abdominal pain and cramps. Examination of a stool sample reveals protozoan trophozoites containing human red blood cells (RBCs) and cysts with four nuclei. What is the genus and species name of the protozoan? How is this protozoan transmitted?

5. Discuss at least two ways to break the chain of infection of humans with *Paragonimus westermani.*

ANSWERS

Matching

I.	1. g	2. a	3. b	4. d	5. h	6. j	7. i	8. k
II.	1. c	2. b	3. c	4. c	5. c	6. a		
III.	1. i	2. f	3. e	4. a	5. b	6. d	7. k	
IV.	1. d	2. d	3. a	4. d	5. c			
V.	1. f	2. c	3. d	4. h	5. b	6. e	7. a	8. g
VI.	1. e	2. b	3. a	4. c	5. d			
VII.	1. a	2. b	3. a	4. c				
VIII.	1. e	2. c	3. d	4. a	5. b			

Fill in the Blanks

1. mycosis 2. Dermatophytes 3. dimorphic 4. yeast infection 5. six 6. eight 7. definitive
8. intermediate 9. hermaphroditic 10. subcutaneous 11. conjugation 12. schizogony 13. stipes
14. mutualistic 15. bladders

Label the Art

1. a. Food vacuole b. Pseudopods c. Nucleus
2. a. Pellicle b. Cytostome c. Food vacuole d. Cilia e. Contractile vacuole f. Macronucleus
g. Micronucleus h. Anal pore

Critical Thinking

1. *Pneumocystis* is difficult to place taxonomically because of its lack of identifiable structures. Until recently, *Pneumocystis* was classified as a protozoan, but comparison of its rRNA to other forms of life suggests that it may be a fungus.

2. a. Algin extracted from the cell walls of algae is used to thicken foods such as ice cream and cake decorations.

   b. Algin is also used in nonfood products such as rubber tires and hand lotions.

   c. Agar is extracted from a red algae and is used to solidify microbiological media.

   d. Diatoms are used as diatomaceous earth.

3. Lichens are a combination of green algae (or cyanobacteria) and a fungus and exist in a mutualistic relationship. The algae or cyanobacteria are photosynthetic and provide nutrients to the relationship. The fungal member provides attachment and protection from desiccation in the harsh, rocky environment in which they live.

4. The protozoan is *Entamoeba histolytica*. It is transmitted by ingestion of the cyst in contaminated food or water.

5. a. Advise infected people not to defecate in the water.

   b. Eliminate the snail host.

   c. Cook crayfish before eating them.

13 Viruses, Viroids, and Prions

In the early days of microbiology, the term **filterable agents** or *filterable virus* (the word *virus* derives from the word *poison*) was used to designate an infectious agent that passed through filters that retained bacteria. Later, the term **virus** alone came into use. At the time, no one was sure of the particulate nature of these submicroscopic agents.

GENERAL CHARACTERISTICS OF VIRUSES

Viruses are **obligatory intracellular parasites** that require a living host cell in order to multiply. The term **host range** refers to the spectrum of host cells the virus can infect. Viruses that infect bacteria are called **bacteriophages** or **phages.** Most viruses are much smaller than bacteria, although some larger ones approach the size of very small bacteria. Viral size ranges from about 20 to 1000 nm (see Figure 13.1 in the text).

VIRAL STRUCTURE

A **virion** is a fully developed complete viral particle.

Nucleic Acid

Viral nucleic acid may be either DNA or RNA in double-stranded or single-stranded forms. It may be linear or even in several separate segments.

Capsid and Envelope

The protein coat is the **capsid;** it is made up of protein subunits, the **capsomeres.** The capsid may be covered by an **envelope** of some combination of lipids, proteins, and carbohydrates. Envelopes may be covered with **spikes** projecting from the surface. Some viruses use these spikes to adhere to red blood cells, causing a clumping called **hemagglutination.** Viruses not covered by an envelope are known as **nonenveloped viruses.**

General Morphology

Viruses may be classified into several morphological types.

    Helical viruses resemble long rods, their capsids a hollow cylinder with a helical structure. Examples are the tobacco mosaic virus or bacteriophage M13.

    Polyhedral viruses usually have a capsid in the shape of an *icosahedron* (a polyhedron of 20 regular triangular faces). Examples are the adenovirus and poliovirus.

    Enveloped viruses have an envelope covering their capsid. They are roughly spherical but pleomorphic (variable in shape). A helical virus (in this case the helical capsid is folded, not extended in rodlike form) such as the influenza virus is referred to as **enveloped helical.** A polyhedral virus such as herpes simplex, with a capsule, is an **enveloped polyhedral virus.**

    Complex viruses, such as the poxviruses, do not contain identifiable capsids. They may have several coats around the nucleic acid or, like many bacteriophages, have a polyhedral head and a helical tail.

TAXONOMY OF VIRUSES

In this text we group viruses according to host range; that is, animal, bacterial, or plant viruses. Current classification systems are based on type of nucleic acid, morphology, presence or absence of an envelope, and so on (see Table 13.1). A **viral species** is a group of viruses sharing the same genetic information and ecological niche. The suffix *-virus* is used for genus names, and family names end in *-viridae*.

Table 13.1 Families of Viruses That Affect Humans

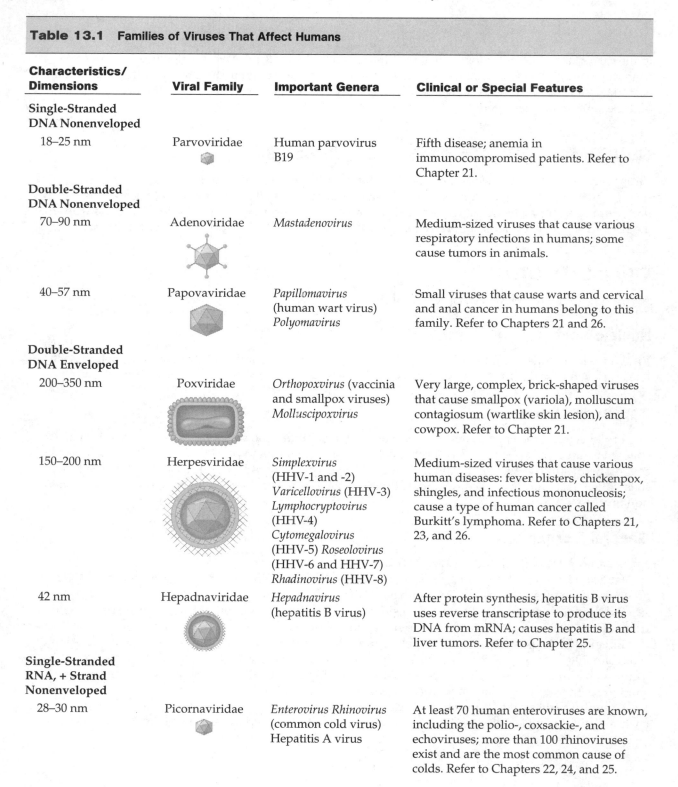

Characteristics/ Dimensions	Viral Family	Important Genera	Clinical or Special Features
Single-Stranded DNA Nonenveloped			
18–25 nm	Parvoviridae	Human parvovirus B19	Fifth disease; anemia in immunocompromised patients. Refer to Chapter 21.
Double-Stranded DNA Nonenveloped			
70–90 nm	Adenoviridae	*Mastadenovirus*	Medium-sized viruses that cause various respiratory infections in humans; some cause tumors in animals.
40–57 nm	Papovaviridae	*Papillomavirus* (human wart virus) *Polyomavirus*	Small viruses that cause warts and cervical and anal cancer in humans belong to this family. Refer to Chapters 21 and 26.
Double-Stranded DNA Enveloped			
200–350 nm	Poxviridae	*Orthopoxvirus* (vaccinia and smallpox viruses) *Molluscipoxvirus*	Very large, complex, brick-shaped viruses that cause smallpox (variola), molluscum contagiosum (wartlike skin lesion), and cowpox. Refer to Chapter 21.
150–200 nm	Herpesviridae	*Simplexvirus* (HHV-1 and -2) *Varicellovirus* (HHV-3) *Lymphocryptovirus* (HHV-4) *Cytomegalovirus* (HHV-5) *Roseolovirus* (HHV-6 and HHV-7) *Rhadinovirus* (HHV-8)	Medium-sized viruses that cause various human diseases: fever blisters, chickenpox, shingles, and infectious mononucleosis; cause a type of human cancer called Burkitt's lymphoma. Refer to Chapters 21, 23, and 26.
42 nm	Hepadnaviridae	*Hepadnavirus* (hepatitis B virus)	After protein synthesis, hepatitis B virus uses reverse transcriptase to produce its DNA from mRNA; causes hepatitis B and liver tumors. Refer to Chapter 25.
Single-Stranded RNA, + Strand Nonenveloped			
28–30 nm	Picornaviridae	*Enterovirus Rhinovirus* (common cold virus) Hepatitis A virus	At least 70 human enteroviruses are known, including the polio-, coxsackie-, and echoviruses; more than 100 rhinoviruses exist and are the most common cause of colds. Refer to Chapters 22, 24, and 25.

Table 13.1 **Families of Viruses That Affect Humans** (continued)

Characteristics/Dimensions	Viral Family	Important Genera	Clinical or Special Features
35–40 nm	Caliciviridae	Hepatitis E virus *Norovirus*	Includes causes of gastroenteritis and one cause of human hepatitis. Refer to Chapter 25.
Single-Stranded RNA, + Strand Enveloped			
60–70 nm	Togaviridae	*Alphavirus Rubivirus* (rubella virus)	Included are many viruses transmitted by arthropods (*Alphavirus*); diseases include eastern equine encephalitis (EEE), western equine encephalitis (WEE), and chikungunya. Rubella virus is transmitted by the respiratory route. Refer to Chapters 21, 22, and 23.
40–50 nm	Flaviviridae	*Flavivirus Pestivirus* Hepatitis C virus	Can replicate in arthropods that transmit them; diseases include yellow fever, dengue, and St. Louis and West Nile encephalitis. Refer to Chapters 22, 23, and 25.
80–160 nm	Coronaviridae	*Coronavirus*	Associated with upper respiratory tract infections and the common cold; SARS virus. Refer to Chapter 24.
– Strand, One Strand of RNA			
70–180 nm	Rhabdoviridae	*Vesiculovirus* (vesicular stomatitis virus) *Lyssavirus* (rabies virus)	Bullet-shaped viruses with a spiked envelope; cause rabies and numerous animal diseases. Refer to Chapter 22.
80–14,000 nm	Filoviridae	*Filovirus*	Enveloped, helical viruses; Ebola and Marburg viruses are filoviruses. Refer to Chapter 23.
150–300 nm	Paramyxoviridae	*Paramyxovirus Morbillivirus* (measles virus)	Paramyxoviruses cause parainfluenza, mumps, and Newcastle disease in chickens. Refer to Chapters 21, 24, and 25.
32 nm	Deltaviridae	Hepatitis D	Depend on coinfection with hepadnavirus. Refer to Chapter 25.
– Strand, Multiple Strands of RNA			
80–200 nm	Orthomyxoviridae	Influenza virus A, B, and C	Envelope spikes can agglutinate red blood cells. Refer to Chapter 24.
90–120 nm	Bunyaviridae	*Bunyavirus* (California encephalitis virus) *Hantavirus*	Hantaviruses cause hemorrhagic fevers such as Korean hemorrhagic fever and *Hantavirus* pulmonary syndrome; associated with rodents. Refer to Chapters 22 and 23.

Table 13.1 Families of Viruses That Affect Humans (*continued*)

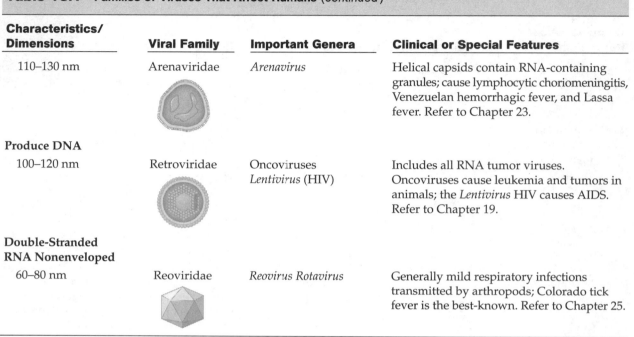

Characteristics/ Dimensions	Viral Family	Important Genera	Clinical or Special Features
110–130 nm	Arenaviridae	*Arenavirus*	Helical capsids contain RNA-containing granules; cause lymphocytic choriomeningitis, Venezuelan hemorrhagic fever, and Lassa fever. Refer to Chapter 23.
Produce DNA			
100–120 nm	Retroviridae	Oncoviruses *Lentivirus* (HIV)	Includes all RNA tumor viruses. Oncoviruses cause leukemia and tumors in animals; the *Lentivirus* HIV causes AIDS. Refer to Chapter 19.
Double-Stranded RNA Nonenveloped			
60–80 nm	Reoviridae	*Reovirus Rotavirus*	Generally mild respiratory infections transmitted by arthropods; Colorado tick fever is the best-known. Refer to Chapter 25.

ISOLATION, CULTIVATION, AND IDENTIFICATION OF VIRUSES

Growing Bacteriophages in the Laboratory

Bacteriophages require a specific host bacterium for growth. The growth medium for the host may be liquid or solid, but solid media are used for detecting and counting viruses by the **plaque method.** For this method, a melted agar suspension of host cells and bacteriophage (phage, for short) are poured in a thin layer over an agar surface on a Petri plate. The bacteria develop into a turbid lawn except where they are destroyed by proliferating phage, forming a circular clearing called a **plaque.** Phage counts are expressed in **plaque-forming units (PFU).**

Growing Animal Viruses in the Laboratory

Viruses can be cultured in suitable *living animals,* and some can be grown only in this way. Because signs of disease in the animal are often significant, this method can be used in diagnosis.

Embryonated eggs can be inoculated by a hole drilled in the shell. Growth may be detected by death of the embryo or formation of pocks or lesions on the membranes.

Most recently, **cell culture** has been the method of choice for viral cultivation. Animal (or plant) cells may be separated and grown as homogeneous collections of cells, not unlike bacterial cultures. In containers, the cells tend to adhere to surfaces and form a **monolayer** of cells. Cell infection by a virus causes observable death or damage known as **cytopathic effects (CPE),** which can be used, much as plaques are, for counting or detecting viruses. **Primary cell lines** are derived directly from tissue and tend to die after a few generations, but a few specialized human cell lines may be cultivated for 100 generations or so. **Diploid cell lines,** developed from embryonic human cells, are used to culture rabies virus for human diploid cell vaccines. They can be maintained for about 100 generations. **Continuous cell lines,** often cancer cells such as the HeLa cells, can be maintained for an indefinite number of generations.

Viral Identification

The most common methods of identification are serological. The virus is detected and identified by its reaction with antibodies, which are specific proteins produced by animals in response to the virus. (Antibodies will be discussed in Chapter 17 and specific methods for viral identification will be discussed in Chapter 18.)

VIRAL MULTIPLICATION

The virus has only a few genes. Most enzymes encoded in viral nucleic acid are not part of the virion but are synthesized and function only within the host cell. The viral enzymes mostly replicate viral nucleic acid and seldom the viral proteins, which are supplied by the host. Thus, the virus must invade a host cell and take over its metabolic machinery.

Multiplication of Bacteriophages

The most familiar example of a viral life cycle is the **lytic cycle** of the **T-even phages** (T2, T4, T6). The tail of the phage is **adsorbed** or **attached** to the host cell. This is a highly specific reaction depending on a complementary receptor site. The phage forms a hole in the cell wall using **phage lysozyme** and drives the tail core through the cell wall (**penetration**); it then injects the DNA of the virus into the cytoplasm. The head (capsid) remains outside. The viral DNA causes transcription of RNA from viral DNA and thus commandeers the metabolic machinery of the host cell for its own biosynthesis. For several minutes following infection, complete phages cannot be found; this is called the **eclipse period.** During the **maturation period** that follows, the phage DNA and capsid, formed separately, are assembled into virions. When complete, the host cell **lyses** and **releases** these virions. The time required from phage adsorption to release is the **burst time** (about 20–40 minutes), and the number released is the **burst size** (about 50–200). The stages of phage multiplication can be demonstrated with the **one-step growth experiment** (see Figure 13.11 in the text).

Lysogeny.   Sometimes the lytic cycle just described does not occur. The phage DNA becomes incorporated as a **prophage** into the host's DNA, a state called lysogeny (Figure 13.12 in the text). Such phages are **lysogenic** or **temperate** phages (such as bacteriophage lambda [λ] and their bacterial host, **lysogenic cells.** The lysogenized cell may exhibit new properties, **phage conversion,** such as toxin production (examples are scarlet fever, diphtheria, and botulism). The prophage is reproduced along with the bacterial chromosome but can be induced to complete the lytic cycle. In **specialized transduction,** a lysogenic phage incorporates small amounts of host DNA along with its own DNA and can confer this DNA to a newly infected cell.

Multiplication of Animal Viruses

Multiplication of animal viruses follows the general pattern just described, but with important differences. Animal viruses have no tail, so attachment is by spikes, small fibers, and so on. Penetration occurs by fusion of the envelope with the host plasma membrane or by the nonenveloped virus somehow entering the cytoplasm. Penetration by **endocytosis** requires the cell's plasma membrane to fold inward as vesicles. The host enfolds the virus into this vesicle, bringing it into the cell. An alternative method of penetration is **fusion.** The viral envelope fuses with the plasma membrane and releases the capsid into the host cell's cytoplasm. Once inside the host cell, the viral nucleic acid separates from the protein coat, a process called **uncoating.**

The Biosynthesis of DNA Viruses

Multiplication of DNA viruses may occur entirely in the cytoplasm (poxviruses). Or, the DNA may be formed in the nucleus and the protein in the cytoplasm, with the final assembly taking place in the nucleus.

Adenoviridae.   Adenoviruses are the cause of some common colds; they are named after adenoids.

Poxviridae.   Poxviruses cause infections such as smallpox. *Pox* are pus-filled sacs on skin.

Herpesviridae.   Herpesviruses are named after the spreading (*herpetic*) appearance of cold sores. Official names are human herpesviruses (HHV) numbered for identification. Most are more commonly known by their vernacular names.

HHV-1   (Herpes simplex 1)

HHV-2   (Herpes simplex 2)

HHV-3   (*Varicella,* or chickenpox virus)

HHV-4   (Epstein-Barr virus)

HHV-5   (*Cytomegalovirus*)

HHV-6   (*Roseolovirus*)

HHV-7   (Mostly infecting infants)

HHV-8   (Cause of Kaposi's sarcoma)

Papovaviridae.   Papovaviruses are named for *pa*pillomas (warts), *po*lyomas (tumors), and *va*cuolation (cytoplasmic vacuoles produced by some of these viruses). DNA viruses, such as papovaviridae, replicate in the nucleus of the host cell (Figure 13.15 in the text). Basically, protein is synthesized by transcription and translation in the conventional manner, similar to the process in bacteria.

Hepadnaviridae.   Named because they cause *hepa*titis and contain *DNA,* hepadnaviruses differ from other DNA viruses because they synthesize DNA by copying RNA with reverse transcriptase, which will be discussed soon with retroviruses.

The Biosynthesis of RNA Viruses

Multiplication of RNA viruses is essentially similar to that of DNA viruses, but it takes place in the cytoplasm. Of course, the transcription of DNA to mRNA is not needed.

Picornaviridae (from *pico*, meaning small, and *RNA*).   Picornaviruses are single-stranded RNA viruses such as poliovirus. The single strand (+ or **sense strand**) acts as mRNA. It serves as a means to make *RNA-dependent RNA polymerase,* which catalyzes the synthesis of another strand of RNA (– or **antisense strand**). The – strand serves as a template for + strands that in turn serve as a means to produce viral RNA or viral protein.

Togaviridae (from *toga*, or covering).   Also containing a single + strand of RNA, togaviruses differ from picornaviruses in that two types of mRNA are transcribed from the – strand. One codes for capsid proteins and the other for envelope proteins.

Rhabdoviridae (from *rhabdo-*, or rod).   Rhabdoviruses are usually bullet-shaped (such as the rabies virus) and contain a single – strand of RNA. Because rhabdoviruses already contain RNA-dependent RNA polymerase, they do not have to synthesize this enzyme. The RNA polymerase produces a + strand, which serves as mRNA and a template for synthesis of viral RNA.

Reoviridae (from the first letters of *respiratory, enteric, and orphan*).   Reoviruses contain double-stranded RNA. One of the capsid proteins of these viruses serves as RNA-dependent RNA polymerase. After the capsid enters a host cell, mRNA is produced inside the capsid and released into the cytoplasm, where it is used to synthesize more viral proteins. One of these proteins acts as RNA-dependent RNA polymerase to produce – strands of RNA. These – strands and the mRNA + strands form the double-stranded RNA in reoviruses.

Retroviridae (from *reverse transcriptase*).   Some retroviruses cause cancers, and one type is the cause of **acquired immunodeficiency syndrome (AIDS)**. These viruses carry a polymerase (**reverse transcriptase**) that uses the RNA of the virus to make a complementary strand of DNA. This DNA becomes integrated into the DNA of a host cell (**provirus**), and transcription into mRNA may then take place normally.

Maturation and Release.   For enveloped cells, the envelope develops around the capsid from the plasma membrane by a process called **budding,** which occurs as the nucleic acid enclosed in the capsid pushes out through the plasma membrane. Budding does not necessarily kill the host cell. Nonenveloped viruses released by host cells usually cause **lysis** and death of the host cell.

VIRUSES AND CANCER

When cells multiply in an uncontrolled way, the excess tissue is called a tumor, which is **malignant** if cancerous and **benign** if not. **Leukemias** are not solid tumors but an excess production of white cells. Chicken **sarcoma** (cancer of connective tissue) and **adenocarcinoma** (cancer of glandular tissue) can be transmitted by viruses.

The Transformation of Normal Cells into Tumor Cells

It is believed that cancer-causing changes in cellular DNA are directed by parts of the genome called **oncogenes.** Mutations can cause oncogenes to bring about cancerous transformations of cells. Oncogenes can be activated by chemicals, oncogenic viruses, and radiation. A tumor cell formed by activation of an oncogene is said to have undergone **transformation** and is distinctly different from normal cells. Viruses that activate oncogenes are called **oncogenic viruses,** or **oncoviruses.** Sometimes the provirus remains latent, much like lysogeny, and replicates only with the host cell; or it may become transcribed and produce new, infective viruses. Finally, it may convert the host cell into a tumor cell. Transformed cells also contain **tumor-specific transplantation antigens (TSTA)** on the surface, or **T antigens** in the nucleus.

DNA Oncogenic Viruses

The adenovirus, herpesvirus, poxvirus, and papovavirus groups all contain **oncogenic viruses.** Among the papovaviruses are the papilloma viruses that cause warts. Herpesviruses, including the Epstein-Barr (EB) virus, may cause Burkitt's lymphoma or nasopharyngeal carcinoma. The hepadnavirus causing hepatitis B also has a role in liver cancer.

RNA Oncogenic Viruses

Only the retroviruses seem to be oncogenic among the RNA types. Their oncogenic activity seems related to the production of reverse transcriptase. The DNA synthesized from viral RNA becomes integrated into the host cell's DNA (provirus). In some cases, this may convert the host cell into a tumor cell.

LATENT VIRAL INFECTIONS

Sometimes the virus remains latent in the nerve cells of the host for long periods without causing disease: a **latent infection.** Stress or some other cause may trigger its reappearance. This is the case with the herpes simplex virus, which causes cold sores, and the chickenpox virus, which causes shingles.

PERSISTENT VIRAL INFECTIONS

The term **persistent** or **chronic viral infection** refers to a disease process that occurs gradually over a long period. Originally called **slow viral infection,** the term refers to the slow progress of the disease. An example may be *subacute sclerosing panencephalitis,* in which the measles virus continues to reproduce slowly, causing this rare encephalitis.

PRIONS

A number of neurological diseases, called *spongiform encephalopathies,* may be caused by **prions** (coined from *proteinaceous infectious particle*), which have characteristics unique to biology. The prototype of these diseases is *scrapie,* a neurological disease of sheep. The prion appears to be pure protein and to lack nucleic acids. Among hypotheses to explain their reproduction is that the prion protein is a gene found in normal host DNA, and that an abnormal form of the protein causes the disease condition. Another possibility is that the prion may contain an undetectably small amount of nucleic acid. Diseases caused by these agents, other than scrapie in sheep, include mad cow disease, Creutzfeldt-Jakob disease, kuru, and Gerstmann-Straüssler-Scheinker syndrome.

These diseases are caused by conversion of the normal host glycoprotein called **PrP^C** (for cellular prion protein) into an infectious form called **PrP^{Sc}** (for scrapie protein). In brief summary of the prion infection, PrP^C produced by cells is secreted on the cell surface. During infection the infectious prion, PrP^{Sc}, reacts with PrP^C on the cell surface and converts it into PrP^{Sc}. The PrP^{Sc} is then taken up by the cell (endocytosis) and accumulates in lysosomes in the cell. Fragments of PrP^{Sc} accumulate in the brain, forming plaques. These plaques are important for diagnosis but do not appear to be the cause of cell damage.

PLANT VIRUSES AND VIROIDS

Some plant diseases are caused by **viroids.** These are very short pieces of nonenveloped RNA with no protein coat.

SELF-TESTS

In the matching section, there is only one answer to each question; however, the lettered options (a, b, c, etc.) may be used more than once or not at all.

I. Matching

_____ 1. A complete, assembled virus.

_____ 2. The subunits making up the protein outer coating of most viruses.

_____ 3. The protein outer coating of most viruses.

_____ 4. A term derived from the word for poison.

_____ 5. A combination of lipids, proteins, and carbohydrates covering the protein coating of a virus.

_____ 6. Infectious prion.

a. Virion

b. Capsid

c. Capsomere

d. Envelope

e. Virus

f. PrP^C

g. PrP^{Sc}

II. Matching

____ 1. Describes the morphology of the capsid of many viruses.

____ 2. A method by which a virus enters an animal host cell.

____ 3. A cell line derived from tissue that normally reproduces for relatively few generations.

____ 4. The HeLa cell line would be placed in this group.

____ 5. A clearing in a "lawn" of susceptible bacterial cells.

____ 6. The number of bacteriophages produced by one bacterial host cell.

____ 7. Presumed agent causing diseases such as sheep scrapie.

____ 8. A bacterial virus.

____ 9. A short strand of RNA virus without a capsid.

____ 10. PrP.

a. Burst size

b. Burst time

c. Primary cell line

d. Continuous cell line

e. Plaque

f. Cytopathic effect

g. Icosahedral

h. Endocytosis

i. Phage

j. Viroid

k. Diploid cell line

l. Prion

III. Matching

____ 1. Describes a method by which an enveloped virus leaves the host cell while acquiring the envelope.

____ 2. Describes growth characteristics of normal cell cultures in glass or plastic containers.

____ 3. A term meaning cancer-causing.

____ 4. Observable changes in a virus-infected cell.

____ 5. The time during which the capsids and DNA of a phage, already formed, are now assembled into complete viruses.

a. Replicative form

b. Maturation period

c. Budding

d. Oncogenic

e. Cytopathic effect

f. Endocytosis

g. Monolayer

h. Eclipse period

IV. Matching

____ 1. Cancer of connective tissue.

____ 2. The clumping of red blood cells due to adherence to spikes on viruses.

____ 3. Equivalent to mRNA in a single-stranded RNA virus.

____ 4. RNA to DNA.

a. Sarcoma

b. + or sense strand

c. Reverse transcription

d. Interferon

e. Hemagglutination

V. Matching

_____ 1. Varicella virus.

_____ 2. Herpes simplex 2.

_____ 3. Epstein-Barr virus.

_____ 4. Cytomegalovirus.

_____ 5. Cause of Kaposi's sarcoma.

a. Human herpesvirus 3

b. Human herpesvirus 4

c. Human herpesvirus 5

d. Human herpesvirus 8

e. Human herpesvirus 2

Fill in the Blanks

1. The virus, once inside the host cell, separates the viral nucleic acid from the capsid; this is called

_____.

2. Another term for a lysogenic phage is _____ phage.

3. _____ are not solid tumors but an excessive production of white blood cells.

4. Many viruses can be grown in _____ eggs.

5. The herpes simplex virus remains _____ in nerve cells of the host for long periods without causing disease.

6. Counts of phage are made in terms of _____ units.

7. An oncogene might become active when placed on the chromosome in a position where normal controls are not active; this is termed _____.

8. The term _____ refers to the spectrum of host cells the virus can infect.

9. When cells multiply in an uncontrolled way, the excess tissue is called a _____.

10. Oncogenic viruses are those that _____ cells into tumor cells.

11. The type of virus implicated as a cause of AIDS is a(n) _____.

12. The abbreviation TSTA stands for tumor-specific _____ antigens.

13. For several minutes following infection by a phage, no complete phages can be found in the host cell; this is called the _____ period.

14. The _____ of the phage is adsorbed to the host cell.

15. The phage forms a hole in the cell wall using phage _____ and drives the tail core through the cell wall.

16. Sometimes the lytic cycle does not occur upon phage infection of a host bacterium. The phage DNA becomes incorporated as a(n) _____ into the host's DNA.

17. When the phage DNA is incorporated into the host's DNA, this state is called _____.

18. Transformed cells lose _____ ; that is, they do not stop reproduction when in contact with neighbor cells.

19. The hepadnavirus has genetic material called _____. (Select DNA or RNA.)

20. Picornaviruses have genetic material called _____. (Select DNA or RNA.)

21. Tumors are malignant when cancerous and _____ when not cancerous.

Critical Thinking

1. What feature of the viral life cycle makes it difficult to produce antiviral drugs?

2. How are viruses able to avoid the action of antibodies?

3. Compare and contrast the lytic and lysogenic cycles of the T-even bacteriophages.

4. By what mechanism may retroviruses induce tumors?

5. During 1993, several deaths caused by a virus occurred in the southwestern United States. Eventually, other cases surfaced in other parts of the country. What method was used to isolate the viral agent? What genus of virus caused the outbreak?

ANSWERS

Matching

I.   1. a   2. c   3. b   4. e   5. d   6. g
II. 1. g   2. h   3. c   4. d   5. e   6. a   7. l   8. i   9. j   10. l
III. 1. c   2. g   3. d   4. e   5. b
IV. 1. a   2. e   3. b   4. c
V.   1. a   2. e   3. b   4. c   5. d

Fill in the Blanks

1. uncoating   2. temperate   3. Leukemias   4. embryonated   5. latent   6. plaque-forming
7. translocation   8. host range   9. tumor   10. transform   11. retrovirus   12. transplantation
13. eclipse   14. tail   15. lysozyme   16. prophage   17. lysogeny   18. contact inhibition   19. DNA
20. RNA   21. benign

Critical Thinking

1. The problem results from the fact that viruses take over the reproductive machinery of host cells to replicate. This means that drugs that inhibit viral replication will also affect reproduction of the host's cells.

2. When viruses infect a host, the host's immune system reacts by producing specific antibodies that act against that virus. Some viruses are able to escape antibodies because proteins on their surface or on their spikes mutate. This means that the antibodies that were originally formed will no longer react with the virus, making them ineffective.

3. The final stage of the lytic cycle involves release of the virions from the host cell. This is accomplished when lysozyme is synthesized within the cell. This enzyme breaks down the cell wall, resulting in lysis and release of the virions. In the lysogenic cycle, the phage remains latent, incorporating its nucleic acid into that of the host. The lytic cycle may be induced by some spontaneous event such as exposure to UV light. Lysogeny also results in the following:

   a.   Lysogenic cells are immune to reinfection by the same phage.

   b.   The infected host cell may have new properties.

   c.   Lysogeny makes specialized transduction possible.

4. Retroviruses induce tumors because some of them contain promoters that turn on oncogenes; others actually contain oncogenes. Also, the fact that the double-stranded DNA of these viruses (produced by reverse transcription of the viral RNA) is incorporated into the DNA of the host and introduces new material to the host's genome can in itself cause problems.

5. A method referred to as PCR was used to amplify RNA from autopsy specimens and eventually helped researchers to identify *Hantavirus* as the cause of the mysterious deaths.

14 Principles of Disease and Epidemiology

PATHOLOGY, INFECTION, AND DISEASE

We are susceptible to **pathogens** (disease-causing microorganisms). **Pathology** is the science that deals with the study of disease. It involves the **etiology** (cause) of the disease, the manner in which a disease develops (**pathogenesis**), the structural and functional changes brought about by the disease, and the final effects on the body. **Infection** means invasion or colonization of the body (the host, in this case) by potentially pathogenic microorganisms. **Disease** itself is any change from a state of health; it is an abnormal state in which the body is not properly adjusted or capable of performing its normal functions.

NORMAL MICROBIOTA

Microorganisms that establish permanent residence without producing disease are known as **normal flora** or **normal microbiota.** Other microorganisms that may be present for a time and then disappear are called **transient microbiota.**

Normal microbiota can benefit the host by preventing the overgrowth of harmful microorganisms, a process called **microbial antagonism** or **competitive exclusion.** An example is the production of **bacteriocins** by *E. coli* cells in the large intestine, which inhibits pathogens such as *Salmonella* and *Shigella.* The relationship between the normal microbiota of a healthy person and that person is called **symbiosis.** If, in the symbiosis, one of the organisms is benefited and the other unaffected, the relationship is known as **commensalism**. If both organisms are benefited, it is called **mutualism,** and if one organism is benefited at the expense of the other, it is **parasitism**.

Under certain conditions these relationships can change, and members of the normal microbiota can become **opportunistic pathogens.**

THE ETIOLOGY OF INFECTIOUS DISEASE

Not all diseases are caused by microorganisms. We have **inherited (genetic)** diseases (like hemophilia) and **degenerative diseases** (osteoarthritis). Here we will discuss **infectious diseases**—those caused by microorganisms.

Koch's Postulates

Koch's postulates must be fulfilled to demonstrate that a specific microorganism is the cause of a specific disease.

1. The same pathogen must be present in every case of the disease.
2. The pathogen must be isolated from the diseased host and grown in pure culture.
3. The pathogen from the pure culture must cause the disease when inoculated into a healthy, susceptible laboratory animal.
4. The pathogen must again be isolated from the inoculated animal and must be shown to be the same pathogen as the original organism.

Exceptions to Koch's Postulates

There are some exceptions to Koch's postulates. For example, a few diseases, such as syphilis, are caused by organisms that cannot be isolated and virulent strains grown on laboratory media. However, evidence based on the first postulate over many years' experience has often been adequate to show that a certain organism is associated with the disease. Some diseases have such clearly defined symptoms—diphtheria or tetanus, for example—that associations between specific microorganisms and the disease are obvious. Other diseases—such as pneumonia, peritonitis, and meningitis—may be caused by a variety of microorganisms, and the specific etiology is not easy to determine from the symptoms. Furthermore, some pathogens may infect a number of different organs or tissues and cause very different diseases or disease symptoms. A good example is *Streptococcus pyogenes*, which can cause sore throats, scarlet fever, erysipelas, and puerperal fever.

CLASSIFYING INFECTIOUS DISEASES

Symptoms are changes in body function felt by the patient, such as pain and **malaise** (a vague feeling of body discomfort), that are *subjective* and not apparent to an observer. The patient also may exhibit **signs,** *objective* changes that the physician can observe and measure. A specific group of symptoms or signs accompanying a particular disease is called a **syndrome.**

Communicable diseases are spread directly or indirectly from one host to another; typhoid fever and tuberculosis are examples. **Noncommunicable diseases** are caused by microorganisms such as *Clostridium tetani,* which only produces tetanus when it is introduced into the body by contamination of wounds. A **contagious disease** is easily spread from one person to another.

Occurrence of a Disease

The **incidence** of a disease is the fraction of a population that contracts it during a particular length of time. The **prevalence** of a disease is the fraction of the population that has the disease at a given time. If a disease occurs only occasionally, it is called **sporadic;** when it is constantly present, as is the common cold, it is termed **endemic.** If many people in a given area acquire a certain disease in a short period of time, it is referred to as an **epidemic** disease, such as influenza. A worldwide epidemic is referred to as a **pandemic** disease.

Severity or Duration of a Disease

An **acute disease** is one that develops rapidly but lasts only a short time—influenza, for example. A **chronic disease** develops more slowly and the body reactions are often less severe, but it is continuous or recurrent for long periods of time. Tuberculosis, syphilis, and leprosy are such diseases. Diseases intermediate between acute and chronic are described as **subacute.** A **latent disease** is one in which the pathogen is inactive for a time but then becomes active to produce the symptoms. Shingles is an example. When many immune people are present in a community, **herd immunity** exists; that is, susceptible persons are so few that a communicable disease does not cause an epidemic.

Extent of Host Involvement

A **local infection** is one in which the invading microorganisms are limited to a relatively small area of the body, as, for example, in boils or abscesses. In a **systemic,** or **generalized,** infection, microorganisms or their products are spread throughout the body by the blood or lymphatic system. A **focal infection** is one in which a local infection, such as infected teeth, tonsils, or sinuses, enters the blood or lymph and spreads to other parts of the body. The presence of bacteria in the blood is known as **bacteremia.** **Sepsis** is a toxic, inflammatory condition arising from the spread of bacteria or bacterial toxins from a focus of infection. **Septicemia** is sepsis that results from the proliferation of bacterial pathogens in the bloodstream. **Toxemia** is the presence of toxins in the blood, and **viremia** is the presence of viruses in the blood.

A **primary infection** is an acute infection that causes the initial illness. A **secondary infection** is one caused by an opportunist only after the primary infection has weakened the body's defenses. A **subclinical (inapparent) infection** is one that does not cause any noticeable illness.

PATTERNS OF DISEASE

Predisposing Factors

Predisposing factors, such as gender, genetic background, climate, age, and nutrition, can greatly affect the occurrence of disease in individuals.

Development of Disease

The development of disease follows a certain sequence of steps. The **incubation period** is the time between actual infection and the first appearance of signs or symptoms. The **prodromal period** follows the incubation period in some diseases and is characterized by mild symptoms of the disease. During the **period of illness,** the overt symptoms of the disease are apparent. During the **period of decline,** the signs and symptoms subside. The patient regains his or her prediseased state during the **period of convalescence.**

THE SPREAD OF INFECTION

Reservoirs of Infection

A continual source of the pathogen, such as an animal or inanimate object, is a **reservoir of infection.**

Human Reservoirs.   Many people harbor pathogens and transmit them to others, directly or indirectly. These people may be diseased and obvious transmitters, but others, called **carriers,** do not exhibit symptoms.

Animal Reservoirs.   Diseases that occur primarily in wild and domestic animals but that can be transmitted to humans are called **zoonoses.** Transmission may be by direct contact with infected animals; contamination of food and water; insect vectors; contact with contaminated hides, fur, or feathers; or consumption of infected animal products.

Nonliving Reservoirs.   Examples of nonliving reservoirs for infectious diseases are soil, which harbors the agent of botulism, and water contaminated by human or animal feces, which transmits gastrointestinal pathogens.

Transmission of Disease

Contact Transmission.   Infections may be spread more or less directly from one host to another by **direct contact transmission,** such as kissing, handshaking, bites, or sexual intercourse. **Droplet infection,** in which agents of disease are spread very short distances (less than a meter) while contained in droplets of saliva or mucus from coughing or sneezing, also is considered a form of contact transmission. **Indirect contact transmission** involves a nonliving object, such as a drinking cup or towel, called a **fomite.**

Vehicle Transmission.   Inanimate reservoirs such as food, water, or blood may transmit diseases (**vehicle transmission**) to large numbers of individuals. Diseases spread by agents of infection traveling on droplets or dust for a distance of more than a meter are considered to occur by *airborne transmission.* Spores produced by fungi also can be transmitted by the airborne route.

Vectors.   Arthropods are the most important group of disease **vectors**—animals that carry pathogens from one host to another. In **mechanical transmission,** insects, such as flies, carry pathogens on their bodies to food that is later swallowed by the host. In **biological transmission**, the arthropod may pass the pathogen in a bite, or it may pass the pathogen in its feces, which later enters the wound caused by the arthropod's bite.

Portals of exit are the routes by which a pathogen leaves the body, often to be transmitted to others. Examples are mouth, nose, genitourinary tract, blood, and feces.

NOSOCOMIAL (HOSPITAL-ACQUIRED) INFECTIONS

A **nosocomial** infection does not show any evidence of being present or incubating at the time of admission to a hospital; it is acquired as a result of a hospital stay. (The word nosocomial is derived from the Greek word for hospital; the term also includes infections acquired in nursing homes and other health care facilities.) In recent years, the term **health care–associated infection (HAI)** has been introduced to include infections acquired in settings other than just hospitals. These include same-day surgical centers, ambulatory outpatient health care clinics, nursing homes, rehabilitation facilities, and in-home health care environments.

Nosocomial infections result from the interaction of several factors:

1. microorganisms in the hospital environment,
2. the compromised (or weakened) status of the host, and
3. the chain of transmission in the hospital.

Microorganisms in the Hospital

Although every effort is made to kill or check the growth of microorganisms in the hospital, the hospital environment is a major reservoir for a variety of pathogens. One reason is that certain normal microbiota of the human body are opportunistic and present a particularly strong danger to hospital patients. In fact, most of the microbes that cause nosocomial infections do not cause disease in healthy people but are pathogenic only for individuals whose defenses have been weakened by illness or therapy. In addition to being opportunistic, some microorganisms in the hospital become resistant to antimicrobial drugs, which are commonly used there.

Compromised Host

A compromised host is one whose resistance to infection is impaired by disease, therapy, or burns. Two principal conditions can compromise the host: broken skin or mucous membranes, and a suppressed immune system.

Chain of Transmission

Given the variety of pathogens (and potential pathogens) in the hospital and the compromised state of the host, routes of transmission are a constant concern. The principal routes of transmission of nosocomial infections are (1) direct contact transmission from hospital staff to patient and from patient to patient and (2) indirect contact transmission through fomites and the hospital's ventilation system (airborne transmission).

Control of Nosocomial Infections

Control measures aimed at preventing nosocomial infections vary from one institution to another, but certain procedures are generally implemented. It is important to reduce the number of pathogens to which patients are exposed by using aseptic techniques, handling contaminated materials carefully, insisting on frequent and thorough handwashing, educating staff members about basic infection control measures, and using isolation rooms and wards.

EMERGING INFECTIOUS DISEASES

Emerging infectious diseases (EIDs) are ones that are new or changing, showing an increase in incidence in the recent past or a potential for increase in the near future.

EPIDEMIOLOGY

The science that deals with the transmission of diseases in the human population, and where and when they occur, is called **epidemiology.** An epidemiologist determines not only the etiology of a disease, but also data such as geographical distribution, and gender, age, and so on, of persons affected. First, an epidemiologist collects data such as the place(s) where a disease occurred and time(s) when it occurred. Persons affected by the disease would provide information on age, sex, personal habits, and so on. This process is known as **descriptive epidemiology.**

These data are then studied (**analytical epidemiology**) to look for common factors among the affected persons that might have preceded the disease outbreak.

Experimental epidemiology tests a hypothesis, such as assumed effectiveness of a drug. Randomly selected groups receive one of two substances, either the selected drug or a *placebo* (a substance that has no effect). The groups are compared to discern the effectiveness of the drug versus the placebo.

The Centers for Disease Control and Prevention (CDC), a branch of the U.S. Public Health Service located in Atlanta, Georgia, are a central source of epidemiological information in the United States. The CDC issues the *Morbidity and Mortality Weekly Report (MMWR),* which contains data on **morbidity** (relative incidence of a disease) and **mortality** (deaths from a disease). **Notifiable infectious diseases** are those for which physicians must report cases to the Public Health Service.

SELF-TESTS

In the matching section, there is only one answer to each question; however, the lettered options (a, b, c, etc.) may be used more than once or not at all.

I. Matching

_____ 1. Invasion or colonization of the body by potentially pathogenic microorganisms.

_____ 2. The cause of a disease.

_____ 3. A change from a state of health, in which the body is not properly adjusted or capable of performing its normal functions.

_____ 4. The manner in which a disease develops.

a. Infection

b. Pathogenesis

c. Disease

d. Etiology

II. Matching

____ 1. One organism is benefited at the expense of another.

____ 2. The general relationship between the normal microbiota and the host.

____ 3. One of the organisms is benefited and the other unaffected.

____ 4. A symbiosis that benefits both organisms.

____ 5. Live bacterial cultures intended to exert a beneficial effect.

a. Symbiosis

b. Opportunistic

c. Commensalism

d. Mutualism

e. Parasitism

f. Prebiotics

g. Probiotics

III. Matching

____ 1. First mild symptoms appear.

____ 2. The individual regains strength, and the body returns to its prediseased state.

____ 3. The time between infection and the first appearance of signs and symptoms.

a. Prodromal period

b. Period of convalescence

c. Period of incubation

d. Period of illness

IV. Matching

____ 1. Easily spread from one person to another person.

____ 2. Tetanus is an example.

a. Contagious disease

b. Communicable disease

c. Noncommunicable disease

V. Matching

____ 1. An inanimate object that may transmit disease.

____ 2. A group of symptoms associated with a disease.

____ 3. Identification of a disease.

____ 4. Objective changes caused by a disease that the physician may observe.

____ 5. An arthropod, for example, that carries malaria.

____ 6. A toxic, inflammatory condition arising from the spread of bacteria or bacterial toxins from a focus of infection.

a. Sepsis

b. Bacteremia

c. Syndrome

d. Diagnosis

e. Signs

f. Vector

g. Fomite

h. Septicemia

VI. Matching

____ 1. People who transmit diseases, but who do not exhibit any symptoms of illness.

____ 2. A disease that occurs only occasionally.

____ 3. A worldwide epidemic.

____ 4. Diseases acquired in a hospital.

____ 5. Diseases that occur in animals and can be transmitted to humans.

a. Nosocomial

b. Zoonoses

c. Carriers

d. Sporadic

e. Pandemic

Fill in the Blanks

1. In _____ transmission of disease, an insect such as a fly carries the pathogen on its body to human food.

2. The _____ of a disease is the fraction of the population that contracts it during a particular period of time.

3. _____ disease is one that develops rapidly but lasts only a short time.

4. A simple presence of bacteria in the blood is known as _____.

5. A(n) _____ infection is one caused by an opportunist after the primary infection has weakened the body's defenses.

6. _____ are changes in body function felt by the patient and subjective in nature, such as pain.

7. The science that deals with transmission of diseases in the human population, and when and where they occur, is called _____.

8. The abbreviation CDC stands for _____ and Prevention.

9. An abscess is an example of a(n) _____ type of infection.

10. An infection in which the microorganisms or their products are spread throughout the body in the blood or lymphatic system is known as a(n) _____ infection.

11. An inapparent, or _____ , disease is one that does not cause any noticeable illness.

12. The _____ of a disease is the fraction of the population having the disease at a given time.

13. Diseases intermediate between acute and chronic are described as _____.

14. A pathogen is found in all cases of a certain disease and is grown in pure culture; then it is inoculated into a laboratory animal. What is the next step in Koch's postulates?

_____.

Critical Thinking

1. What type of symbiotic relationship exists between normal microbiota and the host? Give two examples of contributions made by normal microbiota to the human host.

2. What is microbial antagonism, and how does it contribute to a healthy host? List and briefly discuss three examples of microbial antagonism.

3. Using at least two examples, explain microbial synergism.

4. Under what circumstances is it difficult to use Koch's postulates to determine the etiologic agent of an infectious disease? How has this problem been overcome?

ANSWERS

Matching

    I.   1. a   2. d   3. c   4. b

   II. 1. e   2. a   3. c   4. d   5. g

 III. 1. a   2. b   3. c

 IV. 1. a   2. c

   V. 1. g   2. c   3. d   4. e   5. f   6. a

 VI. 1. c   2. d   3. e   4. a   5. b

Fill in the Blanks

1. mechanical   2. incidence   3. Acute   4. bacteremia   5. secondary   6. symptoms   7. epidemiology
8. Centers for Disease Control   9. local   10. systemic   11. subclinical   12. prevalence   13. subacute
14. Isolate the pathogen from the animal and show that it is the same as the original organism.

Critical Thinking

1. The relationship between normal microbiota and the host may be commensal or mutual. Examples of contributions made by normal microbiota include the synthesis of K and B vitamins by *E. coli*, and microbial antagonism in a healthy host.

2. Microbial antagonism refers to the competition that exists among a host's normal microbiota for space and nutrients. This process protects the host from colonization by potentially pathogenic organisms. Four examples of microbial antagonism include the following:

   - Lactobacilli create an acidic environment that discourages the growth of *Candida albicans*, a common cause of vaginitis.

   - Streptococci living in the mouth prevent the growth of other gram-positive organisms.

   - Bacteriocins produced by *E. coli* inhibit the growth of *Shigella* and *Salmonella*.

   - Normal microbiota of the intestines inhibit growth of the pathogen *Clostridium difficile*.

3. Synergism means that the effect of two microbes acting together is greater than the effect of either acting alone. This definition is demonstrated by the cooperation that exists between oral streptococci and the pathogens that cause periodontal disease and gingivitis, and that between *Mycoplasma* and HIV.

4. Koch's postulates are most easily applied to microbes that can be grown on artificial media. However, some microbes cannot be grown on artificial media. To help overcome this problem, alternate methods of culturing and detecting microbes have been developed. Detection of these organisms might involve culturing in a guinea pig and in the yolk sacs of embryonated eggs.

15 Microbial Mechanisms of Pathogenicity

Pathogenicity is the ability to cause disease by overcoming the defenses of a host. First, however, the pathogen must enter the host's body. **Virulence** is the degree of pathogenicity.

HOW MICROORGANISMS ENTER A HOST

The avenue by which a microbe gains access to the body is a **portal of entry.**

Portals of Entry

Mucous Membranes. To gain access to the body, pathogens can penetrate *mucous membranes* lining the conjunctiva of the eye and the respiratory, gastrointestinal, and genitourinary tracts. The respiratory tract is the easiest, most frequently used route of entry for infectious microorganisms. Diseases contracted by this route are the common cold, pneumonia, tuberculosis, influenza, and measles. Microorganisms contracted from food, water, or fingers enter the body by the gastrointestinal tract. Although many are destroyed by stomach acids and intestinal enzymes, diseases such as poliomyelitis, hepatitis A, typhoid fever, amebic dysentery, shigellosis (bacillary dysentery), and cholera are transmitted in this manner. Examples of pathogens that enter through the mucous membranes of the genitourinary tract are HIV/AIDS, genital warts, chlamydia, syphilis, herpes and gonorrhea.

Skin. With a few exceptions, such as the hookworm, microorganisms cannot penetrate unbroken skin. Some fungi, however, grow on the keratin of the skin, and other microorganisms gain access by penetrating openings such as hair follicles and sweat ducts.

Parenteral Route. When the skin and mucous membranes are punctured or injured (*traumatized*), microorganisms can gain access to body tissues. This route is referred to as the **parenteral route.**

The Preferred Portal of Entry

Whether or not disease results after entry of microorganisms depends on many factors. The organism must enter by a preferred route. *Salmonella typhi,* for example, must enter the gastrointestinal tract to cause typhoid fever, rather than being rubbed onto the skin. *Clostridium tetani* must penetrate the skin to cause tetanus, and *Corynebacterium diphtheriae* must enter the respiratory tract to cause diphtheria.

Numbers of Invading Microbes

A measure of virulence is the **LD_{50} (lethal dose),** which is the dose of pathogen that will kill half of the test animals. If the pathogen causes only a nonfatal disease, the test is referred to as **ID_{50} (infectious dose).** The fewer microorganisms required, the higher the virulence.

Adherence

For most pathogens, **adherence** (or **adhesion**) is necessary for pathogenicity. The attachment between pathogen and host makes use of surface **adhesins (ligands)** and complementary surface **receptors** on the host cells. Adhesins are mostly glycoproteins or lipoproteins, which are frequently associated with structures such as fimbriae. Many microbes have the ability to form **biofilms,** which are masses on surfaces; there they secrete a glycocalyx that aids attachment. Within biofilms they share available nutrients and are relatively protected from antimicrobials and the immune system.

HOW BACTERIAL PATHOGENS PENETRATE HOST DEFENSES

Capsules

For many pathogens, such as *Streptococcus pneumoniae* and *Haemophilus influenzae,* **capsules** confer a resistance to phagocytosis.

Cell Wall Components

Cell surfaces sometimes contribute to invasiveness. *Streptococcus pyogenes,* for example, contains a protein called **M protein** (see Figure 21.6 in the text) on a fuzzy layer of fibrils on the cell surface that helps it resist phagocytosis and improve adherence. The **waxy lipid** (mycolic acid) that makes up the cell wall of *Mycobacterium tuberculosis* also increases virulence by resisting digestion of phagocytes and can even multiply inside phagocytes. *Neisseria gonorrhoeae* attach to host cells by means of **fimbriae** and an outer membrane protein called **Opa.**

Enzymes

Extracellular enzymes (exoenzymes) have the ability to break open cells, dissolve material between cells, and form or dissolve blood clots. These may contribute to invasiveness. **Coagulases** produced by some members of the genus *Staphylococcus* are enzymes that coagulate blood. The fibrin clot formed may protect the bacterium from phagocytosis. Bacterial **kinases** are enzymes that break down fibrin and dissolve clots formed by the body to isolate infections. **Streptokinase (fibrinolysin),** produced by streptococci, and **staphylokinase,** produced by staphylococci, are the best known. **Hyaluronidase** is an enzyme secreted by certain bacteria that digests hyaluronic acid. This compound is a mucopolysaccharide that holds together certain body cells, much like mortar holds together bricks, especially in connective tissue. Both the organisms that produce gas gangrene and a number of streptococci produce this enzyme, which may promote spread of the infections to adjoining tissues. **Collagenase,** produced by several species of *Clostridium,* breaks down the collagen framework of the muscle tissue. As a defense against adherence of pathogens to mucosal surfaces, the body produces a class of antibodies called **IgA antibodies.** Some pathogens have the ability to produce enzymes called **IgA proteases** that can destroy these antibodies.

Antigenic Variation

A major factor in the body's defenses is the specific resistance of the immune system that recognizes antigens by producing antibodies specific against them, as in the IgA antibodies discussed above. Some organisms can continually change their antigenic structure, by means of **antigenic variation,** and evade the immune system.

Penetration into Host Cell Cytoskeleton

Microbes sometimes attach to host cells by **adhesins.** This activates factors in the host cell **cytoskeleton** that results in entrance into the cell by the microbe. A major component of the cytoskeleton is a protein called **actin.** Some pathogens such as salmonellae produce surface proteins called **invasins** that rearrange nearby actin filaments of the cytoskeleton. This causes the pathogen to be engulfed by the cell.

Once inside the host cell, some bacteria can use actin to propel themselves from one host cell to another. Actual movement from one cell to another involves a glycoprotein called **cadherin.**

HOW BACTERIAL PATHOGENS DAMAGE HOST CELLS

Using the Host's Nutrients: Siderophores

Iron is required for the growth of pathogens; however, there is little free iron in the host's body. It is tightly bound to iron-transport proteins such as *lactoferrin, transferrin,* and *ferritin,* as well as *hemoglobin.* Many pathogens secrete **siderophores,** proteins that remove iron from iron-transport proteins by binding it more tightly. Iron can also be released from cells of the host when the cell is killed by toxins (discussed shortly) produced by the pathogen.

Direct Damage

Once pathogens attach to host cells, they can cause direct damage as the pathogens use the host cell for nutrients and produce waste products. Most damage by bacteria, however, is done by toxins.

The Production of Toxins

Toxins are poisonous substances produced by certain microorganisms. The capacity to produce them is called **toxigenicity.** The presence of toxins in the blood and lymph is called **toxemia.** Toxins are of two general types, based on their position relative to the microbial cell: exotoxins and endotoxins.

Exotoxins

Exotoxins are proteins secreted by the bacterium, mostly gram-positive, into the surrounding medium or released following lysis. They are highly specific in their effects on body tissues and are among the most lethal substances known. The body can produce antibodies called **antitoxins** that provide immunity to exotoxins. Exotoxins can be inactivated, called *toxoids,* and used to stimulate antitoxins in the body against diseases such as diphtheria and tetanus.

Types of Exotoxins.   **A-B toxins** are named because they consist of two parts, A and B. Most exotoxins are of this type. After release of the toxin from the pathogen, diphtheria toxin for example, the B part binds to the host cell and the A-B toxin is transported into the cell by endocytosis. There the components separate and the A part inhibits protein synthesis.

Membrane-disrupting toxins cause lysis of host cells by disrupting their plasma membranes. Examples are cell-lysing exotoxins of *Staphylococcus aureus.* These toxins also contribute to virulence by killing host cells, especially phagocytes and macrophages of the body's defense system. Membrane-disrupting toxins such as this are called **leukocidins** (white-cell killers). If the targets of these toxins are red blood cells, they are called **hemolysins.** Streptococci produce hemolysins called *streptolysins,* such as *streptolysin O,* so named because oxygen inactivates it. Streptococcal hemolysins stable in the presence of oxygen are called *streptolysin S.* These hemolysins kill both red blood cells and white blood cells and many other body cells.

Superantigens are antigens that provoke a very intense immune response. They stimulate the proliferation of immune system cells called T cells, an important factor in the body's specific immune defenses discussed more fully in Chapter 17. The T cells in turn release enormous amounts of *cytokines,* which are proteins that stimulate or inhibit many cell functions. Their release by superantigen stimulation can cause many symptoms of disease and even death. Staphylococcal toxic shock syndrome is an example.

Representative Exotoxins.   Exotoxins are named on the basis of the tissues they affect, such as *neurotoxins* (nerve cells), *cardiotoxins* (heart cells), *hepatotoxins* (liver cells), *leukotoxins* (leukocytes), *enterotoxins* (lining of gastrointestinal tract), and *cytotoxins* (a wide variety of body cells).

Diphtheria toxin, an A-B toxin, is produced by *Corynebacterium diphtheriae* when it is infected by a lysogenic phage (*prophage*) carrying the *tox* gene. **Erythrogenic toxin** is produced by *Streptococcus pyogenes* and produces the red skin rash characteristic of scarlet fever. **Botulinum toxin** is produced by *Clostridium botulinum.* It is a neurotoxin that causes flaccid paralysis when ingested. **Tetanus toxin** is produced by *Clostridium tetani* and is known as *tetanospasmin.* It is an A-B neurotoxin that blocks the relaxation pathway of muscle control, producing spasmodic contractions (tetanus). **Vibrio enterotoxin** (cholera toxin) is produced by *Vibrio cholerae.* It is an A-B toxin that causes release of large amounts of fluids and electrolytes (ions). The result is a severe diarrhea. The *heat labile toxin* produced by some strains of *E. coli* has an identical action. **Staphylococcal enterotoxin** is produced by *Staphylococcus aureus* and affects the intestines in the same way as cholera toxin.

Endotoxins

Endotoxins are not secreted by bacteria but are part of the outer portion of the cell wall of gram-negative bacteria. Specifically, this is the lipopolysaccharide (LPS) portion called **lipid A.** (Recall that exotoxins are proteins.) Endotoxins are released upon the death and lysis of the bacteria, as well as during bacterial multiplication. One consequence of endotoxins is *disseminated intravascular coagulation (DIC),* in which the endotoxins activate blood-clotting proteins that block capillaries and thereby cause the death of tissues. Endotoxins also stimulate macrophages to release damaging amounts of cytokines. One symptom is fever caused by **interleukin-1 (IL-1),** formerly called *endogenous pyrogen.* Another symptom is **shock,** or dangerous loss of blood pressure related to release, after phagocytosis of gram-negative bacteria, of tumor necrosis factor alpha (TNF-α). This disease condition is referred to as **septic shock.** A laboratory test for endotoxins is the *Limulus* **amebocyte lysate (LAL) assay.** This uses the amebocytes (white blood cells) from the horseshoe crab, *Limulus polyphemus,* which endotoxins cause to clot.

Plasmids, Lysogeny, and Pathogenicity

Several diseases are caused only when the pathogen is lysogenized by a phage (**lysogenic conversion**) or carries a particular plasmid.

PATHOGENIC PROPERTIES OF VIRUSES

Cytopathic Effects of Viruses

Visible damage to a host cell is known as **cytopathic effects (CPE). Cytocidal** effects kill the cell; **noncytocidal** effects are damaging but not lethal. A virus can produce one or more of the following CPE:

1. Cause macromolecular synthesis within host cell to stop.
2. Cause the cell's lysosomes to release their enzymes, causing cell lysis.
3. **Inclusion bodies,** granules usually composed of viral parts, can be found in virus-infected cells.
4. Adjacent infected cells fuse to form a **syncytium** (plural: *syncytia*) or very large, multinucleated cell.
5. Change in function but otherwise no visible change. For example, production of a hormone may be decreased.
6. Cell produces **interferon** that protects neighboring uninfected cells from infection by a virus.
7. Antigenic changes that elicit host immune response.
8. Chromosomal changes are induced, possibly **oncogenic** (cancer-causing).
9. Transform cell so it loses **contact inhibition,** which leads to unregulated cell growth.

PATHOGENIC PROPERTIES OF FUNGI, PROTOZOA, HELMINTHS, AND ALGAE

Some protozoa and helminths grow on host tissues, causing cellular damage to the host. Their waste products may contribute to symptoms of disease in the host. Some algae produce neurotoxins.

Some toxins are associated with **fungi**—for example, **ergot**, which causes **ergotism.** Ergot is contained in **sclerotia** (resistant mycelia) of the plant pathogen *Claviceps purpurea.* Ergotism can result in hallucinations, or it may constrict blood capillaries and cause gangrene in the extremities. **Aflatoxin** is produced by the mold *Aspergillus flavus* and can cause cancer of the liver in animals. Toxic mushrooms such as *Amanita phalloides* contain dangerous **mycotoxins** such as **phalloidin** and **amanitin.**

Some **algae** produce neurotoxins called **saxitoxin.** For example, people become ill after ingesting mollusks that feed on the algae.

PORTALS OF EXIT

Portals of exit, like portals of entry, are often a characteristic of a disease. Most commonly these are the respiratory or gastrointestinal tract and are important factors in disease transmission. Another example for sexually transmitted diseases is the genitourinary tract.

SELF-TESTS

In the matching section, there is only one answer to each question; however, the lettered options (a, b, c, etc.) may be used more than once or not at all.

I. Matching

____ 1. Produced by some members of the genus *Staphylococcus;* forms a fibrin clot around the bacterium.

____ 2. A substance produced by some bacteria that destroys certain phagocytic cells.

____ 3. Enzymes that cause lysis of red blood cells.

____ 4. Enzymes that break down fibrin and dissolve clots.

____ 5. The fibrinolysins produced by the streptococci.

____ 6. May cause hallucinations or gangrene.

____ 7. Virulence factor that improves adherence of *Neisseria gonorrhoeae.*

____ 8. Tumor necrosis factor.

a. Leukocidins

b. Collagenase

c. Kinases

d. Hyaluronidase

e. Coagulase

f. Hemolysins

g. Cachectin

h. Aflatoxin

i. Amanitin

j. Syncytia

k. Ergot

l. Opa

II. Matching

___ 1. A protein secreted by a bacterium.

___ 2. The tetanus toxin is a good example.

___ 3. A lipopolysaccharide component of the cell wall of many gram-negative bacteria.

___ 4. Released upon lysis of the cell.

___ 5. Detected by *Limulus* amebocyte lysate assay.

a. Exotoxin

b. Endotoxin

III. Matching

___ 1. The capacity to form toxins.

___ 2. Provides immunity to exotoxins.

___ 3. The presence of a toxin in the bloodstream.

a. Pathogenicity

b. Toxemia

c. Antitoxin

d. Toxigenicity

IV. Matching

___ 1. A way to measure virulence.

___ 2. The degree of pathogenicity.

___ 3. The ability of a pathogen to cause disease in a host.

a. Traumatized

b. Virulence

c. Pathogenicity

d. LD_{50}

V. Matching

___ 1. Confers some resistance to phagocytosis.

___ 2. Entrance through skin and mucous membranes to gain access to body tissues.

___ 3. Very large multinucleated cells caused by viral infection.

___ 4. An important component of the cell's cytoskeleton that aids in entrance of pathogens into the cell.

___ 5. Produced by virus-infected cells; helps prevent infection of uninfected neighboring cells.

a. Parenteral

b. Syncytia

c. Actin

d. Interferon

e. Capsule

VI. Matching

___ 1. Portion of the outer wall of gram-negative bacterium that is the endotoxin.

a. Lipid A

___ 2. Formerly called endogenous pyrogen.

b. Streptolysin O

c. Interleukin-1

___ 3. Membrane-disrupting toxin such as a leukocidin.

Fill in the Blanks

1. The term LD_{50} refers to the dose of pathogen that will kill half of the test _____.

2. Hyaluronidase is an enzyme secreted by certain bacteria that digests _____ acid.

3. A disease associated with _____ -toxins is septic shock (supply prefix).

4. The *Limulus* amebocyte lysate assay is intended to detect _____.

5. One effect of a viral infection is unregulated cell growth due to loss of _____ inhibition.

6. Contributing to invasiveness by *Streptococcus pyogenes* is a cell protein called _____ protein.

7. With few exceptions, microorganisms cannot penetrate unbroken skin; however, some fungi grow on the _____ component of the skin.

8. The avenue by which a microorganism gains access to the body is called its _____.

9. If the pathogen causes a nonfatal disease, the equivalent to the LD_{50} test is referred to as _____ (infective dose).

10. The toxin ergot is contained in resistant mycelia called _____.

11. A microbe is pathogenic only if its characteristics are changed due to a prophage. This is called _____.

Critical Thinking

1. Explain the relationship between the waxy cell wall of *Mycobacterium tuberculosis* and virulence.

2. A 14-year-old male patient with a sore throat is examined by his physician. A throat swab cultured on blood agar produces beta-hemolytic streptococci. Antibodies against M protein are detected in the blood.

a. What is the etiologic agent?

b. What is the probable portal of entry?

c. What is the probable route of transmission?

3. Discuss the sequence of events that lead to shock in some gram-negative infections.

ANSWERS

Matching

I.   1. e   2. a   3. f   4. c   5. c   6. k   7. i   8. g
II. 1. a   2. a   3. b   4. b   5. b
III. 1. d   2. c   3. b
IV. 1. d   2. b   3. c
V.   1. e   2. a   3. b   4. c   5. d
VI. 1. a   2. b   3. c

Fill in the Blanks

1. animals   2. hyaluronic   3. endo   4. endotoxins   5. contact   6. M   7. keratin
8. portal of entry   9. ID_{50} 10. sclerotia   11. lysogenic conversion

Critical Thinking

1. The waxy cell wall of *M. tuberculosis* helps it avoid phagocytosis by cells of the immune system, allowing the microorganism to reproduce and become established in the host.

2. a. *Streptococcus pyogenes.*

   b. The respiratory system.

   c. Airborne transmission of cough-produced droplets or vehicle transmission, such as transport of the streptococci on dust particles.

3. Septic shock seen in some gram-negative infections is due to endotoxins produced by the pathogen. The endotoxin causes macrophages to secrete a substance called tumor necrosis factor (TNF) or cachectin. TNF binds to and alters the metabolism of certain tissues in the body. For example, TNF damages and increases permeability of capillaries, causing the loss of large amounts of fluids. Blood pressure drops and results in shock and other serious effects.

16 Innate Immunity: Nonspecific Defenses of the Host

Pathogenic microorganisms are endowed with special properties that, given the right opportunity, enable them to cause disease. Our bodies, however, have defenses against these microorganisms. In general, our ability to ward off disease through our defenses is called **immunity,** or **resistance,** and our vulnerability or lack of resistance is known as **susceptibility.**

THE CONCEPT OF IMMUNITY

Innate immunity refers to defenses that are present at birth. They are always present and available to provide rapid responses to protect us against disease. **Adaptive immunity** is based on a specific response to a specific microbe once a microbe has breached the innate immunity defenses. As noted previously, the innate immune system responds rapidly to invaders by detecting them and then attempting to eliminate them. It has recently been learned that the responses of the innate system are activated by protein receptors in the plasma membranes of defensive cells; among these activators are **Toll-like receptors (TLRs)**. These TLRs attach to various components commonly found on pathogens that are called **pathogen-associated molecular patterns (PAMPs)** (see Figure 16.1). TLRs also attach to components of fungi and parasites. Two of the defensive cells involved in innate immunity are called macrophages and dendritic cells. When the TLRs on these cells encounter the PAMPs of microbes, such as the LPS of gram-negative bacteria, the TLRs induce the defensive cells to release chemicals called cytokines. **Cytokines** (*cyto-* = cell; *-kinesis* = motion) are proteins that regulate the intensity and duration of immune responses.

FIRST LINE OF DEFENSE: SKIN AND MUCOUS MEMBRANES

PHYSICAL FACTORS

The intact skin consists of the **dermis,** an inner, thicker portion composed of connective tissue, and the **epidermis,** an outer, thinner portion consisting of several layers of epithelial cells arranged in continuous sheets. The top layer of epidermal cells contains the protein **keratin.** Intact, these barriers are seldom penetrated by microorganisms. **Mucous membranes** line the body cavities that open to the exterior; these include the digestive, respiratory, urinary, and reproductive tracts. Mucous membranes consist of an **epithelial layer** and an underlying **connective tissue layer.** Cells in mucous membranes secrete **mucus,** which prevents the cavities from drying out. Another mechanical factor involved in protection is the **lacrimal apparatus,** which manufactures and drains away tears. This apparatus has a cleansing effect on the eye surface. **Saliva,** produced by salivary glands, washes microorganisms from the surfaces of the teeth and mucous membrane of the mouth. Mucus tends to trap microorganisms that enter the respiratory and digestive tracts. Cells of the mucous membrane of the lower respiratory tract contain cilia, microscopic hairlike projections that move synchronously (**ciliary escalator**) and propel inhaled microorganisms trapped in mucus out of the respiratory system. The **epiglottis** covers the larynx during swallowing.

The **flow of urine** and vaginal secretions tend to remove microorganisms from the body. **Peristalsis, defecation, vomiting,** and **diarrhea** expel pathogens and toxins.

CHEMICAL FACTORS

Sebaceous oil glands of the skin produce **sebum,** which forms a protective film over the skin surface. The unsaturated fatty acids of sebum inhibit the growth of certain pathogens. The low pH (3 to 5) of the skin is partly due to these secretions. Body odor is the result of the decomposition of both sloughed-off skin cells and sebum secretions by commensal bacteria. **Sweat glands** produce **perspiration,** which flushes microorganisms from the skin surface. Perspiration contains **lysozyme,** an enzyme that breaks down the cell walls of gram-positive bacteria. The enzyme is also found in saliva, tears, nasal secretions, and tissue fluids. **Gastric juice** is a mixture of hydrochloric acid, enzymes, and mucus, with an acidity of pH 1.2 to 3. These stomach secretions destroy many ingested microorganisms and toxins. **Vaginal secretions** have an antibacterial role: glycogen produced by vaginal epithelial cells is converted into lactic acid by *Lactobacillus acidophilus,* which is inhibiting to bacteria. Cervical mucus also has antimicrobial activity. **Urine** contains lysozyme and a number of antimicrobial metabolic by-products. It also has an acidic pH. **Earwax (cerumen)** serves as a physical barrier, and its low pH is also protective.

NORMAL MICROBIOTA AND INNATE IMMUNITY

The normal microbiota, by **competitive exclusion,** can often prevent colonization by pathogens. Live bacterial cultures (**probiotics**) are sometimes added, which are intended to exert a beneficial effect.

———— SECOND LINE OF DEFENSE ————

FORMED ELEMENTS IN BLOOD

Blood consists of fluid, called **plasma,** and **formed elements**—that is, cells and cell fragments suspended in plasma (see Table 16.1). Of the formed elements, those that concern us at present are the **leukocytes,** or white blood cells. Leukocytes are divided into two main categories based on their appearance under a light microscope: **granulocytes** and **agranulocytes.**

Granulocytes owe their name to the presence of large granules in their cytoplasm that can be seen under a light microscope after staining. They are differentiated into three types of cells on the basis of how the granules stain: **neutrophils, basophils,** and **eosinophils.** The granules of neutrophils stain pale lilac with a mixture of acidic and basic dyes. Neutrophils are also commonly called **polymorphonuclear leukocytes (PMNs), or polymorphs.** They have the ability to leave the blood, enter an infected tissue, and destroy microbes and foreign particles. **Basophils** stain blue-purple with the basic dye methylene blue. Basophils release substances, such as **histamine,** that are important in inflammation and allergic responses. **Eosinophils** stain red or orange with the acidic dye eosin. Eosinophils are somewhat phagocytic and also have the ability to leave the blood. Their major function is to produce toxic proteins against certain parasites, such as helminths.

Agranulocytes also have granules in their cytoplasm, but the granules are not visible under the light microscope after staining. There are three different types of granulocytes: **monocytes, dendritic cells**, and **lymphocytes. Monocytes** are not actively phagocytic until they leave circulating blood, enter body tissues, and mature into macrophages. (See pages 172 and 188 for discussions of macrophages.) **Dendritic cells** are also believed to be derived from monocytes. They have long extensions that resemble the dendrites of nerve cells, thus their name. The function of dendritic cells is to destroy microbes by phagocytosis and to initiate adaptive immunity responses (see Chapter 17). **Lymphocytes** include natural killer cells, T cells, and B cells. **Natural killer (NK) cells** are found in blood and in the spleen, lymph nodes, and red bone

Table 16.1 Formed Elements in Blood

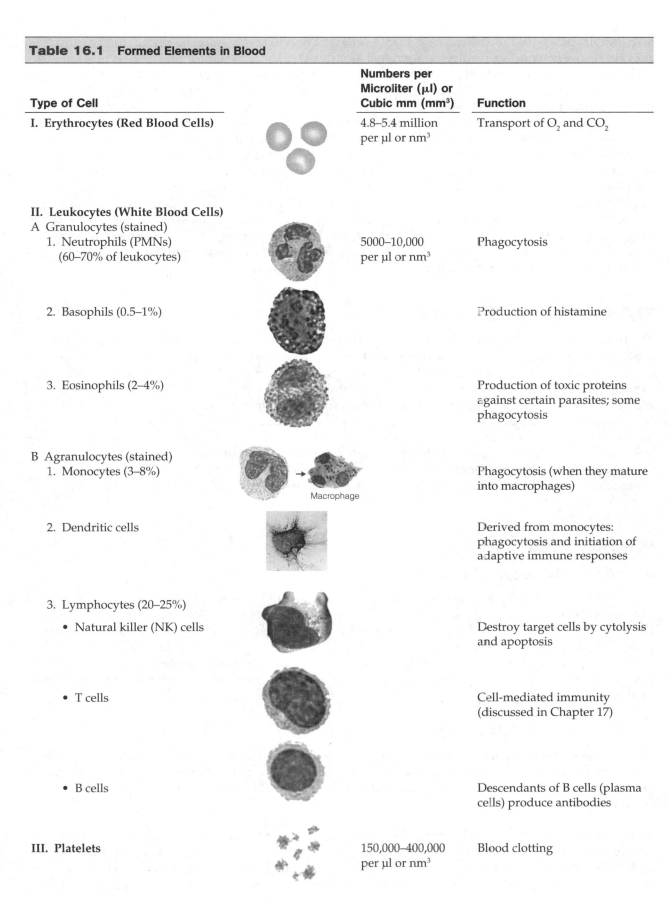

Type of Cell		Numbers per Microliter (µl) or Cubic mm (mm³)	Function
I. Erythrocytes (Red Blood Cells)		4.8–5.4 million per µl or nm³	Transport of O_2 and CO_2
II. Leukocytes (White Blood Cells)			
A Granulocytes (stained)			
1. Neutrophils (PMNs) (60–70% of leukocytes)		5000–10,000 per µl or nm³	Phagocytosis
2. Basophils (0.5–1%)			Production of histamine
3. Eosinophils (2–4%)			Production of toxic proteins against certain parasites; some phagocytosis
B Agranulocytes (stained)			
1. Monocytes (3–8%)	Macrophage		Phagocytosis (when they mature into macrophages)
2. Dendritic cells			Derived from monocytes: phagocytosis and initiation of adaptive immune responses
3. Lymphocytes (20–25%)			
• Natural killer (NK) cells			Destroy target cells by cytolysis and apoptosis
• T cells			Cell-mediated immunity (discussed in Chapter 17)
• B cells			Descendants of B cells (plasma cells) produce antibodies
III. Platelets		150,000–400,000 per µl or nm³	Blood clotting

marrow. NK cells have the ability to kill a wide variety of infected body cells and certain tumor cells. NK cells attack any body cells that display abnormal or unusual plasma membrane proteins. Some granules contain a protein called **perforin,** which inserts into the plasma membrane of the target cell and creates channels (perforations) in the membrane. As a result, extracellular fluid flows into the target cell and the cell bursts, a process called **cytolysis** (sī-tol' i-sis; *cyto-* = cell; *-lysis* = loosening). Other granules of NK cells release **granzymes,** which are protein-digesting enzymes that induce the target cell to undergo apoptosis, or self-destruction. This type of attack kills infected cells but not the microbes inside the cells; the released microbes, which may or may not be intact, can be destroyed by phagocytes. **T cells** and **B cells** are not usually phagocytic but play a key role in adaptive immunity (see Chapter 17). They occur in lymphoid tissues of the lymphatic system and also circulate in blood. During many kinds of infections, especially bacterial infections, the total number of white blood cells increases as a protective response to combat the microbes; this increase is called *leukocytosis*. During the active stage of infection, the leukocyte count might double, triple, or quadruple, depending on the severity of the infection. Leukocyte increase or decrease can be detected by a **differential white blood cell count,** which is a calculation of the percentage of each kind of white cell in a sample of 100 white blood cells.

THE LYMPHATIC SYSTEM

The **lymphatic system** consists of a fluid called **lymph,** vessels called **lymphatic vessels,** a number of structures and organs containing **lymphoid tissue,** and **red bone marrow,** where stem cells develop into blood cells, including lymphocytes. Lymphatic vessels begin as microscopic **lymphatic capillaries** located in spaces between cells. The lymphatic capillaries permit interstitial fluid derived from blood plasma to flow into them, but not out. Within the lymphatic capillaries, the fluid is called lymph. Lymphatic capillaries converge to form larger lymphatic vessels. These vessels, like veins, have one-way valves to keep lymph flowing in one direction only. At intervals along the lymphatic vessels, lymph flows through bean-shaped **lymph nodes,** which are the sites of activation of T cells and B cells, which destroy microbes by immune responses (see Chapter 17). Lymphoid tissues and organs are scattered throughout the mucous membranes that line the gastrointestinal, respiratory, urinary, and reproductive tracts.

PHAGOCYTES

Actions of Phagocytic Cells

Granulocytes and monocytes migrate to sites of infection. Granulocytes are mostly neutrophils that wander in the blood and can pass through capillary walls to reach trauma sites. Macrophages are highly phagocytic cells called **free (wandering) macrophages** because of their ability to migrate. **Fixed macrophages** (*histiocytes*) enter tissues and organs and remain there. Examples are Kupffer cells in the liver. The fixed macrophages are referred to as the **mononuclear phagocytic (reticuloendothelial) system.** Granulocytes predominate early in infections, as indicated by differential counts; as infections subside, monocytes predominate.

The Mechanism of Phagocytosis

Chemotaxis and Adherence. **Adherence,** or attachment between the cell membrane of the phagocyte and the organism, is facilitated by **chemotaxis,** which is the attraction of microorganisms to chemicals. (See Figure 16.1.) **Opsonization**—coating a microorganism with plasma proteins such as antibodies and complement—promotes phagocytosis.

Ingestion and Digestion. Following adherence, projections of the cell membrane of the phagocyte (**pseudopods**) engulf the microorganism and then fold inward, forming a sac around it called a **phagosome** or **phagocytic vesicle.** This process is called **ingestion.** Within the cell, enzyme-containing phagosomes and lysosomes of the cell fuse to form a larger structure, the **phagolysosome,**

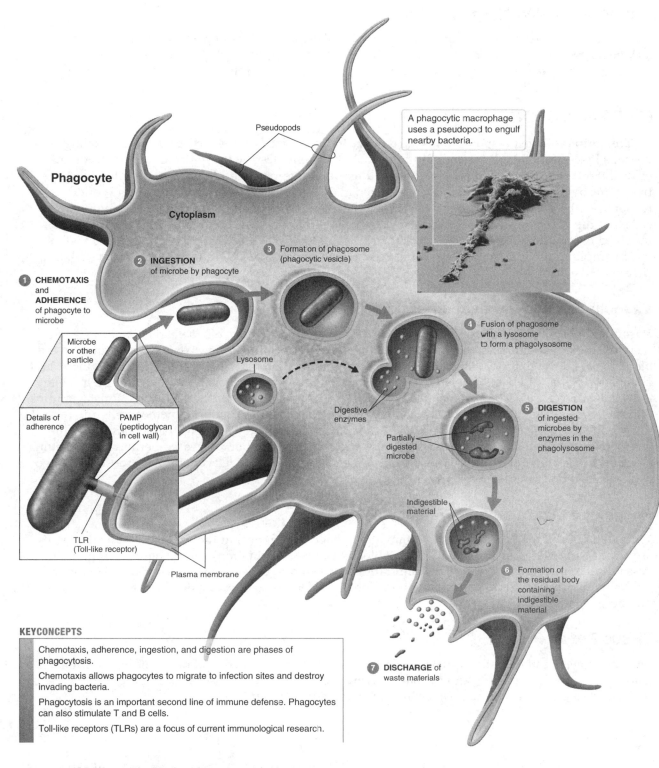

Pseudopods

A phagocytic macrophage uses a pseudopod to engulf nearby bacteria.

Phagocyte

Cytoplasm

3 Formation of phagosome (phagocytic vesicle)

2 **INGESTION** of microbe by phagocyte

1 **CHEMOTAXIS** and **ADHERENCE** of phagocyte to microbe

4 Fusion of phagosome with a lysosome to form a phagolysosome

Microbe or other particle

Lysosome

Digestive enzymes

5 **DIGESTION** of ingested microbes by enzymes in the phagolysosome

Details of adherence

PAMP (peptidoglycan in cell wall)

Partially digested microbe

TLR (Toll-like receptor)

Indigestible material

Plasma membrane

6 Formation of the residual body containing indigestible material

KEYCONCEPTS

Chemotaxis, adherence, ingestion, and digestion are phases of phagocytosis.

Chemotaxis allows phagocytes to migrate to infection sites and destroy invading bacteria.

Phagocytosis is an important second line of immune defense. Phagocytes can also stimulate T and B cells.

Toll-like receptors (TLRs) are a focus of current immunological research.

7 **DISCHARGE** of waste materials

Figure 16.1 The mechanism of phagocytosis in a phagocyte.

within which the bacteria are usually quickly killed. Enzymes in lysosomes include lysozyme, various hydrolytic enzymes, and myeloperoxidases. Their action is a process called an **oxidative burst.** Indigestible material in the phagolysosome is called the **residual body** and moves to the cell boundary, where it is discharged.

Microbial Evasion of Phagocytosis

Some microbes evade adherence by phagocytes with structures such as M proteins or capsules. Others may be ingested but not killed. For example, they produce *leukocidins* and other complexes that kill the phagocyte. Still others can survive inside phagocytes and even multiply.

INFLAMMATION

Inflammation is a host response to tissue damage, characterized by redness, pain, heat, swelling, and perhaps loss of function. It is generally beneficial, serving to destroy the injurious agent, to confine or wall it off, and to repair or replace damaged tissue. **Acute-phase proteins** are converted to active forms by inflammation and induce localized and systemic responses. Inflammation is often in response to the action of *cytokines,* which, you will recall, are small proteins released from human cells that regulate immune responses. An example is **tumor necrosis factor alpha (TNF-α)**. All cells involved in inflammation have receptors for TNF-α and are activated by it to produce more TNF-α, which amplifies the inflammatory response. Excessive amounts of TNF-α lead to disorders such as rheumatoid arthritis.

Vasodilation and Increased Permeability of Blood Vessels

Vasodilation is the first stage of inflammation; it involves an increase in blood vessel diameters, and therefore more blood flow, in the injured area. **Increased permeability** allows defense substances in the blood to pass through the walls of the blood vessels. Vasodilation is also responsible for the redness, heat, **edema** (swelling), and pain of inflammation. **Histamine** is released by injury and increases permeability as well. **Kinins** cause vasodilation and also increase permeability and attract phagocytes. **Prostaglandins** (substances released by damaged cells) and **leukotrienes** (substances produced by most cells and related to prostaglandins) cause increased permeability of blood vessels and help attach pathogens to phagocytes. Activated fixed macrophages secrete cytokines, which bring about increased permeability. Clotting elements in the blood help prevent the spread of the infection. A localized collection of **pus** (dead phagocytic cells and body fluids) in a cavity is called an **abscess.**

Phagocyte Migration and Phagocytosis

Within an hour of the beginning of inflammation, phagocytes appear at the site. Blood flow decreases as the phagocytes stick to the inner lining of blood vessels (**margination**) and then pass through the vessel wall to the damaged area. This **emigration** is called **diapedesis.** Monocytes appear later in the inflammatory response and mature into macrophages. Macrophages ingest dead tissue and invading microorganisms.

Tissue Repair

The final stage of inflammation is **tissue repair,** which involves production of new cells by the **stroma** (supporting connective tissue) and the **parenchyma** (functioning part of tissue). Scar tissue (aggregations of fibers, by a process called *fibrosis*) results from stroma-type repair.

FEVER

The hypothalamus controls body temperature (normally 37°C or 98.6°F). The setting can be altered by ingestion of gram-negative bacteria by phagocytes. The resulting release of endotoxins causes release of a cytokine called *interleukin-1 (endogenous pyrogen)*. A **chill** (cold skin and **shivering**) is a sign of **fever** (rising temperature). **Crisis** refers to a very rapid fall in temperature. Fevers are often beneficial as aids to body tissue repair and inhibitors of microbial growth.

ANTIMICROBIAL SUBSTANCES

The Complement System

Complement consists of a group of over 30 different proteins found in blood serum. **Blood serum** is the liquid portion of blood that remains after it is drawn and clotting proteins form a clot with the formed elements. Complement participates in lysis of foreign cells, inflammation, and phagocytosis. The system can be activated by an immune reaction in the classical pathway, the alternative pathway, or the newly discovered lectin pathway.

The complement proteins act in an ordered sequence, or **cascade;** one protein activates another. Protein C3, as shown in Figure 16.2, plays a central role in both the classical and the alternative pathways.

The Classical, Alternative, and Lectin Pathways.   In the **classical pathway,** activity is initiated when antibody molecules bind to the antigen—a bacterial cell, for example. In the **alternative pathway,** which does not involve antibodies, complement proteins, and proteins called factors B, D, and P (*properdin*), combine with certain microbial polysaccharides. Especially affected are the lipopolysaccharide cell wall portions (endotoxins) of gram-negative enteric bacteria. In the **lectin pathway,** macrophages stimulate the liver to release **lectins.** These enhance opsonization by binding to cell carbohydrates.

The Result of Complement Activation.   **Cytolysis:** Complement protein then binds to two adjacent antibodies and initiates a sequence known as the **membrane attack complex.** Circular lesions called **transmembrane channels (membrane pores)** are formed and cause the eventual lysis of the cell, to which the antibodies are attached. **Inflammation** also can develop from complement. Other complement proteins combine with mast cells and trigger the release of histamine, which increases blood vessel permeability. One protein chemotactically attracts phagocytes to the site. **Opsonization,** or *immune adherence,* promotes attachment of a phagocyte to the microbe. Complement involvement results from interaction with special receptors on the phagocytes.

Interferons (IFNs)

Interferons of three types (interferon alpha and interferon beta, which are produced by virus-infected host cells; and interferon gamma, which is produced by lymphocytes and kills bacteria) are proteins. They tend to interfere with viral multiplication by inducing the uninfected cell to manufacture mRNA for synthesis of **antiviral proteins (AVPs).** One example is *oligoadenylate synthetase,* which degrades viral mRNA; another is *protein kinase,* which inhibits protein synthesis. Interferons are most important in protection against acute virus-caused infections, such as influenza. Interferons can be produced by biotechnological methods and have been tested to determine their antiviral and anticancer effects. The three types have different effects.

Iron-Binding Proteins

Most pathogenic bacteria require iron for their growth. Free iron is scarce in the human body because most of it is bound to molecules such as transferrin, lactoferrin, ferritin, and hemoglobin, collectively called **iron-binding proteins,** whose function is to transport and store iron. **Transferrin** is found in blood and tissue fluids. **Lactoferrin** is found in milk, saliva, and mucus. **Ferritin** is located in the liver, spleen, and red bone marrow, and **hemoglobin** is located within red blood cells. Pathogenic bacteria obtain iron by secreting proteins called **siderophores** (see Figure 15.3 in the text). Once the iron-siderophore complex is formed, it is taken into the bacterium, and the iron is split from the siderophore and utilized. There are a few exceptions to this process.

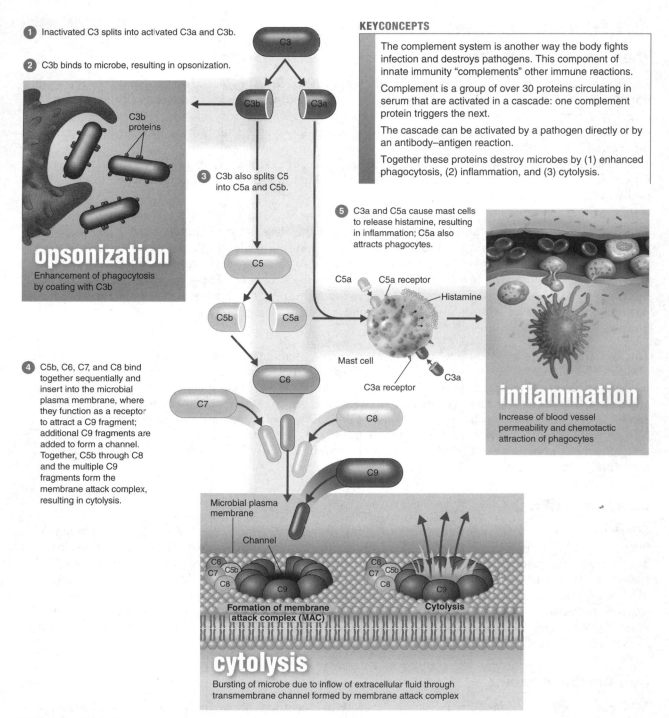

① Inactivated C3 splits into activated C3a and C3b.

② C3b binds to microbe, resulting in opsonization.

C3b proteins

opsonization
Enhancement of phagocytosis by coating with C3b

③ C3b also splits C5 into C5a and C5b.

④ C5b, C6, C7, and C8 bind together sequentially and insert into the microbial plasma membrane, where they function as a receptor to attract a C9 fragment; additional C9 fragments are added to form a channel. Together, C5b through C8 and the multiple C9 fragments form the membrane attack complex, resulting in cytolysis.

KEYCONCEPTS

The complement system is another way the body fights infection and destroys pathogens. This component of innate immunity "complements" other immune reactions.

Complement is a group of over 30 proteins circulating in serum that are activated in a cascade: one complement protein triggers the next.

The cascade can be activated by a pathogen directly or by an antibody–antigen reaction.

Together these proteins destroy microbes by (1) enhanced phagocytosis, (2) inflammation, and (3) cytolysis.

⑤ C3a and C5a cause mast cells to release histamine, resulting in inflammation; C5a also attracts phagocytes.

C5a   C5a receptor

Histamine

Mast cell

C3a receptor   C3a

inflammation
Increase of blood vessel permeability and chemotactic attraction of phagocytes

Microbial plasma membrane

Channel

C6 C5b C7 C8 C9

C6 C5b C7 C8 C9

Formation of membrane attack complex (MAC)   **Cytolysis**

cytolysis
Bursting of microbe due to inflow of extracellular fluid through transmembrane channel formed by membrane attack complex

Figure 16.2 Outcomes of complement activation.

Antimicrobial Peptides

A relatively recent discovery that confers resistance to microbial infections are the **antimicrobial peptides (AMPs),** which are short peptides synthesized on ribosomes. Their synthesis is triggered by certain protein and sugar molecules on the microbial surface, including Toll-like receptors. AMPs inhibit cell wall synthesis, form pores in the bacterial plasma membrane, and destroy DNA and RNA of target cells. Examples of AMPs are *dermicidin, defensins, cathelicidins,* and *thrombocidin.*

SELF-TESTS

In the matching section, there is only one answer to each question; however, the lettered options (a, b, c, etc.) may be used more than once or not at all.

I. Matching

____ 1. Produces tears.

____ 2. The outer layer of the skin.

____ 3. An oily substance forming a protective film over the skin surface.

____ 4. Secreted by cells in mucous membrane; prevents the cavities from drying.

____ 5. Covers larynx during swallowing.

____ 6. The inner portion of the skin, composed of connective tissue.

____ 7. Live bacteria intended for beneficial effect.

a. Dermis

b. Epidermis

c. Mucus

d. Mucous membrane

e. Lacrimal apparatus

f. Sebum

g. Epiglottis

h. Probiotics

II. Matching

____ 1. The blood fluid.

____ 2. Cells and cell fragments of the blood.

____ 3. Immunity based on antibodies.

____ 4. Movement by a microorganism toward an attractant chemical.

____ 5. An increase in the diameter of blood vessels.

____ 6. A collection of dead phagocytic cells and fluids.

____ 7. Vulnerability to a pathogen.

a. Serum

b. Plasma

c. Formed elements

d. Susceptibility

e. Specific resistance

f. Chemotaxis

g. Vasodilation

h. Pus

i. Opsonization

III. Matching

B 1. Neutrophils.

C 2. Monocytes.

A 3. Lymphocytes.

a. No granules in cellular cytoplasm; important to specific immunity

b. Granulocytes

c. Mature into macrophages

IV. Matching

___ 1. An increase in the number of white blood cells.

___ 2. Projections of the cell membrane of a phagocyte.

B 3. A larger structure formed when lysosome and phagosome fuse.

G 4. A decrease in the number of white blood cells.

a. Pseudopods

b. Phagolysosome

c. Phagosome

d. Lysozyme

e. Lysosome

f. Leukocytosis

g. Leukopenia

V. Matching

___ 1. Blood flow decreases as phagocytes stick to the inner lining of blood vessels.

___ 2. Complement reacts with mast cells and attached antibodies to release this compound.

___ 3. A protein in blood that inhibits microbial growth by reducing the amount of available iron.

B 4. Controls body temperature.

___ 5. Emigration of phagocytes through the vessel wall to damaged tissue.

F 6. Protein secreted by bacteria to obtain iron.

a. Diapedesis

b. Hypothalamus

c. Margination

d. Transferrin

e. Histamine

f. Siderophore

VI. Matching

____ 1. Polymorphonuclear leukocytes.

____ 2. Most numerous granulocytes in blood.

____ 3. Stain red or orange with the acidic dye eosin.

____ 4. Attach externally to large parasites such as worms and lyse them by discharge of peroxides.

____ 5. Granulocytes that stain with basic methylene blue dyes.

____ 6. Become macrophages.

____ 7. Kupffer cells in the liver, for example.

a. Neutrophils

b. Basophils

c. Eosinophils

d. Monocytes

e. Macrophages

VII. Matching

____ 1. Innate immunity.

____ 2. Adaptive immunity.

____ 3. Toll-like receptors.

____ 4. Cytokines.

____ 5. Dermicidin.

a. Something to do with cookies

b. Refers to defenses that tend to protect us from any kind of pathogen

c. Immunity based on antibody production, for example

d. Attach to components of the outer membrane of gram-negative bacteria

e. Proteins that regulate the intensity and duration of immune responses

f. Example of an antimicrobial peptide

Fill in the Blanks

1. Some cells of the mucous membrane of the lower respiratory tract contain

    _____, which are microscopic, hairlike projections.

2. The _____ glands produce perspiration.

3. Complement acts in a sequence called a(n) _____.

4. In the membrane attack complex associated with the action of complement, circular lesions called

    _____ channels are formed.

5. _____ is a group of more than 30 proteins found in blood serum.

6. Lymphocytes and monocytes do not have _____ in their cytoplasm.

7. The coating of a microorganism with plasma proteins such as antibodies and complement is called _____ and promotes phagocytosis.

8. Scar tissue results from _____ -type repair.

9. The complement pathway that does not involve antibodies is called the _____ pathway.

10. Another name for cellular self-destruction is _____.

Label the Art

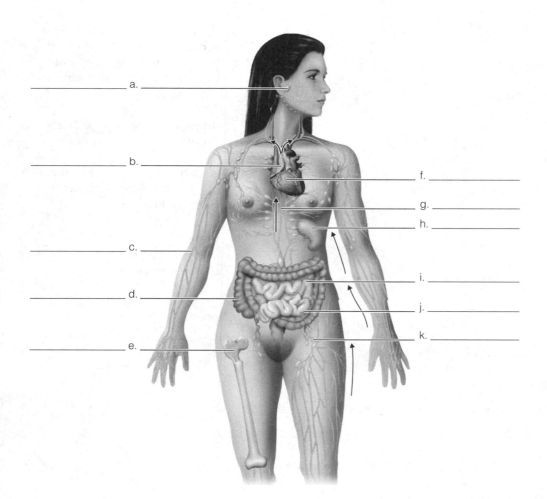

_____ a. _____

_____ b. _____

_____ c. _____

_____ d. _____

_____ e. _____

f. _____

g. _____

h. _____

i. _____

j. _____

k. _____

Critical Thinking

1. List and briefly discuss the nonspecific defense mechanisms that help prevent infection of the respiratory tract.

2. How do some bacteria overcome the strongly acidic environment of the stomach to cause disease?

3. Define *phagocytosis*. How do some bacteria avoid or survive the action of phagocytes?

4. Compare and contrast the roles of wandering macrophages and fixed macrophages.

5. Define *opsonization*. How does opsonization help enhance phagocytosis?

ANSWERS

Matching

I.	1. e	2. b	3. f	4. c	5. g	6. a	7. h
II.	1. b	2. c	3. e	4. f	5. g	6. h	7. d
III.	1. b	2. c	3. a				
IV.	1. f	2. a	3. b	4. g			
V.	1. c	2. e	3. d	4. b	5. a	6. f	
VI.	1. a	2. a	3. c	4. c	5. b	6. d	7. e
VII.	1. b	2. c	3. d	4. e	5. f		

Fill in the Blanks

1. cilia   2. sweat   3. cascade   4. transmembrane   5. Complement   6. granules   7. opsonization
8. stroma   9. alternative   10. apoptosis

Label the Art

a. Tonsil   b. Thymus   c. Lymphatic vessel   d. Large intestine   e. Red bone marrow   f. Heart
g. Thoracic duct   h. Spleen   i. Small intestine   j. Peyer's patch   k. Lymph node

Critical Thinking

1. • Mucous membranes lining the respiratory tract help prevent the penetration of pathogens.

   • Mucus-coated hairs of the nose trap microbes, dust, and so on.

   • The ciliary escalator helps remove microbes from the lower respiratory tract.

   • The epiglottis covers the larynx during swallowing, preventing microbes from entering the lower
   respiratory tract.

2. Bacteria such as *Clostridium botulinum* and *Staphylococcus aureus* don't actually survive stomach acids.
   Instead, they produce acid-resistant toxins that produce the symptoms of the diseases that they cause.
   Other microorganisms are protected from the effects of acid by food particles. Finally, the bacterium
   Helicobacter pylori is able to neutralize the acid and grow in the stomach lining.

3. Phagocytosis refers to the ingestion of microorganisms or particulate matter by a cell. Phagocytosis is
   a means by which the body counters infection.

   • Lipids in the cell wall of *Mycobacterium* help it to avoid phagocytosis.

   • Capsules help bacterial cells avoid phagocytosis.

   • Toxins produced by staphylococci and *Actinobacillus* kill phagocytes.

   • Some cells evade the immune system by entering and reproducing inside phagocytes.

   • Some microorganisms prevent fusion of the phagosome with a lysosome, preventing digestion.

   • M proteins of streptococci inhibit attachment of phagocytes.

4. Wandering macrophages and fixed macrophages make up the mononuclear phagocytic system. Wandering macrophages develop from monocytes that migrate to the infected area. They leave the blood
   and migrate to the infected area, where they scavenge and phagocytize bacteria and debris. Fixed
   macrophages are located in certain tissues such as the liver, lungs, and nervous system. Fixed macrophages remove microorganisms from blood or lymph as these substances pass through the organs.

5. *Opsonization* refers to the coating of microorganisms with certain plasma proteins to promote attachment and phagocytosis.

17 Adaptive Immunity: Specific Defenses of the Host

As discussed in previous chapters, the body's defenses are termed **innate (nonspecific) immunity.** **Adaptive (specific) immunity** is the subject of this chapter.

THE ADAPTIVE IMMUNE SYSTEM

The adaptive immune system is so called because it adapts to experience with conditions it encounters, such as the immunity to smallpox that occurs after recovery from, or *vaccination* against, smallpox.

DUAL NATURE OF THE ADAPTIVE IMMUNE SYSTEM

Humoral immunity describes immunity brought about by **antibodies,** which are proteins formed in the host after exposure to a pathogen. They are highly specific in neutralizing the pathogen when it is encountered again.

Cellular immunity is based on certain specialized cells called **T cells** (because they mature under the influence of the thymus). Antibodies, in contrast, are produced by the stimulation of **B cells** (named for the **bursa of Fabricius,** a structure in birds similar to a lymph node, in which they are produced).

T cells, like B cells, respond to antigens by means of receptors on their surface called **T-cell receptors (TCRs).** When stimulated by an antigen complementary to the TCR, the T cell proliferates and secretes, not antibodies, but *cytokines.* Cytokines are chemical messengers that impart instructions to other cells to perform certain functions.

ANTIGENS AND ANTIBODIES

The Nature of Antigens

Most **antigens** (sometimes called **immunogens**) are mostly proteins or large polysaccharides. These may be part of microorganisms or antigens such as pollen, egg white, blood cells, or transplanted tissues or organs. Antibodies usually recognize and interact with **antigenic determinants,** or **epitopes,** on the antigen, rather than an entire antigen. A bacterium or virus may have numerous antigenic determinants. **Haptens** are low-molecular-weight antigens that are not antigenic unless first attached to a carrier molecule. Once an antibody against the hapten has formed, the hapten will react independently of its carrier. Penicillin is a good example of a hapten.

The Nature of Antibodies

Antibodies are proteins called **immunoglobulins (Ig).** (**Globulins** are proteins of certain solubility characteristics.) They are highly specific and react with only one type of antigenic determinant. Each antibody has at least two **antigen-binding sites.** The **valence** is the number of such sites on the antibody.

Antibody Structure. Figure 17.1a shows a typical *monomer-type antibody.* It has four protein chains—two identical **light (L) chains** and two identical **heavy (H) chains.** At the ends of the arms of Y-shaped molecules are **variable (V) regions,** with a structure that accounts for the ability of different antibodies to recognize and bind with different antigens (Figure 17.1b)—the **antigen-binding sites.**

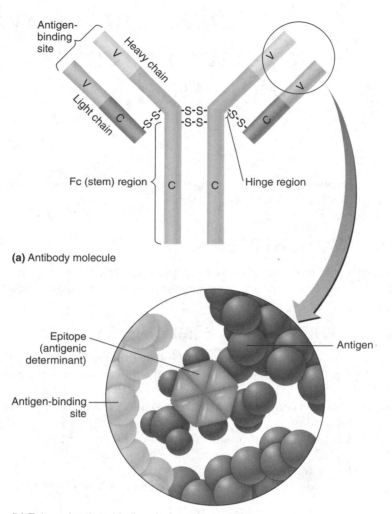

(a) Antibody molecule

(b) Enlarged antigen-binding site bound to an epitope

Figure 17.1 **Structure of a typical antibody molecule.** The Y-shaped molecule is composed of two light chains and two heavy chains linked by disulfide bridges (S-S). Most of the molecule is made up of constant regions (C), which are the same for all antibodies of the same class. The amino acid sequences of the variable regions (V), which form the two antigen-binding sites, differ from molecule to molecule.

The stem of the antibody monomer and lower parts of the Y arms are called **constant (C) regions.** The stem of the Y-shaped monomer is the **Fc** (stem) **region.** The antibody molecule can attach to a host cell by the Fc region. Complement can bind to the Fc region.

Immunoglobulin Classes.   (Refer to Table 17.1 in the text.) **IgG** antibodies account for 80% of all antibodies in serum. They are **monomers** and readily cross blood vessel walls to enter tissue fluids. Maternal IgG crosses the placenta to confer immunity to the fetus. IgG antibodies protect against circulating bacteria and viruses, neutralize bacterial toxins, trigger the complement system, and bind to antigens to enhance action of phagocytic cells. IgG is so long-lived that its presence may indicate immunity against a disease condition in the more distant past.

IgM antibodies constitute 5–10% of antibodies in serum. IgM has a pentamer structure of five Y-shaped monomers held together by a **J (joining) chain.** IgM antibodies are the first to appear in response to an antigen, but their concentration declines rapidly. The presence of IgM in high concentrations in a patient makes it likely that the antibodies are associated with the disease pathogen. IgM antibodies generally do not enter surrounding tissue. IgM molecules are especially effective at cross-linking particulate antigens, causing their aggregation. It is the predominant antibody in the ABO blood group antigen reactions.

IgA antibodies account for about 10–15% of the antibodies in serum. IgA circulates in serum as a monomer, **serum IgA.** IgA may be joined by a J chain into **dimers** of two Y-shaped monomers called **secretory IgA.** IgA is found on mucosal surfaces and in body secretions such as colostrum. A **secretory component** protects the IgA from enzymes and may help it enter secretory tissues. The main function of IgA is preventing attachment of viruses and certain bacteria to mucosal surfaces.

IgD antibodies are only about 0.2% of the total serum antibodies. They are monomers and are found in blood and lymph cells and on B-cell surfaces. IgD functions in serum are little known, but IgD antibodies are on the surface of B cells.

IgE antibodies are monomers slightly larger than IgG and constitute only 0.002% of the total serum antibodies. They bind by their **Fc** (stem) sites to mast cells and basophils. When an antigen reacts with IgE antibodies, the mast cell or basophil releases histamine and other chemical mediators involved in allergic reactions. These inflammatory reactions can be protective, attracting IgG and phagocytic cells.

B CELLS AND HUMORAL IMMUNITY

Antibodies are produced by a special group of lymphocytes called B cells, which develop from stem cell precursors. The process that leads to the production of antibodies starts when B cells are exposed to *free,* or *extracellular, antigens.*

Clonal Selection of Antibody-Producing Cells

Each B cell carries immunoglobulins on its surface, mostly IgM and IgD—all of which are specific for the recognition of the same epitope. When a B cell's immunoglobulins bind to the epitope for which they become specific, the B cell is *activated.* An activated B cell undergoes *clonal expansion,* or proliferation. B cells usually require the assistance of a *T helper cell (T_H)* as shown in Figure 17.2 (T cells will be discussed in detail shortly).

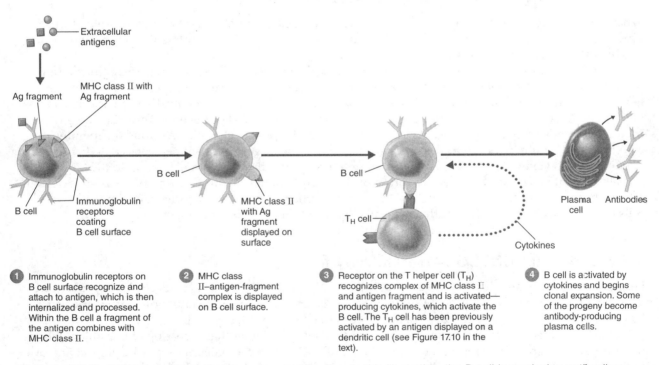

Figure 17.2 Activation of B cells to produce antibodies. In this illustration, the B cell is producing antibodies against a T-dependent antigen.

An antigen that requires a T_H cell for antibody production is known as a **T-dependent antigen.** These antigens are mainly proteins. The antigen contacts the surface immunoglobulins on the B cell and is enzymatically processed within the B cell—fragments of it are combined with membrane proteins called the major histocompatibility complex (MHC). The **major histocompatibility complex (MHC)** is a collection of genes that encode molecules of genetically diverse glycoproteins (part carbohydrate, part protein) found on plasma membranes on nucleated cells. In humans it is also called the *human leukocyte antigen (HLA) system,* which will be discussed in Chapter 19. The MHC identifies the host and prevents the immune system from making antibodies that would be harmful. The complex of the antigenic fragments and the MHC are then displayed on the B cell's surface for identification by the TCR of the T_H.

In this instance the MHC is of class II, which is found only on the surface of *antigen-presenting cells (APCs)*—in this case, a B cell. The T_H becomes activated by this contact and begins producing cytokines that cause the activation of the B cell. An activated B cell proliferates into a large clone of cells, some of which differentiate into antibody-producing **plasma cells.** Other clones of the activated B cell become long-lived **memory cells.** This is called **clonal selection** (a similar process occurs with T cells, as will be seen later in the chapter). B cells that might produce antibodies harmful to the host (self) are eliminated at the immature lymphocyte stage by **clonal deletion.**

Antigens that stimulate B cells directly without the help of T cells are called **T-independent antigens.** These are characterized by repeating subunits such as found in polysaccharides or lipopolysaccharides; bacterial capsules, for example. T-independent antigens generally provoke a relatively weak immune response, primarily of IgM, and no memory cells.

The Diversity of Antibodies

The human immune system can recognize a huge number of antigens. Simplistically, the mechanism is analogous to the generation of huge numbers of words from a limited alphabet. The "alphabet" is part of the genetic makeup of the variable (V) region of the immunoglobulin molecule, which can be linked to various genes in the antibody's constant (C) region.

ANTIGEN–ANTIBODY BINDING AND ITS RESULTS

When an antibody encounters an antigen for which it is specific, an **antigen–antibody complex** forms. The strength of the bond is called **affinity.** The capability of recognizing an antigen's epitope and distinguishing it from another is the **specificity.**

Foreign organisms and toxins are rendered harmless by only a few mechanisms. In **agglutination,** the antibodies cause antigens to clump together; these clumps are more easily ingested by phagocytes. IgM is more effective at agglutination because it has more binding sites. In **neutralization,** the antibodies inactivate viruses or toxins by blocking their attachment to host cells. For **opsonization,** the antigen is coated with antibodies that enhance its ingestion and lysis by phagocytic cells. **Antibody-dependent cell-mediated cytotoxicity** resembles opsonization in that the target organism becomes coated with antibodies—however, destruction is done by immune system cells that remain external to the target cell.

Finally, antibodies may trigger **activation of the complement system.** This part of the body's defenses was discussed in Chapter 16. Sometimes this attracts phagocytes and other defensive immune system cells to the infected area and other times it leads to lysis of the microbial pathogen.

T CELLS AND CELLULAR IMMUNITY

Humoral antibodies are effective against freely circulating antibodies. T cells and cellular immunity are directed against intracellular antigens, which are not exposed to circulating antibodies. They are also important in the recognition and destruction of cells that are nonself, especially cancer cells.

Like B cells, each T cell is specific for only a certain antigen. Rather than a coating of immunoglobulins that provides the specificity for B cells, a T cell has TCRs. The stem cell precursors of a T cell reach maturity in the thymus. Most are eliminated there, a process called **thymic selection,** which is analogous to clonal deletion in B cells. They then migrate to lymphoid tissues where they are most likely to encounter antigens.

Most pathogens that the cellular immune system combats enter the gastrointestinal tract or lungs, where they encounter a barrier of epithelial cells. They normally pass this barrier in the gastrointestinal tract by way of **microfold cells** or **M cells.** M cells are located over **Peyer's patches,** which are secondary lymphoid organs located on the intestinal wall. The recognition of an antigen by a T cell requires that the antigen first be processed by a specialized **antigen-presenting cell (APC).** This resembles the situation in humoral immunity in which the B cells served as the APC. After processing by the APC the antigen is presented on the APC surface together with a molecule of MHC. (APCs will be described in detail later in the chapter.)

Classes of T Cells

Primarily, there are two populations of T cells that concern us: T helper cells (T_H) and T cytotoxic cells (T_C). T_C can differentiate into an effector cell called a **cytotoxic T lymphocyte (CTL).** T cells are also classified by glycoproteins on their surface called **clusters of differentiation (CDs).** The CDs of greatest interest are CD4 and CD8; T cells that carry these molecules are named CD4$^+$ T cells (which are T_H cells that bind to MHC class II molecules on B cells and APCs, as shown in Figures 17.4 and 17.10 in the text) and CD8$^+$ T cells (T_C cells that bind to MHC class I molecules).

T helper cells (CD4$^+$ cells) present T-dependent antigens to B cells. They also help other T cells respond to antigens. T_H cells differentiate into two major subpopulations. **T_H1 cells** mostly activate cells related to cellular immunity, especially delayed-type hypersensitivity (see Chapter 19). They also enhance the activity of complement in opsonization and inflammation (see Figure 17.11 in the textbook). **T_H2 cells** are especially important in allergic reactions such as hypersensitivity (see Chapter 19) and defense against helminthic infections. A recently discovered third subset is the **T_H17** T cells, which are named for their production of large quantities of cytokine IL-17. They are effective against certain infections not affected by T_H1 and T_H2 cells. Another recently described subset is follicular helper T cells (T_{FH}), which stimulate B cells to produce plasma cells and are also involved in class switching.

T regulatory cells (T_{reg} cells) were formerly called *suppressor cells.* They are a subset of CD4$^+$ T helper cells and carry an additional CD25 molecule. They combat autoimmunity by suppressing T cells that escape deletion in the thymus without the necessary "education" to avoid reacting against the body's self. They also protect against intestinal bacteria required for digestion.

T cytotoxic cells (CD8$^+$ cells) are precursors to cytotoxic T lymphocytes (CTLs). It is only the CTL that can attack a target cell considered to be nonself (see Figure 17.11 in the text). The target cells can be self-cells that have been altered by infection by a pathogen, especially a virus. On their surface they carry fragments of **endogenous antigens** that are generally synthesized within the cell. Other important target cells are tumor cells and transplanted tissue. Rather than reacting with antigenic fragments presented by an APC in a complex with MHC class II molecules, the CD8$^+$ T cell recognizes endogenous antigens on the target cell's surface that are in combination with and MHC class I molecule. Such MHC molecules are found on nucleated cells; therefore, a CTL can attack almost any cell of the host that has been altered. In its attack a CTL attaches to the target cell and releases **perforin.** This protein forms a pore that contributes to the death of the cell. **Granzymes,** proteases that induce apoptosis, are now able to enter through the pore. **Apoptosis** is also called *programmed cell death* and results in cell disintegration that attracts phagocytes that digest the remains.

Apoptosis, or *programmed cell death,* causes the death and disintegration of cells. One benefit of apoptosis is that it clears the cells of pathogens and prevents the spread of infectious viruses into other cells. It also limits uncontrolled growth that would be harmful in the long run.

ANTIGEN-PRESENTING CELLS (APCs)

In cellular immunity, the APCs are not B cells, but dendritic cells and macrophages.

Dendritic cells are characterized by long extensions called dendrites. They are plentiful in the lymph nodes, spleen, and skin but not the brain. There are several different populations of dendritic cells, which are named for their derivation or location. Those located in the skin or genital tract, for example, are called *Langerhans cells* after their discoverer. Compared to macrophages, dendritic cells are poorly phagocytic but much more important as APCs.

Macrophages are important for innate immunity and ridding the body of cellular debris. Their phagocytic capabilities are increased when stimulated by ingestion of antigenic material to become **activated macrophages.** When activated their appearance changes, becoming larger and ruffled—they are then important factors in the control of cancer cells and intracellular pathogens.

EXTRACELLULAR KILLING BY THE IMMUNE SYSTEM

Certain granular leukocytes called **natural killer (NK) cells** can attack relatively large parasites (compared to bacteria). They are not immunologically specific and do not require stimulation by an antigen. They contact a target cell to determine whether it expresses MHC class I self-antigens. If it does not, they kill the target cell by mechanisms similar to that of a CTL. Tumor cells are also targets.

The functions of NK cells and the other principal cells involved in cellular immunity are summarized in Table 17.2 in the textbook.

ANTIBODY-DEPENDENT CELL-MEDIATED CYTOTOXICITY

In addition to attacks by NK cells, relatively large parasites can be targeted by **antibody-dependent cell-mediated cytotoxicity.** The target cell is first coated with antibodies, and then a variety of cells of the immune system bind to the F_C regions of these antibodies and thus to the target cell. The target cell is then lysed by substances secreted by the attacking cells.

CYTOKINES: CHEMICAL MESSENGERS OF IMMUNE CELLS

The immune response requires complex interactions between different cells, which are mediated by chemical messengers called **cytokines.** When they communicate between leukocytes, they are termed **interleukins.** A family of cytokines that induces migration of leukocytes into areas of infection is called **chemokines.** The family of cytokines called **interferons** protects cells from viral infections. **Tumor necrosis factor** cytokines are a factor in inflammatory diseases such as rheumatoid arthritis. A family of cytokines, **hematopoietic cytokines,** control pathways by which stem cells develop into different red or white blood cells. Cytokines may stimulate cells to produce more cytokines. When this feedback loop gets out of control, a **cytokine storm** may damage tissues, an important factor in certain diseases (also see the discussion of superantigens in the text).

IMMUNOLOGICAL MEMORY

The intensity of the antibody-mediated humoral response is reflected by the relative amount of antibody in the serum, the **antibody titer.** After contact with the antigen there is a slow rise in antibody titer; first IgM class antibodies, then IgG antibodies (see Figure 17.17 in the text), which is the **primary response.** A second exposure to the antigen results in a **secondary response** (the **memory** or **anamnestic response**). This is more rapid and of greater magnitude than the primary response. By way of explanation, some activated B cells become long-lived **memory cells,** which even years later will differentiate into antibody-producing plasma cells when stimulated by the same antigen.

TYPES OF ADAPTIVE IMMUNITY

Adaptive immunity refers to the protection an animal develops against certain specific microbes or foreign substances. **Naturally acquired active immunity** develops when a person is exposed to antigens, becomes ill, and then recovers. *Subclinical* or *inapparent* infections that produce no noticeable signs or symptoms can also confer immunity.

Naturally acquired passive immunity involves the natural transfer of antibodies from a mother to her infant by *transplacental transfer* (across the placenta) to the fetus or by the first secretions of breast milk, the *colostrum*. This type of immunity usually persists only for a few weeks or months. **Artificially acquired active immunity** is the result of **vaccination (immunization)** by introduction of **vaccines** into the body, which are antigens such as killed or living microorganisms or inactivated toxins (**toxoids**).

    Artificially acquired passive immunity involves the injection of antibodies (rather than antigens) into the body. These antibodies come from an animal or person already immune to the disease.

    Antiserum is the generic term for blood-derived fluids containing antibodies. **Serology** is the study of reactions between antibodies and antigens (which are found in blood serum). Most antibodies in blood serum are in the globular proteins of the gamma fraction when separated by electrophoresis—therefore, the serum containing antibodies for passive immunity is called **gamma globulin.** The half-life of such injected antibodies is typically about 3 weeks.

SELF-TESTS

In the matching section, there is only one answer to each question; however, the lettered options (a, b, c, etc.) may be used more than once or not at all.

I. Matching

_____ 1. Antigen converts these into plasma cells.

_____ 2. Involved in cell-mediated immunity.

_____ 3. Directed against transplanted tissue cells and cancer cells.

_____ 4. Have been influenced by the thymus.

_____ 5. Defend mainly against bacteria and viruses circulating in blood and lymph.

_____ 6. Responsible for rejection of foreign tissue transplants.

a. B cells

b. T cells

II. Matching

_____ 1. Based on antibodies produced as a result of recovery from a disease.

_____ 2. Passed to a fetus by transplacental transfer.

_____ 3. Passed to an infant in human colostrum.

_____ 4. Passed to a recipient by injection of gamma globulin blood fraction from other people.

_____ 5. Based on production of antibodies by vaccination.

a. Naturally acquired active immunity

b. Artificially acquired passive immunity

c. Naturally acquired passive immunity

d. Artificially acquired active immunity

III. Matching

____ 1. An incomplete antigen that will react with antibodies but will not, by itself, stimulate their formation.

____ 2. The number of determinant sites on an antigen or antibody.

____ 3. The source of B cells and T cells.

____ 4. Chemical messengers by which cells of the immune system communicate with each other.

____ 5. The relative strength of the antigen–antibody bond.

a. Hapten

b. Valence

c. Stem cells

d. Cytokines

e. Affinity

f. Memory cells

g. Specificity

IV. Matching

____ 1. A pentamer; the first antibody class to appear, though comparatively short-lived.

____ 2. The most abundant immunoglobulin in serum.

____ 3. Functions of this immunoglobulin class are not well defined, but it is found on the surface of B cells.

____ 4. Involved in allergic reactions, such as hay fever.

____ 5. Often forms dimers of two immunoglobulin monomers.

a. IgA

b. IgG

c. IgD

d. IgE

e. IgM

V. Matching

____ 1. Synonym for antigens.

____ 2. B cells that interact with self-antigens are destroyed.

____ 3. Protein bound to IgA immunoglobulins.

____ 4. Blood fraction that contains most of the serum immunoglobulins.

____ 5. Antigenic; will stimulate the production of antitoxins.

____ 6. Activated B cell proliferates into a large clone of cells, some of which will differentiate into plasma cells.

a. Gamma globulin

b. Immunogens

c. Clonal deletion

d. Toxoid

e. Secretory component

f. Clonal selection

g. Thymic selection

VI. Matching

___ 1. CD4+.

___ 2. CD8+.

___ 3. Can differentiate into CTLs.

___ 4. Present T-dependent antigens to B cells.

___ 5. Recognize and target cells that carry endogenous antigens.

a. T helper cells (T_H)

b. T cytotoxic cells (T_C)

c. Regulatory T cells (T_{reg})

VII. Matching

___ 1. Requires assistance of a T helper cell to form antibodies.

___ 2. Typically a protein.

___ 3. Typically a polysaccharide such as a bacterial capsule.

a. T-independent antigen

b. T-dependent antigen

VIII. Matching

___ 1. Cytokine that inhibits viral infections.

___ 2. Released by a cytotoxic T lymphocyte to lyse a target cell.

___ 3. Stem region of an antibody molecule.

___ 4. Programmed cell death.

a. Perforin

b. Apoptosis

c. F_C

d. Interferon

IX. Matching

___ 1. First breast milk secretions of a mammal.

___ 2. Adjective applied to a component in IgA that protects it from enzyme activity.

___ 3. Adjective applied to the cells that actually produce antibodies after a B cell is stimulated by an antigen.

___ 4. Usual configuration of IgA.

a. Dimer

b. Colostrum

c. Secretory

d. Monomer

e. Plasma

X. Matching

____ 1. Clumping of antigens when binding with antibodies.

____ 2. Coating of target cell with antibody that enhances phagocytosis.

____ 3. Coating of target cell with antibody that leads to lysis by substances secreted by immune cells external to the target cell.

____ 4. Relative amount of antibody in the serum.

a. Antibody-dependent cell-mediated cytotoxicity

b. Antibody titer

c. Agglutination

d. Opsonization

XI. Matching

____ 1. Communicate between leukocytes.

____ 2. A factor in inflammatory diseases such as rheumatoid arthritis.

____ 3. Control pathways by which stem cells develop into different red or white blood cells.

____ 4. Induce migration of leukocytes into areas of infection.

a. Interleukins

b. Hematopoietic cytokines

c. Tumor necrosis factor

d. Chemokines

XII. Matching

____ 1. Especially important in allergic reactions such as hypersensitivity and defense against helminthic infections.

____ 2. Stimulate B cells to produce plasma cells.

____ 3. Enhance the activity of complement in opsonization and inflammation.

a. T_H1 T cells

b. T_H2 T cells

c. T_H17 T cells

d. T_{FH} T cells

Fill in the Blanks

1. Resistance present at birth that does not involve humoral or cell-mediated immunity is

   _____ immunity.

2. A(n) _____ site is a specific chemical group on an antigen that combines with the

   antibody.

3. The five monomers that constitute the IgM molecule are held together by a _____.

4. The antibody _____ is the measured amount of antibody in the serum.

5. Certain lymphocytes called _____ cells kill virus-infected cells and tumor cells, but

   are not immunologically specific. They contact and kill the target cells.

6. B cells derive their name from an organ in poultry, the _____.

7. CD is short for clusters of _____.

8. Low-molecular-weight substances such as penicillin that do not (by themselves) cause formation of

   antibodies are known immunologically as _____.

9. The second time we encounter an antigen, our immune response is faster and more intense; this is

   termed the _____ response.

10. Some antibodies are poorer matches for an antigen than others; they are said to have less

    _____.

11. The subpopulation of T cells that mostly activate cells related to cell-mediated immunity such as

    macrophages, CD8$^+$ T cells, and natural killer cells is _____.

12. An antigen-presenting cell (APC) that is not efficient at phagocytosis, but is the most imporant in

    APC, is called a _____.

Critical Thinking

1. An infant's mother had diphtheria prior to pregnancy. Is the infant born with an immunity to
   diphtheria? If so, why would we need vaccination for diphtheria?

2. What are cytokines, and why are they necessary?

ANSWERS

Matching

I.	1. a	2. b	3. b	4. b	5. a	6. b
II.	1. a	2. c	3. c	4. b	5. d	
III.	1. a	2. b	3. c	4. d	5. e	
IV.	1. e	2. b	3. c	4. d	5. a	
V.	1. b	2. c	3. e	4. a	5. d	6. f
VI.	1. a	2. b	3. b	4. a	5. b	
VII.	1. b	2. b	3. a			
VIII.	1. d	2. a	3. c	4. b		
IX.	1. b	2. c	3. e	4. a		
X.	1. c	2. d	3. a	4. b		
XI.	1. a	2. c	3. b	4. d		
XII.	1. b	2. d	3. a			

Fill in the Blanks

1. innate   2. antigenic determinant   3. J chain   4. titer   5. natural killer (NK)   6. bursa of Fabricius
7. differentiation   8. haptens   9. secondary or anamnestic   10. affinity   11. T_H1 12. dendritic cell

Critical Thinking

1. The infant does have immunity to diphtheria by maternal antibodies acquired from the mother, but this immunity is short-lived. The purpose of maternally acquired immunity is to protect the infant until its own immune system matures.

2. Cytokines are soluble chemical messengers produced by immune system cells such as lymphocytes and macrophages. Cytokines are needed to allow cells of the immune system to communicate with each other—for example, to warn of the presence of pathogens that must be dealt with.

18 Practical Applications of Immunology

VACCINES

Principles and Effects of Vaccination

A **vaccine** is a suspension of microorganisms, or some part or product of them, that will induce immunity when it is administered to the host. Vaccines had their origin in the practice of **variolation,** in which material from smallpox scabs was inoculated into the bloodstream to provide immunity to the disease. We now call this strategy **vaccination** or **immunization.**

A disease can be controlled if most, but not all, of the population is immune. This is called **herd immunity.**

Types of Vaccines and Their Characteristics

A **live, attenuated vaccine** uses pathogens that have been weakened by age or extended laboratory culture, in the case of viruses, cell culture. An **inactivated, killed vaccine** uses microbes that have been killed, usually by formalin or phenol. Attenuated vaccines tend to mimic actual infection and usually provide better immunity.

A **subunit vaccine** uses only those antigenic fragments of a microorganism that are best suited to stimulate an immune response. These subunits may be produced by bacteria or yeasts, for example, by use of genetic engineering techniques; such vaccines are called **recombinant vaccines. Toxoids** are inactivated bacterial toxins. An *antitoxin* is serum that contains antibodies against a toxin.

Vaccines can be fragmented and the desired antigens separated out, rather than using the complete cell; these are called **acellular vaccines.** Some polysaccharide vaccines have enhanced effectiveness when combined with tetanus or diphtheria toxoids. These are **conjugated vaccines.**

Nucleic-acid vaccines (DNA vaccines) inject "naked" DNA into muscle. This results in production of proteins encoded in the DNA, which persist and stimulate an immune response. RNA, which could replicate in the recipient, might be a more effective agent. **Adjuvants** are chemicals—alum, for example— added to vaccines to improve their effectiveness.

DIAGNOSTIC IMMUNOLOGY

We cannot see antibodies and must infer their presence indirectly by a variety of reactions. Essential elements of diagnostic tests are **sensitivity** (the probability that the test is reactive if the specimen is a true positive) and **specificity** (the probability that a positive test will not be reactive if a specimen is a true negative).

Monoclonal Antibodies

Techniques have been developed recently that allow large volumes of antibodies to be produced in vitro— an important factor in many new diagnostic tests (see Figure 18.2 in the text). In these techniques, a cancerous cell (considered immortal, in the sense that it can be propagated indefinitely) and an antibody-secreting plasma cell (B cell), taken from a mouse that has been immunized with a particular antigen, are fused. This hybrid cell is called a **hybridoma.** Grown in culture, such a hybridoma will continue to produce the type of antibody characteristic of the ancestral B cell. Because the antibodies produced by such a hybridoma are identical, they are called **monoclonal antibodies** (Mabs). A Mab programmed to react with a cancer can theoretically be combined with a toxin (**immunotoxin** or **conjugated Mabs**) against the cancer cells and kill them. A problem in therapeutic use of Mabs is that they are produced from mouse cells; these are considered foreign, and human immune systems react against their presence.

Monoclonal antibodies constructed with variable regions from mouse cells and constant regions from human sources (**chimeric Mabs**) would be more compatible with the human immune system. Even better would be **humanized Mabs** constructed so that the murine portion is limited to the antigen-binding sites. The eventual goal is to develop **fully human Mabs** that could be an exact match to the patient.

The source of particular Mabs can be recognized from the spelling of the final letters of the name, such as human (*u*), mouse (*o*), chimera (*xi*), or humanized (*zu*). For example, a Mab name ending in *–umab* shows it to be derived from a human source. The general disease state can also be included in a similar manner. For example, from the spelling of the Mab *biciromab*, it can be seen to be derived from a mouse (*-omab*) and directed at treating a cardiovascular condition (known by the letters *cir*).

Precipitation Reactions

Precipitation reactions involve the reaction of *soluble* antigens with IgG or IgM antibodies to form large interlocking aggregates called **lattices.** Precipitation can be easily arranged by allowing the antigen and antibody to diffuse toward each other. A cloudy line of precipitation forms in the area in which the **optimal ratio** has been reached (the **zone of equivalence**). In a capillary tube, this is called a **precipitin ring test.** In a Petri dish, the antigens and antibodies can be placed into wells cut into the agar medium, called **immunodiffusion tests.** If passive diffusion of the reagents is too slow, the test can be speeded up by applying an electric current, a procedure called **immunoelectrophoresis,** used in research to separate proteins in human serum.

Agglutination Reactions

Agglutination reactions involve either *particulate* antigens (particles such as cells that carry antigenic molecules) or soluble antigens adhering to particles. These antigens can be linked together by antibodies to form visible aggregates, a reaction called **agglutination**.

Direct Agglutination Tests.   **Direct agglutination tests** detect antibodies against relatively large *cellular* antigens, such as red blood cells, bacteria, and fungi. The amount of antigen in each well of a microtiter plate is the same. The serum containing the antibodies is sequentially diluted out in a series of wells. The higher the concentration of antibodies in the serum, the more dilutions are required to dilute it to the point at which no reaction occurs with the antigen. This is a measure of **titer,** or concentration of antibody. For diagnostic purposes a **rise in titer** during the course of a disease indicates that the antibodies are associated with the disease. If it can be shown that a person's blood has no antibody titer before illness but develops one while the disease is progressing, this is called **seroconversion,** and it is diagnostic.

Indirect (Passive) Agglutination Tests.   Soluble antigens can respond to agglutination tests if the antigens are adsorbed onto particles, especially minute latex spheres. In such **indirect (passive) agglutination tests** (usually called **latex agglutination tests**), the antibody reacts with the soluble antigen adhering to the latex particle. This is the basis of many new diagnostic tests.

Hemagglutination.   When agglutination reactions involve the clumping of red blood cells, the reaction is called **hemagglutination**. Certain viruses, such as those causing mumps, measles, and influenza, have the ability to agglutination red blood cells without an antigen–antibody reaction; this process is called **viral hemagglutination**.

Neutralization Reactions

The harmful effects of a bacterial exotoxin or a virus can be neutralized by specific antibodies. One such substance is called an **antitoxin,** as used in the treatment of diphtheria. Antitoxins produced in other animals can be injected to provide passive immunity against a toxin. Similarly, if an antibody binds to a virus, it prevents it from attaching to a host cell and infecting it. If a person's serum contains antibodies against viruses that cause hemagglutination, these antibodies will react with the viruses and neutralize them. This is the basis of the **viral hemagglutination inhibition test,** used in the diagnosis of influenza, measles, and mumps.

Complement-Fixation Reactions

Complement (see Chapter 16 in the text) is a group of serum proteins. During most antigen–antibody reactions, the complement binds to the antigen–antibody complex and is fixed. This **complement fixation** can be used to detect antibody presence. Antibodies that do not produce a visible reaction by other means can sometimes be demonstrated by the fixing of complement during the antigen–antibody reaction (see Figure 18.10 in the text).

Fluorescent-Antibody Techniques

Fluorescent-antibody (FA) techniques use fluorescent dyes (fluorescein) that are combined with antibodies to make them fluoresce when exposed to ultraviolet light. This requires a fluorescence microscope with ultraviolet light illumination. In a **direct FA test,** the antigen to be identified is fixed onto a slide. The fluorescein-labeled antibodies are added. The unbound antigen is washed free, and the slide is examined under the ultraviolet microscope. In an **indirect FA test,** a known antigen is fixed onto a slide. A test serum with an unknown antibody is added. If the serum contains antibodies against the test antigen, it will bind to it, but this is not visible. So that the antigen–antibody complex can be seen, fluorescein-labeled **antihuman immune serum globulin (anti-HISG),** an antibody that reacts specifically with *any* human antibody, is added to the slide. Unbound reagents are washed from the slide, which is examined under the fluorescence microscope. If the known antigen on the slide appears fluorescent, antibody specific to it was in the test serum. A **fluorescence-activated cell sorter** (see Figure 18.12 in the text) uses FA markers to isolate and identify cells such as T cells.

Enzyme-Linked Immunosorbent Assay (ELISA)

A widely used serological test is the **enzyme-linked immunosorbent assay (ELISA),** one of the group of tests called **enzyme immunoassays (EIAs).** A *microtiter plate* with numerous shallow wells is used in the two basic methods we will describe. Variations of the test may use reagents bound to tiny latex particles rather than to the surfaces of the microtiter plates.

In the **direct ELISA,** the goal is to identify an *unknown antigen,* such as a drug in a serum sample. In the **indirect ELISA,** the object is to determine the presence of certain *antibodies* in the serum. This assay currently is used to test for antibodies against the AIDS virus. In both of these variations of the ELISA, a positive reaction is indicated by a color change in the well due to action of an enzyme linked to the last antibody added and allowed to react with a substrate added in the final step. Figure 18.1 shows the mechanism of ELISAs.

Western Blotting (Immunoblotting)

Western blotting, often called *immunoblotting,* can be used to identify a specific protein in a mixture. When this specific protein is an antibody, the technique is valuable in diagnosing a disease. The components of the mixture are separated by electrophoresis in a gel and then transferred to a protein-binding sheet (blotter). There the protein/antigen is flooded with enzyme-linked antibody. The location of the antigen and the enzyme-linked antibody reactant can be visualized, usually with a color-reacting label (see Figure 10.12 in the text). Western blotting is often used as a confirmatory test for HIV infection.

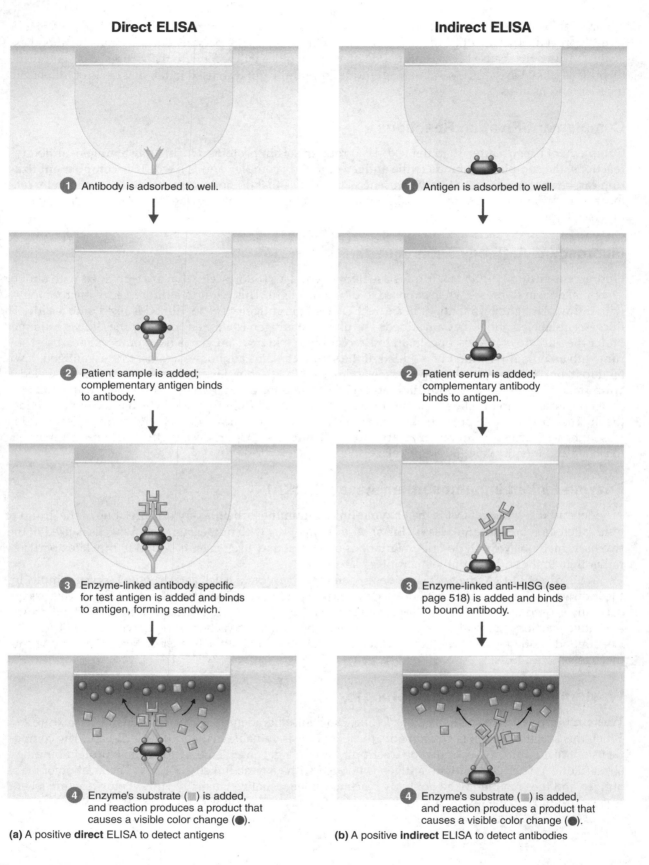

Direct ELISA

1 Antibody is adsorbed to well.

2 Patient sample is added; complementary antigen binds to antibody.

3 Enzyme-linked antibody specific for test antigen is added and binds to antigen, forming sandwich.

4 Enzyme's substrate (■) is added, and reaction produces a product that causes a visible color change (●).

(a) A positive **direct** ELISA to detect antigens

Indirect ELISA

1 Antigen is adsorbed to well.

2 Patient serum is added; complementary antibody binds to antigen.

3 Enzyme-linked anti-HISG (see page 518) is added and binds to bound antibody.

4 Enzyme's substrate (■) is added, and reaction produces a product that causes a visible color change (●).

(b) A positive **indirect** ELISA to detect antibodies

Figure 18.1   The ELISA method. The components are usually contained in small wells of a microtiter plate.

SELF-TESTS

In the matching section, there is only one answer to each question; however, the lettered options (a, b, c, etc.) may be used more than once or not at all.

I. Matching

___ 1. Makes use of the fact that certain viruses will cause agglutination of red blood cells.

___ 2. The absence of complement is indicated by hemolysis.

___ 3. A precipitation-type test in which wells are cut into the agar on a Petri dish.

___ 4. Soluble antigens are detected by binding them to small latex particles, for example, and causing their agglutination.

___ 5. The ELISA used to screen for AIDS antibodies in serum.

___ 6. Often used as a confirmatory test for HIV infection.

a. Immunodiffusion test

b. Indirect agglutination test

c. Complement-fixation test

d. Direct ELISA

e. Indirect ELISA

f. Hemagglutination inhibition test

g. Western blotting

II. Matching

___ 1. Subunit vaccine using genetically engineered organisms to produce it.

___ 2. Unwanted components are removed from a whole-cell vaccine.

___ 3. An inactivated toxin.

___ 4. "Naked" DNA or RNA injected into muscle.

___ 5. Chemical additive that improves effectiveness of a vaccine.

a. Recombinant vaccine

b. Toxoid

c. Acellular vaccine

d. Antitoxin

e. Nucleic-acid vaccine

f. Adjuvant

III. Matching

____ 1. The probability that a positive diagnostic test will not be reactive if a specimen is a true negative.

____ 2. The probability that a diagnostic test is reactive if the specimen is a true positive.

____ 3. A monoclonal antibody combined with a toxin and programmed to react with a cancer cell.

____ 4. An antibody-producing plasma cell fused with a cancerous cell.

____ 5. A monoclonal antibody in which the variable regions are from mouse cells and the constant regions are from human sources.

a. Chimeric Mab

b. Hybridoma

c. Specificity

d. Sensitivity

e. Conjugated Mab

f. Humanized Mab

Fill in the Blanks

1. Before the invention of modern vaccines, material from smallpox scabs was inoculated into the bloodstream to give immunity to the disease; this was called _____.

2. The measure of the concentration of antibody in serum is called _____.

3. Fluorescein-labeled antihuman gamma globulin would be used in the _____ fluorescent antibody test.

4. A vaccine using a living, weakened organism is called _____.

5. For diagnostic purposes, a rise in _____ during the course of a disease is very significant.

6. A disease can be controlled if most, but not all, of the population is immune; this is called _____ immunity.

7. Polysaccharide vaccines can be enhanced in effectiveness by adding toxoids such as diphtheria; these are so-called _____ vaccines.

Label the Art

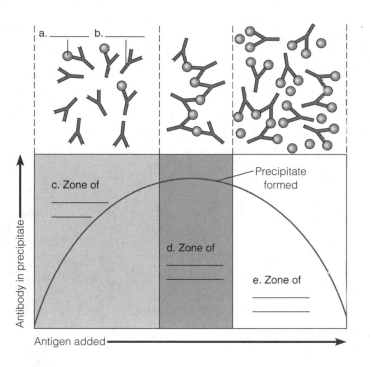

a. _____   b. _____

c. Zone of
_____
_____

Precipitate formed

Antibody in precipitate

d. Zone of
_____
_____

e. Zone of
_____
_____

Antigen added ⟶

Critical Thinking

1. Discuss herd immunity as it relates to the control of disease.

2. Compare and contrast attenuated and inactivated vaccines. Which type of vaccination usually provides better immunity? Why?

3. Discuss two advantages of recombinant vaccines. Are there any disadvantages?

4. How is the measured antibody titer used to diagnose disease? Which of the following titers indicates greater immunity, 1:94 or 1:312? Why?

ANSWERS

Matching

I.   1. f   2. c   3. a   4. b   5. e
II. 1. a   2. c   3. b   4. e   5. f
III. 1. c   2. d   3. e   4. b   5. a   6. g

Fill in the Blanks

1. variolation   2. titer   3. indirect   4. attenuated   5. titer   6. herd   7. conjugate

Label the Art

a. antigen   b. antibody   c. antibody excess   d. equivalence   e. antigen excess

Critical Thinking

1. Herd immunity refers to a situation in which most individuals of a population are immune to a disease. If an outbreak of the disease occurs, there will not be enough susceptible individuals to support an epidemic.

2. Attenuated and inactivated vaccines are both examples of whole-agent vaccines. The process for making the vaccines and the results achieved by using the vaccines differ. Attenuated vaccines use weakened microorganisms to confer immunity to the recipient. Very often viruses used for this type of vaccine are derived from viruses that have mutated during long-term cell culture. Many attenuated vaccines provide lifelong immunity and may be 95% effective. This is due to the fact that the attenuated virus replicates in the body, increasing the original dose. Inactivated vaccines use killed viruses. The viruses are killed by treatment with formalin or other chemicals. Even though the viruses are dead, they still stimulate an immune response. Inactivated vaccines are usually used in situations in which live vaccines are considered too risky.

3. Recombinant vaccines, or subunit vaccines, are produced by genetically altered organisms such as a yeast. The genetically engineered yeast produces a portion of the virus, such as a viral coat protein. Subunit vaccines are safer than whole-agent vaccines because they cannot cause infection in the recipient under any circumstances.

4. Titer refers to the concentration of serum antibody and is the lowest dilution of the antibody that will result in agglutination. In general, the higher the serum antibody titer, the greater the immunity to the disease. This means that a titer of 1:312 shows greater immunity to a disease than a titer of 1:94.

19 Disorders Associated with the Immune System

Antigens such as the staphylococcal enterotoxin are called **superantigens.** They indiscriminately activate many T cells at once, causing a harmful immune response.

HYPERSENSITIVITY

The term **hypersensitivity (allergy)** refers to sensitivity beyond what is considered normal. It occurs in people who have been previously sensitized by exposure to an antigen, called in this context an **allergen.** Once the person is sensitized, another exposure to the antigen triggers an immune response that damages host tissue. There are four principal types of hypersensitivity reactions.

TYPE I (ANAPHYLACTIC) REACTIONS

Anaphylaxis means "the opposite of protected," from the prefix *ana-,* meaning against, and the Greek *phylaxis,* meaning protection. Anaphylactic responses can be **systemic reactions** which produce shock and breathing difficulties and are sometimes fatal, or **localized reactions** which include common allergic conditions such as hay fever, asthma, and hives.

IgE antibodies bind to the surfaces of mast cells and basophils. **Mast cells** are prevalent in the connective tissue of skin, the respiratory tract, and surrounding blood vessels. **Basophils** circulate in the blood. When an antigen combines with antigen-combining sites on two adjacent IgE antibodies and bridges the space between them, the mast cell or basophil undergoes **degranulation,** releasing chemicals called *mediators.* The best-known mediator is **histamine,** which affects the blood vessels, causing edema (swelling), erythema (redness), increased mucus secretion, and smooth-muscle contractions resulting in breathing difficulty. Other mediators are **leukotrienes** (which tend to cause contractions, such as the spasms of asthmatic attacks) and **prostaglandins** (which tend to increase secretions of mucus). Collectively, mediators attract neutrophils and eosinophils to the site and cause inflammatory symptoms.

Systemic Anaphylaxis

When an individual sensitized to an injected antigen, such as an insect sting or penicillin, receives a subsequent injection, the release of mediators can result in a drop in blood pressure (shock) that can be fatal in a few minutes. This is termed **systemic anaphylaxis,** or **anaphylactic shock.** Injections of **epinephrine** are used to treat anaphylactic shock and severe asthma attacks.

Localized Anaphylaxis

Localized anaphylaxis usually is associated with antigens that are ingested or inhaled, rather than injected. Examples are hay fever, for which **antihistamine** drugs often are useful to treat symptoms. Allergens that enter the gastrointestinal tract can also sensitize a person. However, many so-called food allergies are not related to hypersensitivity and are more accurately described as food intolerances. It is sometimes difficult to distinguish food hypersensitivity from food intolerance.

Preventing Anaphylactic Reactions

If a person cannot avoid contact with the allergen, **desensitization** might be attempted. This consists of injections of a series of small doses of the antigen. The idea is to induce IgG antibodies to serve as **blocking antibodies** that intercept and neutralize antigens before they can react with cell-bound IgE.

TYPE II (CYTOTOXIC) REACTIONS

Immunological injury resulting from type II reactions is caused by antibodies that are directed at antigens on the host's blood cells or tissue cells. The host cell plasma membrane may be damaged by antibody and complement, or macrophages may attack antibody-coated cells. *Transfusion reactions*, such as those involving the ABO and Rh blood group systems, are of this type.

The ABO Blood Group System

Human blood is grouped into four principal types: A, B, AB, and O—a classification called the **ABO blood group system.** People with type A, for example, have antigens designated A on their red blood cells. People with blood type O lack both A and B surface antigens. The main features of the ABO blood group system are summarized in Table 19.1. In about 80% of the population, called *secretors*, antigens of the ABO type appear in saliva, semen, and other bodily fluids.

The Rh Blood Group System

Roughly 85% of the human population has an antigen named the **Rh factor,** and they are called Rh+. Giving blood from an Rh+ donor to an Rh− recipient will stimulate the production of anti-Rh antibodies and cause an adverse reaction.

Table 19.1   The ABO Blood Group System

Blood Group	Erythrocyte, or Red Blood Cell, Antigens	Illustration	Plasma Antibodies	Blood That Can Be Received	Frequency (% U.S. Population) White	Black	Asian
AB	A and B		Neither anti-A nor anti-B antibodies	A, B, AB, O	3	4	5
B	B		Anti-A	B, O	9	20	27
A	A		Anti-B	A, O	41	27	28
O	Neither A nor B		Anti-A and anti-B	O	47	49	40

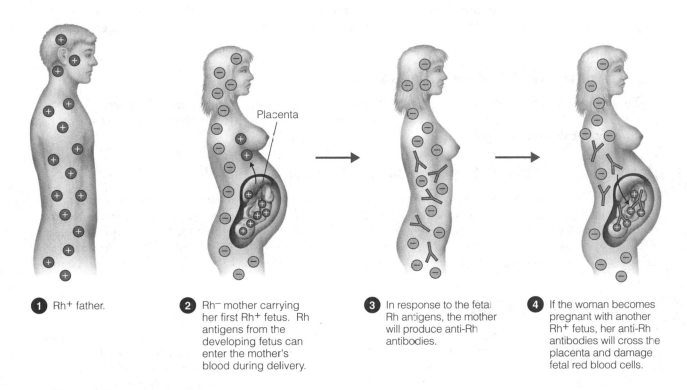

① Rh+ father.

② Rh⁻ mother carrying her first Rh+ fetus. Rh antigens from the developing fetus can enter the mother's blood during delivery.

Placenta

③ In response to the fetal Rh antigens, the mother will produce anti-Rh antibodies.

④ If the woman becomes pregnant with another Rh+ fetus, her anti-Rh antibodies will cross the placenta and damage fetal red blood cells.

Figure 19.1 Hemolytic disease of the newborn.

If an Rh⁻ woman and an Rh+ man produce a child, the child may be Rh+ and sensitize the mother at birth. If the fetus in a later pregnancy is Rh+, the antibodies developed in the mother may attack the fetal red blood cells. This results in **hemolytic disease of the newborn,** once called *erythroblastosis fetalis* (Figure 19.1). This may be prevented by passive immunization with anti-Rh antibodies or even replacement transfusion of the fetus's blood.

Drug-Induced Cytotoxic Reactions

Thrombocytopenic purpura occurs when blood platelets, which are essential for clotting, are coated with drug molecules that function as haptens. If antibodies developed against these haptens cause destruction of the platelets, a disease condition may result. In **hemolytic anemia,** the body may form antibodies against its own red blood cells. Immune-caused destruction of white blood cells is called **agranulocytosis.**

TYPE III (IMMUNE COMPLEX) REACTIONS

Immune complexes form when certain ratios of antigen and antibody occur. When there is a slight excess of antigen, the soluble complexes that form are small and escape phagocytosis. Circulating in the blood, they may locate in the basement membrane beneath endothelial cells of blood vessels. This can set up an inflammatory, tissue-damaging reaction. **Glomerulonephritis** is an immune complex condition that causes inflammatory damage to kidney glomeruli.

TYPE IV (DELAYED CELL-MEDIATED) REACTIONS

Up to this point, our discussion of hypersensitivity has involved IgE, IgG, or IgM. Type IV reactions involve cell-mediated immune responses caused mainly by T cells, but sometimes by macrophages.

These reactions are often not apparent for a day or more (**delayed cell-mediated reactions** or **delayed hypersensitivity**), during which time participating cells migrate to and accumulate near the foreign antigens.

Causes of Delayed Cell-Mediated Reactions

Usually in delayed hypersensitivity reactions, the foreign antigens are phagocytized by macrophages and then presented to receptors on the T-cell surface. A principal factor is the release of cytokines by T cells reacting with the target antigen.

Delayed Cell-Mediated Hypersensitivity Reactions of the Skin

The skin test for tuberculosis is a reaction by a sensitized individual to protein components of tuberculosis bacteria injected into the skin. A day or two is required for the reaction to appear. Cases of **allergic contact dermatitis** are usually caused by haptens that combine with proteins in the skin. Typical foreign antigens are poison ivy, cosmetics, latex, and metals such as nickel in jewelry. The *patch test,* in which samples of suspected material are taped to the skin, may determine the offending environmental factor.

AUTOIMMUNE DISEASES

In our discussion of the different types of hypersensitivity reactions, we have mentioned several **autoimmune diseases.** These occur when there is a loss of **self-tolerance,** the immune system's ability to discriminate between self and nonself. It is believed that some clones of lymphocytes (forbidden clones) having the potential to respond to self-antigens may be produced during fetal life but are destroyed (**clonal deletion**) or inactivated.

Cytotoxic Autoimmune Reactions

Myasthenia gravis is a disease in which antibodies coat the acetylcholine receptor junctions, preventing nerve impulses from reaching the muscles. In **Graves' disease,** antibodies attach to receptors on the thyroid gland and cause excessive production of thyroid-stimulating hormones.

Immune Complex Autoimmune Reactions

Systemic lupus erythematosus is a systemic autoimmune disease in which individuals produce antibodies directed at components of their own cells, including DNA. Immune complexes damage the kidney glomeruli. **Rheumatoid arthritis** results when immune complexes are deposited in the joints. Immune complexes called *rheumatoid factors* may be formed by IgM binding to the Fc region of normal IgG. Chronic inflammation causes joint damage.

Cell-Mediated Autoimmune Reactions

Multiple sclerosis is a neurological disease in which T cells and macrophages attack the myelin sheath of nerves. **Insulin-dependent diabetes mellitus** is caused by immunological destruction of insulin-secreting cells of the pancreas by the cell-mediated immune system. The skin condition **psoriasis** is an autoimmune disorder characterized by itchy-red patches of thickened skin. As many as 25% of patients develop **psoriatic arthritis.**

REACTIONS RELATED TO THE HUMAN LEUKOCYTE ANTIGEN (HLA) COMPLEX

One inherited genetic characteristic is differences in **histocompatibility antigens** on cell surfaces. The genes controlling these antigens are the **major histocompatibility complex (MHC) antigens;** in humans these genes are also called the **human leukocyte antigen (HLA) complex.** For successful transplant surgery, **tissue typing** is used to match donor and recipient. Matching for MHC class I antigens (HLA-A, -B, and -C) has long been standard procedure, but matching for MHC class II antigens (HLA-DR, -DP, and -DQ) might be more important. The donor and recipient must be of the same ABO blood type.

Reactions to Transplantation

Privileged Sites and Privileged Tissue.   The cornea and brain are examples of **privileged sites;** antibodies do not circulate to these regions. **Privileged tissue,** such as pig heart valves, is not antigenic and does not stimulate an immune response.

Stem Cells.   Transplantation medicine may be transformed by the use of **stem cells,** which are master cells capable of generating any of the cell types of the body. Most interest is centered on **embryonic stem cells (ESCs).** These cells can be isolated from the earliest stage of an embryo. ESCs are *pluripotent,* meaning they are capable of generating many different types of tissue cells, such as muscle, nerve, or blood cells. Possible alternatives to the use of ESCs are *adult stem cells (ASCs),* which exist in some tissues, such as the blood or skin. These produce only a few cell types, mostly of the tissue type of origin. A new avenue of research is to genetically reprogram ASCs to convert them into *induced pluripotent stem cells.* Other nonembryonic sources of stem cells are cord blood cells, considered ASCs, harvested from umbilical cords. These are primarily *hematopoietic stem cells,* which are progenitors of blood and lymphatic (immune system) cells.

Grafts and Bone Marrow Transplants.   The transfer of tissue such as skin from one part of an individual to another on the same individual is an **autograft. Isografts** are transplants between identical twins. Such transplants are not rejected. **Allografts**—transplants between related people—represent most transplants. **Xenotransplantation products,** formerly known as **xenografts,** are transplants of tissue from animals other than humans, which tend to be strongly rejected. To be successful, xenotransplantation must overcome **hyperacute rejection** caused by the development in early infancy of antibodies against all distantly related animals such as pigs. Such antibodies attack the transplanted animal tissue and destroy it within an hour.

Xenotransplants—and, under some conditions, human-to-human transplants—are subject to **hyperacute rejection,** caused by antibodies that humans develop in early infancy against distantly related animals such as pigs. Antibodies and complement destroy xenotransplants within hours. When bone marrow is transplanted to people, now often called *hemapoietic stem cell transplants,* the transplanted tissue may carry cells capable of mounting an immune response. This is called **graft-versus-host (GVH) disease.** Umbilical cord blood, which contains many stem cells, can be a substitute for bone marrow transplantation (which is a form of stem cell transplantation).

Immunosuppression

People receiving transplants require suppression of their immune system (**immunosuppression**) to prevent rejection of the new tissue. The drug *cyclosporine* suppresses cell-mediated immunity, but at the cost of some liver and kidney toxicity. Other drugs that block rejection are *tacrolimus (FK506)* and *sirolimus (Rapamune). Mycophenolate mofectil* inhibits proliferation of T cells and B cells. Chimeric monoclonal antibodies such as *basiliximab* and *daclizumab* block IL-2.

THE IMMUNE SYSTEM AND CANCER

The immune system's patrol of the body for cancer cells is called **immunological surveillance.** Individual cancer cells, before they become established, are recognized as foreign and are destroyed by an effective immune system.

Immunotherapy for Cancer

The hypothesis that cancer represents a failure of the immune system has led to the thought that the immune system might be used to prevent or cure cancer—that is, **immunotherapy.** Treating or preventing cancer by immunological means will probably be an increasingly important approach. An attractive aspect is that this approach avoids the damage to healthy cells caused by chemotherapy and radiation treatments. Cancer vaccines might be either *therapeutic* (used to treat existing cancers) or *prophylactic* (to prevent the development of cancer). Prophylactic vaccines already exist; hepatitis B virus is a common cause of liver cancer, and a vaccine against infection by this virus is widely used. Also, a vaccine recommended for young girls, Gardasil, minimizes the chance of later development of cervical cancer caused by strains of a virus that also causes genital warts. The world's first therapeutic cancer vaccine was approved by the FDA in 2010. Used to treat men with advanced prostate cancer, it is able to extend lives for only a few months—about the same as chemotherapy. However, it has fewer uncomfortable side effects and is considered to represent a proof of concept. **Monoclonal antibodies** are a promising tool for delivering cancer treatment. A humanized monoclonal antibody, *Herceptin* is currently being used to treat a form of breast cancer. Another approach is to combine a monoclonal antibody with a toxic agent, forming an **immunotoxin.** Theoretically, an immunotoxin might be used to specifically target and kill cells of a tumor with little damage to healthy cells.

IMMUNODEFICIENCIES

Occasionally people are born with defective immune systems (**congenital immunodeficiencies**). Hodgkin's disease, a form of cancer, lowers cell-mediated immunity, as does removal of the spleen. Several drugs, cancers, or infectious agents can result in such **acquired immunodeficiencies.**

ACQUIRED IMMUNODEFICIENCY SYNDROME (AIDS)

Acquired immunodeficiency syndrome (AIDS) is a type of immunodeficiency disease. It is caused by the **human immunodeficiency virus (HIV),** which destroys helper T cells. AIDS is the final stage of a lengthy HIV infection. At this time the loss of an effective immune system leaves the victim susceptible to many opportunistic infections.

HIV is a *retrovirus* and requires the enzyme reverse transcriptase to form DNA from its RNA genome. The envelope of HIV has spikes of gp120 that allow the virus to attach to the CD4 receptors found on helper T cells. Coreceptors such as *CXCR4* and *CCR5* may also be required. *Attachment* is followed by *fusion,* then by *entry* into the cell, where the viral DNA is integrated into the DNA of the host cell. It may cause new HIVs to bud from the T cell; it may remain *latent* or as a *provirus* in the cell. HIV is capable of very rapid antigenic changes. Worldwide, HIV is beginning to separate into groups called *clades,* or *subtypes.* There are now thirteen clades of **HIV-1,** the most common major type; **HIV-2** is rare in the United States, occurring mostly in West Africa.

HIV Infection

A period of several weeks or months passes before **seroconversion,** when antibodies to HIV appear.

HIV infection typically progresses through three successive phases:

Phase 1. In the first week or so, enormous numbers of viral RNA molecules per milliliter of blood plasma appear. At the same time, billions of CD4$^+$ T cells are infected within a couple of weeks. The patient may be asymptomatic or show lymphadenopathy (swollen lymph nodes).

Phase 2. The numbers of CD4$^+$ T cells declines steadily. HIV replication is controlled by CD8$^+$ T cells. HIVs replicate at a low level but are present in host cells in latent or proviral form. The patient expresses few serious disease symptoms but evidence of declining immunity; for example, *Candida albicans* (yeast) infections may appear on mucous membranes, and symptoms may also include fever and persistent diarrhea.

Phase 3. Clinical AIDS emerges, usually within 10 years of infection. CD4$^+$ T-cell counts are below 350 cells/µl (a count of 200 cells/µl defines AIDS). Important AIDS indicator conditions, such as cytomegalovirus eye infection, tuberculosis, *Pneumocystis* pneumonia, toxoplasmosis of the brain, and Kaposi's sarcoma, appear.

The CDC classifies, mainly for clinical guidance such as for drug administration, the progress of HIV infection based on T-cell populations. The normal population is 800 to 1000 CD4$^+$ T cells/µl.

Diagnostic Methods

The most commonly used screening tests are versions of the ELISA test (see Figure 18.1 in this study guide), which are confirmed with the Western blot test (see Figure 10.12 in the text) or by the APTIMA assay (which detects the RNA of the HIV-1 virus). A problem with antibody-type tests is the window of time between infection and seroconversion. In contrast, **plasma viral load tests** detect and quantify HIV circulating in the blood and minimize the window during which HIV infection cannot be detected.

HIV Transmission

HIV transmission requires transfer of, or direct contact with, infected body fluids. Routes include sexual contact, blood-contaminated needles, organ transplants, and blood transfusions.

AIDS Worldwide

Approximately 33 million people are infected with, and living with, HIV today (see Figure 19.17 in text). An estimated 67% of these are in sub-Saharan Africa; the prevalence in adults (aged 15 to 49) is about 5.2% of the population. South and southeast Asia, with their dense populations, also has a high number of cases, an estimated 3.8 million, although the adult prevalence is only about 0.3%. As the disease becomes established in the huge populations of China and India, the incidence of HIV could exceed more than 1 million new cases a year. Eastern Europe, Russia, and central Asia are also areas reporting a steep rise in HIV infections. In western Europe and the United States, the mortality from AIDS has decreased because of the availability of effective antiviral drugs.

Preventing and Treating AIDS

HIV Vaccines. The immune system has not shown much capability in coping with natural infections, which is an important consideration. A vaccine effective against HIV infection will be a formidable problem, primarily because of the high mutation rate of the virus. A partially effective vaccine that assists the body's own defenses might be a useful goal.

Chemotherapy. Research into the reproductive mechanisms of HIV has increased the number of potential targets for chemical intervention. **Reverse transcriptase inhibitors** were the first target of anti-HIV drugs. In fact, the term **antiretroviral** implies that a drug is used to treat HIV infections. These include drugs *nucleoside reverse transcriptase inhibitors* that terminate viral DNA through competitive inhibition. *Non-nucleoside reverse transcriptase inhibitors* do this but are not analogs of nucleic acids. Therapy using current treatment by multiple drugs is termed **highly active antiretroviral therapy (HAART)**. **Protease inhibitors** target HIV enzymes that cleave lengthy viral precursor proteins into smaller structural and functional proteins. **Cell entry inhibitors** block entry of the virus into the cell. They may block attachment or fusion with the cell. Once fusion is completed, the viral DNA must be integrated into the host chromosome to form the HIV provirus. This requires an enzyme HIV integrase, which is a target for **integrase inhibitors**. Other drugs are **maturation inhibitors** that affect the conversion of a precursor of capsid protein to mature capsid protein—making the virus noninfectious. Other potential drugs are **tetherins,** which tether the newly formed virus to the cell and prevent its release and spread.

SELF-TESTS

In the matching section, there is only one answer to each question; however, the lettered options (a, b, c, etc.) may be used more than once or not at all.

I. Matching

____ 1. Hypersensitivity.

____ 2. Hypersensitivity specifically involving the interaction of humoral antibodies of the IgE class with mast cells.

____ 3. A skin graft from a brother to a sister.

____ 4. The heart of a baboon transplanted to a human.

____ 5. A term used for an antigen causing hypersensitivity reactions.

____ 6. A skin graft transferred from the thigh to the nose of the same person.

a. Allergen

b. Anaphylaxis

c. Xenotransplantation product

d. Allergy

e. Autograft

f. Allograft

g. Autoimmunity

h. Degranulation

II. Matching

____ 1. A drug used for transplantation surgery.

____ 2. A drug that suppresses cell-mediated immunity.

____ 3. The reason why transplantation of a cornea is usually successful.

____ 4. The mediator of a type I reaction that affects the blood capillaries and results in swelling and reddening.

____ 5. The development of blocking antibodies by repeated exposure to small doses of the antigen.

a. Histamine

b. Leukotrienes

c. Prostaglandins

d. Cyclosporine

e. Privileged site

f. Privileged tissue

g. Desensitization

III. Matching

____ 1. The naturally learned ability of the body not to respond immunologically against its own antigens.

____ 2. Destruction of a transplant—especially a xenograft—by antibodies and complement, usually within hours.

____ 3. Inhibition of the immune response by drugs, radiation, and so on.

____ 4. The treatment of cancer or other disease conditions by using monoclonal antibodies with which toxic compounds have been combined.

a. Hyperacute rejection

b. Immunological surveillance

c. Immunosuppression

d. Immunological tolerance

e. Immunotherapy

IV. Matching

____ 1. A mediator released from an antigen-triggered mast cell.

____ 2. Sirolimus.

____ 3. The release of mediators from mast cells or basophils during an anaphylactic reaction.

____ 4. The destruction of Rh⁺ red blood cells by antibodies of maternal origin in a newborn infant; the antibodies are derived from the mother.

____ 5. Individuals in whom ABO antigens are present in body fluids such as saliva and semen.

____ 6. Hematopoietic.

a. Leukotrienes

b. Erythroblastosis fetalis

c. Degranulation

d. Drug used for immunosuppression

e. Secretors

f. Blood-forming

g. Pluripotent

V. Matching

____ 1. Tuberculin test.

____ 2. Asthma.

____ 3. Glomerulonephritis.

____ 4. Poison ivy dermatitis.

____ 5. Graves' disease.

____ 6. Reaction to an insect sting.

a. Type I (anaphylaxis) reaction

b. Type II (cytotoxic) reaction

c. Type III (immune complex) reaction

d. Type IV (cell-mediated) reaction

VI. Matching (HIV Categories)

___ 1. Persistent lymphadenopathy.

___ 2. Full-blown AIDS.

a. Phase 1

b. Phase 2

c. Phase 3

VII. Matching

___ 1. Autoimmune condition in which antibodies coat the receptor sites at which nerve impulses reach the muscles.

___ 2. An immune reaction against the thyroid gland receptor sites that causes excessive production of thyroid hormones.

___ 3. Immune response against M protein of streptococci causes damage to kidneys.

___ 4. Antibodies formed against the body's own DNA; damage to kidney glomeruli is the most damaging factor in the disease.

___ 5. T cells destroy the thyroid gland.

___ 6. T cells attack the myelin sheath of the nervous system.

a. Graves' disease

b. Myasthenia gravis

c. Hashimoto's thyroiditis

d. Systemic lupus erythematosus

e. Glomerulonephritis

f. Multiple sclerosis

VIII. Matching (Stem Cells)

___ 1. Isolated from earliest stage of an embryo.

___ 2. Progenitors of blood and lymphatic cells.

___ 3. Genetically reprogrammed adult stem cells.

a. Induced pluripotent stem cells

b. Hematopoietic stem cells

c. Embryonic stem cells

Fill in the Blanks

1. Endotoxins from gram-negative bacteria stimulate macrophages to produce the cancer-inhibiting _____ factor.

2. The type of anaphylaxis that develops very rapidly after an antigen is presented to a sensitized host, and that may result in life-threatening shock, is _____ anaphylaxis.

3. In the ABO system, absence of antigens makes a person blood type _____.

4. A graft between identical twins is a(n) _____.

5. MHC stands for _____.

6. HLA stands for _____.

7. One result of immunosuppression could be development of graft _____ disease.

8. The treatment for systemic anaphylaxis is to administer an injection of _____ promptly.

9. Destruction of some clones of lymphocytes having the potential to respond to self-antigens during fetal life is called _____.

10. The cornea does not usually reject transplants; it is an example of a(n) _____ site.

11. Pig heart valves are not antigenic and are an example of _____ tissue.

12. About 85% of the population is Rh _____.

13. Immune-caused destruction of white blood cells is called _____.

14. Supply the missing word: highly active _____ therapy.

Critical Thinking

1. What are superantigens? Give an example of a superantigen, and explain the reaction that it causes.

2. The following substances are active in causing some of the signs and symptoms of hypersensitivity. What particular effects do they have on the body, and what sort of symptoms result?

   a. Histamine

   b. Prostaglandins

   c. Leukotrienes

3. Why would a person with type A blood have a reaction against type B blood upon the first exposure, whereas an Rh⁻ person wouldn't have a reaction to Rh⁺ blood until the second exposure?

4. From an inspection of Table 19.1, which shows the ABO blood group system, it is easy to see why persons of blood group AB are considered universal recipients; they do not have antibodies against type A or B blood. However, persons of blood group O are considered universal donors, although their blood contains antibodies against both A- and B-type blood. Why do you think that this type of transfusion, that is, type O blood transfused into patients with blood type AB, A, or B, is not considered harmful?

ANSWERS

Matching

I. 1. d   2. b   3. f   4. c   5. a   6. e
II. 1. d   2. d   3. e   4. a   5. g
III. 1. d   2. a   3. c   4. e
IV. 1. a   2. d   3. c   4. b   5. e   6. f
V. 1. d   2. a   3. c   4. d   5. b   6. a
VI. 1. a   2. c
VII. 1. b   2. a   3. e   4. d   5. c   6. f
VIII. 1. c   2. b   3. a

Fill in the Blanks

1. tumor necrosis   2. systemic   3. O   4. isograft   5. major histocompatibility complex   6. human lymphocyte antigens   7. versus host   8. epinephrine   9. clonal deletion   10. privileged   11. privileged   12. positive   13. agranulocytosis   14. antiretroviral

Critical Thinking

1. Superantigens are antigens that cause a drastic immune response. They act as nonspecific antigens, indiscriminately activating many T-cell receptors at once. This causes the release of large amounts of cytokines and in turn the production of a flood of T cells. Enterotoxins produced by some staphylococci act as superantigens.

2. a. Histamine increases the dilation and permeability of blood capillaries, resulting in edema, erythema, runny nose, and difficulty in breathing.

   b. Prostaglandins affect the smooth muscles of the respiratory system and increase mucus secretion.

   c. Leukotrienes usually cause prolonged contractions of certain smooth muscles, contributing to spasms of the bronchial tubes associated with asthma attacks.

3. People with type A blood have anti-B antibodies in their plasma, so they will react to type B blood on the first exposure. Rh$^-$ people do not have anti-Rh antibodies in their plasma, so they won't react to Rh$^+$ blood until the second exposure.

4. The anti-A and -B antibodies in type O blood do react with the antigens of the recipient, but the relative amounts are small, and the reaction is not damaging.

20 Antimicrobial Drugs

Chemotherapy is the treatment of disease with chemicals (drugs) taken into the body. The class of chemotherapeutic agents used to treat infectious diseases is **antimicrobial drugs;** unlike disinfectants, they must act within the host, where they kill the harmful organism without damaging the host, called **selective toxicity. Antibiotics** are produced by microorganisms and, in small amounts, inhibit another microorganism.

THE HISTORY OF CHEMOTHERAPY

During the early part of the twentieth century, Dr. Paul Ehrlich of Germany speculated about a "magic bullet" that would destroy pathogens but not harm the host. Eventually, he found an arsenic derivative, *salvarsan,* that was useful against syphilis. Prior to this discovery, the only chemotherapeutic agent available was *quinine,* for the treatment of malaria. *Sulfa drugs* were discovered during the 1930s. The drugs are wholly synthetic and technically are not antibiotics. *Penicillin,* an antibiotic, was first discovered in 1928 but was not available in a useful form until after 1940. Today, most antibiotics are discovered by so-called *high-throughput methods* that screen very high numbers of microbes, generally from soil or aquatic samples.

THE SPECTRUM OF ANTIMICROBIAL ACTIVITY

It is comparatively easy to find antimicrobials against prokaryotes (bacteria) because prokaryotes differ substantially from the eukaryotic cells of humans. Fungi, protozoa, and helminths are eukaryotic, which makes selective toxicity for the pathogen (without affecting the host) more difficult. It is difficult to find antimicrobials against viruses, which exist inside a host cell and interact with the host cell to synthesize new viruses.

If an antimicrobial drug affects relatively few bacteria, it has a narrow **spectrum of microbial activity,** as opposed to **broad-spectrum antibiotics.** Antibiotics may eliminate normal microbiota and allow opportunistic pathogens to flourish (**superinfection**).

THE ACTION OF ANTIMICROBIAL DRUGS

See the summary in Figure 20.1. Antimicrobial drugs are either **bactericidal** (they kill microbes directly) or **bacteriostatic** (they prevent microbes from growing).

Inhibiting Cell Wall Synthesis

The cell walls of bacteria consist of a layer of peptidoglycan, which is found only in bacterial cells. Therefore, interference with the synthesis of bacterial cell walls usually does not harm the host. Antibiotics using this mode of action include *penicillins, cephalosporins, bacitracin,* and *vancomycin.* Because the peptidoglycan layer of gram-positive bacteria is more accessible than that of gram-negative ones, these bacteria are the most susceptible to such agents.

Inhibiting Protein Synthesis

Ribosome structure differs greatly between prokaryotic and eukaryotic cells. Many antibiotics such as *chloramphenicol, gentamicin, erythromycin, tetracyclines,* and *streptomycin* interfere with protein synthesis by reacting with the ribosomes of bacteria.

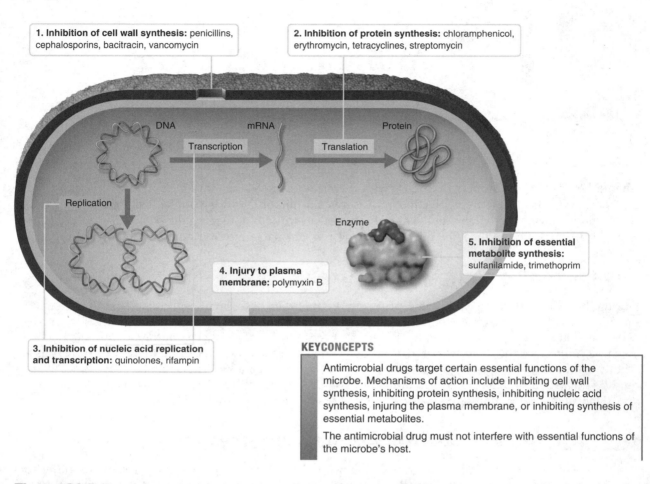

1. Inhibition of cell wall synthesis: penicillins, cephalosporins, bacitracin, vancomycin

2. Inhibition of protein synthesis: chloramphenicol, erythromycin, tetracyclines, streptomycin

DNA · mRNA · Protein

Transcription · Translation

Replication

Enzyme

5. Inhibition of essential metabolite synthesis: sulfanilamide, trimethoprim

4. Injury to plasma membrane: polymyxin B

3. Inhibition of nucleic acid replication and transcription: quinolones, rifampin

KEYCONCEPTS

Antimicrobial drugs target certain essential functions of the microbe. Mechanisms of action include inhibiting cell wall synthesis, inhibiting protein synthesis, inhibiting nucleic acid synthesis, injuring the plasma membrane, or inhibiting synthesis of essential metabolites.

The antimicrobial drug must not interfere with essential functions of the microbe's host.

Figure 20.1 A summary of the major modes of action of antimicrobial drugs. This illustration shows these actions as they might affect a highly diagrammatic representation of a bacterial cell.

Injuring the Plasma Membrane

Antibiotics, especially such polypeptides as *polymyxin B,* can adversely affect the membrane permeability of microbial cells. Loss of important metabolites occurs from these changes in permeability. Similarly, the effectiveness of *nystatin, miconazole, ketoconazole,* and *amphotericin B* against fungi is based on their combining with sterols to disrupt fungal plasma membranes.

Inhibiting Nucleic Acid Synthesis

Similarities between microbial and host cell DNA and RNA are so close that drugs that act by interfering with the nucleic acid synthesis of microbial cells have only limited clinical application. Drugs acting on this principle are *rifampin* and the *quinolones.*

Inhibiting the Synthesis of Essential Metabolites

Sulfa drugs, sulfanilamide for example, competitively inhibit the synthesis of folic acid, which is a vitamin that is synthesized by bacteria but not humans. The drug resembles the metabolite **para-aminobenzoic acid,** which is required to synthesize folic acid.

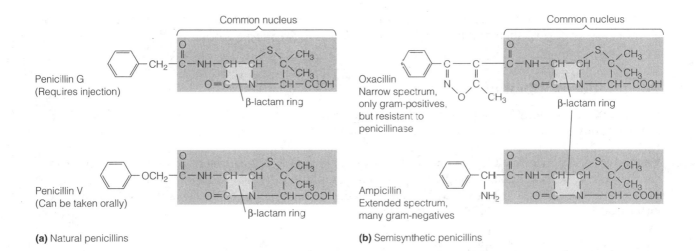

Penicillin G
(Requires injection)

Penicillin V
(Can be taken orally)

β-lactam ring

Oxacillin
Narrow spectrum,
only gram-positives,
but resistant to
penicillinase

β-lactam ring

Ampicillin
Extended spectrum,
many gram-negatives

(a) Natural penicillins

(b) Semisynthetic penicillins

Common nucleus

Figure 20.2   The structure of penicillins, antibacterial antibiotics. The portion that all penicillins have in common—which contains the β-lactam ring—is shaded. The unshaded portions represent the side chains that distinguish one penicillin from another.

A SURVEY OF COMMONLY USED ANTIMICROBIAL DRUGS

Antibacterial Antibiotics: Inhibitors of Cell Wall Synthesis

Penicillins.   The term **penicillin** refers to a group of related antibiotics (Figure 20.2).

Natural Penicillins.   **Natural penicillins,** such as *penicillin G or V,* are products of *Penicillium* mold growth. *Procaine penicillin* and *benzathine penicillin* combine penicillin G with other drugs to prolong the antibiotic's retention in the body. **Penicillinases** (β-*lactamases*) are enzymes that cleave the β-*lactam* ring of penicillins, causing resistance.

Semisynthetic Penicillins.   A large number of **semisynthetic penicillins** have been developed to overcome the disadvantages of natural penicillins. Side chains of natural penicillins are removed and other side chains added to extend their spectrum or make them resistant to penicillinases.

Penicillinase-Resistant Penicillins.   The first semisynthetic penicillin designed to evade the action of penicillinases was *methicillin.* Eventually, so many staphylococcal strains became resistant that the abbreviation **MRSA (methicillin-resistant** Staphylococcus aureus) made its appearance. Methicillin has been discontinued; however, *oxacillin* and *nafcillin* are still in use.

Extended-Spectrum Penicillins.   Certain semisynthetic penicillins, such as *ampicillin, amoxicillin* (both **aminopenicillins**), *carbenicillin,* and *ticarcillin* (both **carboxypenicillins**), have a broader spectrum of activity than do natural penicillins. Semisynthetics such as the **ureidopenicillins,** which include *mezlocillin* and *azlocillin,* also have a broader spectrum of activity.

Penicillins Plus β-Lactamase Inhibitors.   Another approach to penicillinase resistance is to combine penicillins with *potassium clavulanate (clavulanic acid),* which is a noncompetitive inhibitor of penicillinase. Augmentin is the trade name of such a combination.

Carbapenems.   The **carbapenems** are a class of β-lactam antibiotics that have an extremely broad spectrum of activity. An example is Primaxin, a combination of *imipenem* and *cilastin.* A very recent addition, useful against *Pseudomonas aeruginosa* infections, is *doripenem.*

Monobactams.   Another penicillin variant, **monobactams** have only a single-ring structure. One of these, *aztreonam,* affects only gram-negative bacteria.

Cephalosporins. The structural nucleus and mode of action of **cephalosporins** resemble those of penicillin. Cephalosporins, such as *cephalothin, cefamandole,* and *cefotaxime,* are often used as substitutes for penicillin.

Polypeptide Antibiotics

Bacitracin. **Bacitracin** is a polypeptide antibiotic effective primarily against gram-positive bacteria. It inhibits synthesis of cell walls and is used only topically.

Vancomycin. Vancomycinis a member of the small glycopeptide group that inhibits peptidoglycan synthesis. It is used against penicillinase-producing staphylococci that cause life-threatening infections. Vancomycin is used to treat MRSA and has led to the selection of **VRE (vancomycin-resistant enterococci).**

Antimycobacterial Antibiotics

Isoniazid (INH). **Isoniazid,** used in treating tuberculosis, is believed to inhibit synthesis of mycolic acids, which are part of the cell wall of mycobacteria.

Ethambutol. **Ethambutol** is effective only against mycobacteria and is used in chemotherapeutic treatment of tuberculosis. It inhibits the incorporation of mycolic acid into the cell wall.

Inhibitors of Protein Synthesis

Chloramphenicol. *Chloramphenicol* is a broad-spectrum antibiotic that affects protein synthesis. Structurally simple, it is often synthesized chemically. It is the drug of choice for typhoid fever and certain types of meningitis, for which the risk of its side effects is considered justified.

Aminoglycosides. **Aminoglycosides** are a group of antibiotics with amino sugars and an aminocyclitol ring. Examples are *streptomycin* (used for tuberculosis treatment), *neomycin* (used in topical ointment with bacitracin and polymyxin B), and *gentamicin* (effective against most gram-negatives, especially *Pseudomonas*). *Tobramycin* is administered by aerosol to treat cystic fibrosis patients infected with pseudomonads. Aminoglycosides sometimes are toxic to the auditory nerve or the kidneys.

Tetracyclines. **Tetracyclines** are broad-spectrum antibiotics that are also effective against chlamydias and rickettsias. They inhibit protein synthesis. They produce such side effects as tooth discoloration and liver damage. Commonly encountered are *tetracycline, oxytetracycline (Terramycin),* and *chlortetracycline (Aureomycin).* Some newer semisynthetic versions, such as *doxycycline* and *minocycline,* are retained in the body longer.

Glycylcyclines. The **glycylcyclines** are a new class of antibiotics structurally similar to the tetracyclines. They are broad-spectrum antibiotics, administered intravenously, that avoid antibiotic resistance based on rapid efflux. The best known example is *tygecycline (Tygacil).*

Macrolides. **Macrolides** are named for their macrocyclic lactone ring and are especially effective against gram-positive bacteria. *Erythromycin* inhibits protein synthesis and is used in treating infections resistant to penicillins, as well as legionellosis and mycoplasmal pneumonia. Other macrolides are *azithromycin* and *clarithromycin;* compared to erythromycin they have a broader spectrum and penetrate tissues better. **Ketolides** are new semisynthetic macrolides developed to combat microbial resistance. An example is *telithromycin (Ketek).*

Streptogramins. The **streptogramins** are a unique group of antibiotics developed to combat resistance to vancomycin. Synercid is a combination of two cyclic peptides *quinupristin* and *dalfopristin,* which are distantly related to macrolides.

Oxazolidinones.   The **oxazolidinones** are a new class of totally synthetic antibiotics. They have a unique target, binding to the 50S ribosomal subunit close to the point where it interfaces with the 30S subunit. *Linozolid (Zyvox)*, a member of the group, is used mainly to combat MRSA.

Pleuromutinlins.   Two of the best known examples of this newly introduced class of antibiotics against gram-positive bacteria, the **pleuromutinlins,** are *mutilin* and *retpamulin*. Their action is similar to that of macrolides but they are unaffected by resistance to macrolides.

Injury to the Plasma Membrane

Polymyxin B is effective against gram-negative bacteria, even *Pseudomonas.* It is available for topical use in the antiseptic ointment that also contains *bacitracin* and *neomycin.*

Lipopeptides.   The **lipopeptides** are a new class of antibiotics, active against gram-positive bacteria, which attack the membrane of the bacterial cell. An example is *daptomycin*.

Inhibitors of Nucleic Acid (DNA/RNA) Synthesis

Rifamycins.   *Rifampin,* the best known of the **rifamycin** family, is used in tuberculosis therapy. These drugs inhibit the synthesis of mRNA.

Quinolones and Fluoroquinolones.   The first of the **quinolone** group was *nalidixic acid,* which selectively inhibits the enzyme DNA gyrase needed for DNA replication. This led to the development of a prolific group of synthetic quinolones, the **fluoroquinolones.** *Norfloxacin* and *ciprofloxacin (Cipro)* are the most widely used. A third generation of flucroquinolones that have a broader spectrum and can be taken orally include *moaxifloxacin* and *gatifloxacin.*

Competitive Inhibitors of the Synthesis of Essential Metabolites

Sulfonamides.   The **sulfonamides (sulfa drugs)** act by competitive inhibition of folic acid, a precursor to nucleic acids. *Silver sulfadiazine* is used on burn patients. The most widely used sulfa-containing preparation is the combination of *trimethoprim* and *sulfamethoxazole.* These are structural analogs that inhibit synthesis of DNA at different stages.

Antifungal Drugs

Agents Affecting Fungal Sterols.   In fungal membranes, the principal sterol is ergosterol; in animal membranes, it is cholesterol. This forms a basis for selective toxicity.

Polyenes.   *Amphotericin B* is the most commonly used of the **polyene antibiotics.** Their activity is based on damage to fungal plasma membranes by combining with the membrane sterols. *Amphotericin B* is used for systemic fungal infections.

Azoles.   **Imidazole** antifungals such as *miconazole* and *clotrimazole* are **azole antibiotics** used topically against cutaneous fungal infections. *Ketoconazole,* taken orally, is a substitute for amphotericin B for many systemic fungal infections. **Triazoles** such as *fluconazole* and *itraconazole* are used for systemic fungal infections. *Voriconazole* and *posaconazole* are new azoles.

Allylamines.   The **allylamines** are a recently developed class of antifungals that inhibit ergosterol synthesis in a different manner and are often used when resistance to azoles appears. *Terbinafine* and *naftifine* are examples.

Agents Affecting Fungal Cell Walls.   A primary target for selective toxicity is the β-glucans that are unique to fungal cell walls. A new class of antifungals, **echinocandins,** interferes with synthesis of glucans. An example is *caspofungin (Cancidas).*

Agents Inhibiting Nucleic Acids.   *Flucytosine*, an analog of cytosine, interferes with synthesis of RNA, and therefore protein synthesis.

Other Antifungal Drugs.   *Griseofulvin* is a fungistatic drug that interferes with mitosis. Although taken orally, this drug binds selectively to keratin in skin, hair, and nails, preventing fungal growth at these sites. *Tolnaftate* is a topical agent used as an alternative to miconazole for athlete's foot infections. *Undecylenic acid* is a fatty acid with antifungal activity. *Pentamidine isethionate* is used in treating *Pneumocystis* pneumonia.

Antiviral Drugs

The number of antiviral drugs, compared to that of antibacterial drugs, is very limited. A drug used to treat influenza, *amantadine,* was the first to be licensed, though its mode of action is unknown. Most new antivirals are directed at control of HIV.

Nucleoside and Nucleotide Analogs.   Among the nucleoside analogs, *acyclovir* is widely used for many herpesvirus infections. Others are *famciclovir, genciclovir, cidoflovir,* and *ribavirin.* The nucleoside analog *lamivudine* is used to treat hepatitis B, and the nucleotide analog, *adefovir dipivoxil (Hepsera)* has been recently introduced to counter resistance to lamivudine.

Interferons.   Cells infected with a virus often produce interferon, which inhibits further spread of the infection. Interferons are cytokines; *alpha interferon* is used for viral hepatitis infections. A new drug, *imiquimod*, stimulates the production of interferons.

Antivirals for Treating HIV/AIDS

Antiretrovirals.   RNA viruses depend upon the enzyme reverse transcriptase, which is an enzyme humans do not have. HIV is an RNA virus, and the antiretroviral group is largely directed at it; in fact, the term *antiretroviral* implies that it is used to treat HIV. Most antiretrovirals are nucleoside analogs such as *zidovudine (AZT),* which is an analog of thymidine. Another example is *cidofovir,* which is used to treat cytomegalovirus eye infections and is considered a possible treatment for smallpox. The only example of a nucleotide analog is *tenovir.* An antiretroviral that is a non-nucleoside agent is *nevirapine.*

Other Enzyme Inhibitors.   The inhibitors of the enzyme neuraminidase, *zanamivir (Relenza)* and *oseltamivir phosphate (Tamiflu),* are used for treating influenza. Another approach to controlling HIV is to inhibit enzymes that control the last stage of viral reproduction, which requires protease enzymes. The **protease inhibitors** *atazanavir, indinavir,* and *saquinavir* are examples. Other targets for antivirals treating HIV/AIDS are those that inhibit an enzyme that integrates viral DNA into the DNA of the infected cell. Examples are **integrase inhibitors**, such as *raltegravir.* Other approaches are to block entry into the cell by targeting CCR5 coreceptors. The first of a new class of **entry inhibitors** of this type includes *maraviroc.* Entry can also be blocked by **fusion inhibitors** such as *enfuvirtide.*

Antiprotozoan and Antihelminthic Drugs

Antiprotozoan Drugs.   *Quinine* still has limited use against malaria, but it has generally been replaced with synthetic derivatives, such as *chloroquine* and *mefloquine.* The latest groups of drugs against malaria are products of a Chinese shrub, artemisinin. They kill the asexual stages of *Plasmodium* spp. They are known as *artemisinin-based combination therapies (ACTs).* *Quinacrine,* used against giardiasis, functions similarly. *Diiodohydroxyquin (iodoquinol)* is an amebicide. *Metronidazole* is used for treating many protozoan diseases and also is effective against certain anaerobic bacteria. It probably causes disruption of DNA under anaerobic conditions. Newer antiprotozoan drugs are *tinidazole (Fasigyn)* and *nitazoxanide.*

Antihelminthic Drugs.   *Niclosamide* inhibits ATP production in tapeworms. *Praziquantel* also is effective against tapeworms and several fluke-caused diseases. *Mebendazole* and *albendazole* are used to treat ascariasis. *Ivermectin* is widely used in the livestock industry for helminth control. It also is useful in the treatment of some infestations by mites (scabies), ticks, and head lice.

TESTS TO GUIDE CHEMOTHERAPY

The Diffusion Methods

The **disk-diffusion method (Kirby-Bauer test)** uses a dish of agar medium seeded uniformly with a test organism. Filter paper disks impregnated with known concentrations of chemotherapeutic agents are placed on the agar surface. If the chemotherapeutic agent is effective, a zone cf inhibition (no growth) is observed around the disk. The diameter of the zone can be used to calculate the susceptibility of the organisms to the agent.

A more advanced diffusion method, the **E test,** includes an estimate of the **minimum inhibitory concentration (MIC).**

Broth Dilution Tests

A series of dilutions of an antibiotic can be placed in tubes (shallow wells in a plastic plate usually are used in practice) and inoculated with test bacteria. After incubation they are examined for turbidity. The **minimum inhibitory concentration (MIC)** of the antimicrobial is defined as the lowest concentration that prevents growth. Subculturing from the tubes that show no growth will determine whether the bacteria have been killed or only inhibited. The lethal concentration that actually kills the bacteria is called the **minimum bactericidal concentration (MBC).** Many of these tests are highly automated and use light scattering to determine bacterial growth. Hospital personnel responsible for infection control prepare periodic reports, called **antibiograms,** on the susceptibility to antibiotics encountered clinically.

RESISTANCE TO ANTIMICROBIAL DRUGS

Resistance to antimicrobial drugs, a threat to the usefulness of antibiotics, arises from random mutations. These can spread *horizontally* by conjugation or transduction, for example. The resistance is often carried by plasmids, such as resistance (R) factors, or transposons (small segments of DNA). Once acquired, the mutation is transmitted by normal reproduction. Bacteria resistant to large numbers of antibiotics are called **superbugs.**

There are only a few major mechanisms by which bacteria become resistant:

- *Enzymatic destruction or inactivation of the drug.* The best known example of this is the action of β-lactamase, which inactivates penicillins. The notorious pathogen MRSA is a well-known result of this mechanism. MRSA infections that are associated with hospitals are referred to as *health-care associated MRSA.* Outbreaks caused by *community-associated MRSA* occur in the general community, affecting otherwise healthy persons.

- *Prevention of penetration to the target site within the microbe.* This is most often seen in gram-negative bacteria.

- *Alteration of the drug's target site* and *rapid efflux (ejection) of the drug* tend to be common with the tetracycline-type antibiotics.

EFFECTS OF COMBINATIONS OF DRUGS

Two drugs given simultaneously may be more effective than either given alone; this is called **synergism.** Other combinations can show **antagonism,** in which the two drugs are less effective than either used alone.

THE FUTURE OF CHEMOTHERAPEUTIC AGENTS

The most pressing concern currently is the spread of resistance to antibiotics. A truly new approach to controlling pathogens has been to target their virulence factors rather than the microbe producing them. Similarly, targeting the iron needed for growth would limit the reproduction of pathogens. New, exotic ecological niches such as deep-sea settlements need to be explored. Many birds, amphibians, plants, and mammals often produce *antimicrobial peptides*. These are found as part of the defense systems of most forms of life. The best known of these are the *magainins*. A steroid named *squalamine* has been isolated from sharks. Perhaps there will be renewed interest in *phage therapy*.

SELF-TESTS

In the matching section, there is only one answer to each question; however, the lettered options (a, b, c, etc.) may be used more than once or not at all.

I. Matching

_____ 1. Plasmids that carry antibiotic resistance.

_____ 2. Chemotherapeutic agents produced by microorganisms.

_____ 3. Disk-diffusion test for antibiotic sensitivity.

_____ 4. Diffusion test that also measures minimum inhibitory concentration of an antibiotic.

_____ 5. Periodic reports on antibiotic sensitivity in hospitals.

a. Antibiotics

b. E test

c. Chemotherapy

d. Kirby-Bauer test

e. R factors

f. Antibiograms

II. Matching

_____ 1. Activity based on damage to the sterols in plasma membrane of fungi.

_____ 2. Inhibition of protein synthesis.

_____ 3. Inhibition of RNA synthesis.

_____ 4. Inhibition of synthesis of cell wall peptidoglycans.

_____ 5. Inhibition of DNA synthesis.

_____ 6. Inhibition of synthesis of cell wall mycolic acids.

a. Cephalosporins

b. Chloramphenicol

c. Amphotericin B

d. Isoniazid

e. Sulfonamides

f. Rifampin

III. Matching

____ 1. Similar structurally to penicillin.

____ 2. A synthetic drug used in tuberculosis chemotherapy.

____ 3. Antifungal, a polyene.

____ 4. Causes plasma membrane leakage; useful against *Pseudomonas*.

____ 5. An antiviral drug; a nucleoside analog.

____ 6. Antifungal; an allylamine.

____ 7. An antiviral drug; a nucleotide analog.

a. Cephalosporin

b. Polymyxin B

c. Ethambutol

d. Idoxuridine

e. Voriconazole

f. Terbinafine

g. Adefovir dipivoxil

IV. Matching

____ 1. Used in treating diseases caused by protozoa.

____ 2. An antifungal drug taken orally that concentrates in keratin.

____ 3. Useful against tapeworms.

____ 4. A drug that is useful against symptoms of genital herpes.

____ 5. An antifungal drug of the polyene type.

____ 6. A macrolide antibiotic.

____ 7. A streptogramin-type antibiotic.

____ 8. Avoid resistance based on rapid efflux.

a. Erythromycin

b. Griseofulvin

c. Amphotericin B

d. Niclosamide

e. Metronidazole

f. Acyclovir

g. Synercid

h. Tygecycline

V. Matching

____ 1. Inhibits ATP production in tapeworms.

____ 2. Used in the treatment of malaria.

____ 3. Used in the treatment of *Pneumocystis* pneumonia.

____ 4. Stimulates production of interferons.

a. Chloroquine

b. Niclosamide

c. Pentamidine isethionate

d. Imiquimod

VI. Matching

___ 1. Used in treating HIV infections, a nucleoside analog.

___ 2. Acts by competitive inhibition of folic acid, usually in combination with sulfamethoxazole.

___ 3. A derivative of penicillin G designed to be retained for a longer time in the body.

___ 4. A penicillin designed to be resistant to penicillinase; no longer in use.

___ 5. A synthetic fluoroquinolone that acts against the gyrase enzyme.

___ 6. An amebicide.

___ 7. Very broad spectrum; carbapenem group.

___ 8. An integrase inhibitor.

a. Zidovudine

b. Diiodohydroxyquin

c. Methicillin

d. Ampicillin

e. Procaine penicillin

f. Trimethoprim

g. Norfloxacin

h. Primaxin

i. Raltegravir

VII. Matching

___ 1. Zidovudine, an analog of thymidine.

___ 2. Inhibits DNA synthesis of bacteria.

___ 3. An arsenic derivative used against syphilis before the development of modern antibiotics.

___ 4. A rifamycin-type drug used for therapy of tuberculosis; inhibits synthesis of mRNA.

___ 5. An antifungal drug; interferes with synthesis of RNA.

___ 6. A possibility for use against smallpox.

a. Nalidixic acid

b. Flucytosine

c. Salvarsan

d. Rifampin

e. AZT

f. Cidofovir

VIII. Matching

___ 1. Inhibitor of protein synthesis; inexpensive; may cause aplastic anemia.

___ 2. An aminoglycoside antibiotic used in tuberculosis treatment; may cause deafness or kidney damage.

___ 3. A cytokine.

___ 4. An antiviral protease inhibitor.

___ 5. Used mainly against life-threatening staphylococcal infections resistant to penicillin.

a. Chloramphenicol

b. Vancomycin

c. Streptomycin

d. Indinavir

e. Interferon

Fill in the Blanks

1. Cell walls of most bacteria contain _____, the target of activity by penicillins.

2. The treatment of disease with chemicals taken into the body by injection or ingestion is called

   _____.

3. Many bacteria develop resistance to penicillin by producing the enzyme _____.

4. The usual principle of antibiotic activity is _____, meaning it kills the harmful

   organism without damaging the host.

5. The term *penicillin* is applied to a group of antibiotics that all have a(n) _____ in

   their structure.

6. The aminocyclitol ring and amino sugars are found in the _____ group of antibiotics.

7. The lowest concentration of a chemotherapeutic agent that will prevent growth is the

   _____.

8. The lowest concentration of a chemotherapeutic agent that will kill the pathogen (as contrasted to

   inhibition) is called the _____.

9. Miconazole and ketoconazole are examples of the _____ group of antifungals.

10. Many antibiotics inhibit protein synthesis by reacting with the _____ of the

    bacterium, which differ greatly between prokaryotic and eukaryotic cells.

11. When antibiotics eliminate much of the natural microbiota, there may be an overgrowth of resistant

    pathogens; this is called _____.

12. Aztreonam is a variant of penicillin with a single-ring structure; this group of antibiotics is called

    _____.

13. Magainin, squalamine, and cecropin are examples of the new antimicrobial

    _____ agents isolated from animals.

14. A new antiviral drug, envuvirtide, is intended to block entry into the host cell; it is a

    _____ inhibitor.

15. The M in MRSA refers to _____.

Critical Thinking

1. List and briefly discuss four criteria used to evaluate antimicrobial drugs.

2. Using antibiotics as a supplement in animal feed has been linked to *Salmonella* infections in humans. Why did this practice begin? Why has it continued? What risks are associated with the continued use of antibiotics as an animal feed supplement?

3. What advantages and disadvantages are associated with the use of broad-spectrum antibiotics?

4. Define *synergism*. For what purposes should combinations of antimicrobial drugs be used?

ANSWERS

Matching

I.   1. e   2. a   3. d   4. b   5. f
II. 1. c   2. b   3. f   4. a   5. e   6. d
III. 1. a   2. c   3. e   4. b   5. d   6. f   7. g
IV. 1. e   2. b   3. d   4. f   5. c   6. a   7. g   8. h
V.   1. b   2. a   3. c   4. d
VI. 1. a   2. f   3. e   4. c   5. g   6. b   7. h   8. i
VII. 1. e   2. a   3. c   4. d   5. b   6. f
VIII. 1. a   2. c   3. e   4. d   5. b

Fill in the Blanks

1. peptidoglycan   2. chemotherapy   3. penicillinase   4. selective toxicity   5. β-lactam ring
6. aminoglycoside   7. minimum inhibitory concentration (MIC)   8. minimum bactericidal concentration (MBC)   9. azole   10. ribosomes   11. superinfection   12. monobactams   13. peptides   14. fusion
15. methicillin

Critical Thinking

1. a. Selective toxicity—The drug should be toxic to the pathogen but not to the host.

   b. The drug should not produce hypersensitivity in most hosts.

   c. The drug must be soluble in body fluids so that it can rapidly penetrate body tissues. It must also remain in the body long enough to be effective.

   d. Microorganisms should not become readily resistant to the drug.

2. Antibiotics were first added to animal feed to lower the incidence of infection in closely penned animals. Another reason that antibiotics are still added to animal feed is that they accelerate the growth of the animal. The practice has been linked with *Salmonella* infections in humans from meat and milk. FDA testing has shown that most milk and meat have little or no detectable antibiotics, but many people still consider the practice undesirable. Continued use of antibiotics in animal feed will result in the development of antibiotic-resistant strains of bacteria.

3. Advantages:

   • Broad-spectrum activity means that the identity of the pathogen need not necessarily be known; this saves valuable time.

   • Broad-spectrum antibiotics eliminate many important pathogens and opportunistic organisms.

   Disadvantages:

   • Normal microbiota are killed, allowing opportunistic microbiota (for example, yeasts) to proliferate.

   • The indiscriminant use of broad-spectrum drugs leads to the development of resistant strains of bacteria.

4. Synergism refers to the use of two or more antimicrobial drugs simultaneously. It has been found that their combined effect is greater than the effect of either given alone.

   Combinations of antimicrobial drugs should be given for the following purposes:

   a. To prevent or minimize development of resistant strains.

   b. To take advantage of the synergistic effect.

   c. To provide optimal therapy in life-threatening, time-critical situations.

   d. To lessen drug toxicity by reducing the necessary drug concentration.

21

Microbial Diseases of the Skin and Eyes

The first lines of defense of the body are the skin and the mucous membranes. Recently, it was found that the skin and mucous membranes contain antimicrobial peptides, called *defensins,* that have a wide spectrum of activity.

STRUCTURE AND FUNCTION OF THE SKIN

The **epidermis** is the outer, thinner portion of the skin; it is composed of several layers of epithelial cells. The outermost epidermal layer, the **stratum corneum,** consists mainly of dead cells containing the protein **keratin.** The inner layer of the skin is the **dermis,** composed of connective tissue with numerous blood and lymph vessels, nerves, hair follicles, and sweat and oil glands. *Perspiration* provides the moisture necessary for microbial growth; however, salts interfere with the growth of many organisms, and the enzyme lysozyme in perspiration digests the cell walls of many bacteria.

Sebum from oil glands is mainly a mixture of unsaturated fatty acids, proteins, and salts. The fatty acids inhibit the growth of certain pathogens.

Mucous Membranes

In the lining of body cavities such as the mouth, nasal passages, urinary and genital tracts, and gastro-intestinal tract, there is a layer of specialized *epithelial cells,* which are attached at their bases to a layer of extracellular material called the *basement membrane.* Some cells produce **mucus,** which traps particles, including microbes, hence the name **mucous membrane (mucosa).** Other cells are ciliated, functioning to sweep particles and mucus from the body.

NORMAL MICROBIOTA OF THE SKIN

Many skin bacteria are gram-positive spherical bacteria such as staphylococci and micrococci. Gram-positive pleomorphic rods, **diphtheroids,** such as *Propionibacterium acnes,* metabolize the sebum secretions in hair follicles. Acid produced by them keeps the skin pH between 3 and 5, a condition that tends to be bacteriostatic. Other diphtheroids, such as *Corynebacterium xerosis,* are aerobic. The yeast *Malassezia furfur* grows on oily skin secretions. High-moisture regions such as the armpit and the area between the legs are moist enough to support the growth of microorganisms, and populations there are high compared to dry regions such as the scalp.

MICROBIAL DISEASES OF THE SKIN

Vesicles are small, fluid-filled lesions on the skin. Skin vesicles larger than about 1 cm in diameter are **bullae.** Flat, reddened lesions are **macules.** Raised lesions are called **papules,** or **pustules** when they contain pus. A skin rash that arises from disease conditions is called an **exanthem;** on mucous membranes such as the interior of the mouth, such a rash is called an **enanthem.**

Bacterial Diseases of the Skin

Staphylococcal Skin Infections.   **Staphylococci** are spherical gram-positive bacteria that form irregular grapelike clusters of cells. *Staphylococcus aureus* is the most pathogenic of the staphylococci. It forms golden yellow colonies on agar, and almost all pathogenic strains produce **coagulase,** an enzyme that coagulates the fibrin in blood. The coagulase test is used to distinguish *S. aureus* from other species of *Staphylococcus.* The predominant species of coagulase-negative staphylococci is *S. epidermidis. S. aureus* produces a number of toxins, such as enterotoxins, that affect the gastrointestinal tract (which will be discussed in Chapter 25). *S. aureus* is found primarily in nasal passages and is a common hospital problem. It is universally present, and constant exposure to antibiotics causes rapid development of resistance. The species, therefore, causes many infections of surgical wounds and other artificial breaches of the skin barrier.

Pimples are infections of hair follicles, which are natural openings in the skin barrier; an eyelash follicle infection is a **sty. Furuncles (boils)** are a type of **abscess** usually arising from a hair follicle infection. In furuncles, a region of pus is surrounded by inflamed tissue; more extensive invasion of tissue is termed a **carbuncle.**

Staphylococci are the most important causative organism for **impetigo,** which mostly affects children 2 to 5 years of age. (A few cases are caused by *Streptococcus pyogenes.*) The pathogen usually enters through a skin break but can spread to surrounding areas, a process called *autoinoculation.* The disease takes the forms of *nonbullous impetigo* and *bullous impetigo.* This latter type is caused by a staphylococcal toxin and is a localized form of staphylococcal **scalded skin syndrome.** (This is also a symptom of **toxic shock syndrome,** also caused by growth of staphylococci.) Outbreaks of bullous impetigo are a frequent problem in hospital nurseries, known there as **pemphigus neonatorum,** or *impetigo of the newborn.*

Streptococcal Skin Infections.   Streptococci are gram-positive spherical bacteria that tend to grow in chains. They are aerotolerant anaerobes that do not use oxygen and are catalase-negative. They produce several toxins, as well as enzymes such as **hemolysins.** Group A streptococci (GAS), which is synonymous with *S. pyogenes,* is the most important of the beta-hemolytic streptococci. The beta-hemolytic streptococci are divided into immunological groups by the antigenic **M protein.** The M protein is carried externally on a fuzzy layer of fibrils, has antiphagocytic properties, and is an aid in adherence. Also produced by these cells are *erythrogenic toxins* (scarlet fever rash), *deoxyribonucleases* (enzymes degrading DNA), *streptokinases* (enzymes dissolving blood clots), and *hyaluronidase* (an enzyme that dissolves hyaluronic acid, the cementing substance of connective tissue).

S. pyogenes is the most common cause of **erysipelas,** a disease in which the dermis is affected and the skin erupts into reddish patches. Usually it is preceded by a streptococcal infection, such as a sore throat, elsewhere in the body.

Streptococci may attack solid tissue or muscle covering. An *exotoxin,* exotoxin A, acts as a superantigen, causing the immune system to contribute to damage (see Chapter 19).

Necrotizing fasciitis, is often associated with *streptococcal toxic shock syndrome (streptococcal TSS)* which resembles staphylococcal TSS. In streptococcal TSS, a rash is less likely, but bacteremia is more likely to occur. The immediate cause is M proteins that form a complex with blood fibrinogen that binds to neutrophils, precipating shock and organ damage.

Infections by Pseudomonads.   Pseudomonads, particularly *Pseudomonas aeruginosa,* may cause opportunistic skin infections. These bacteria are common in soil, water, and plants, and they grow on minimal organic material. Resistance to antibiotics often makes them a hospital problem. *P. aeruginosa* produces several exotoxins and endotoxins. It may cause respiratory infections, growing in dense biofilms, in compromised hosts (especially people with cystic fibrosis) or in burn patients, in whom infection may produce a characteristic blue-green pus from **pyocyanin** pigments. **Otitis externa,** or swimmer's ear, is a pseudomonad infection of the outer ear. Skin infections, collectively called ***Pseudomonas* dermatitis,** are contracted in such places as whirlpool baths; the organisms are relatively resistant to chlorines. Gentamicin and carbenicillin, often used in combination, are effective; silver sulfadiazine is used to treat *P. aeruginosa* infections of burn patients.

Buruli Ulcer.   Buruli ulcer is a disease of localized tropical and temperate areas including Mexico, Australia, and areas of South America. The disease is caused by *Mycobacterium ulcerans* and progresses slowly from an introductory site on the skin. Eventually, it causes a deep, damaging ulcer. The tissue damage is attributed to a toxin, *mycolactone.* It is treated by antimycobacterial drugs, such as streptomycin-rifampin combinations.

Acne.   Acne can be classified into three categories: *comedonal acne, inflammatory acne*, and *nodular cystic acne*. Acne develops when channels for the passage of sebum to the skin surface are blocked. As sebum accumulates, *whiteheads (comedos)* form; if the blockage protrudes through the skin a *blackhead (comedone or open comedo)* forms. **Comedonal (mild) acne** is usally treated with topical agents such as azelaic acid (Azelex), salicylic acid preparations, or retinoids. These are derivatives of vitamin A such as tretinoin, tazarotene (Tazorac), or adapalene (Differin). They do not affect sebum formation.

   Inflammatory (moderate) acne arises from action of the anaerobic bacterium *Propionibacterium acnes.* It metabolizes sebum and causes an inflammatory response resulting in pustules and papules. Drugs that reduce sebum formation such as isotretinoin (Accutane) are very effective. However, isotretinoin is teratogenic, meaning it causes serious birth defects. Also used are nonprescription benzoyl peroxide compounds and prescription antibiotics. A recent development in the treatment of moderate inflammatotory acne is the Clear Light system. The affected skin is exposed to blue light that destroys *Propionibacterium acnes* bacteria. Also approved recently is ThermaClear, which delivers a brief pulse of heat to the lesions.

   Some patients progress to **nodular cystic (severe) acne** that leaves prominent scars. Isotretinoin is the usual treatment.

Viral Diseases of the Skin

Warts.   More than 50 types of *Papillomavirus* cause the uncontrolled nonmalignant growth of skin cells called **warts.** Warts can be spread by contact; the incubation period is several weeks. Treatment of warts includes burning them with acids or the use of liquid nitrogen (cryotherapy), or electrodesiccation. Genital warts have been successfully treated with injected interferon and lasers.

Smallpox (Variola).   Smallpox is caused by a poxvirus known as the smallpox (variola) virus. **Variola major** smallpox has a mortality rate of 20% or more, and **variola minor** has a mortality rate of less than 1%. Recovery from one form gives immunity to both. Transmission is by the respiratory route, and the viruses eventually move through the bloodstream to infect the skin. Vaccination has resulted in the elimination of smallpox in the world. **Monkeypox** closely resembles smallpox and has been mistaken for it.

Chickenpox (Varicella) and Shingles (Herpes Zoster).   Chickenpox (varicella) is a relatively mild disease, the second most common reportable disease in the United States. Infection is by the respiratory route, and after about 2 weeks of incubation it localizes in the skin. The skin is vesicular for 3 or 4 days. The herpesvirus varicella-zoster (official name, human herpesvirus 3) that causes chickenpox may remain latent in nerve cells and later in life be reactivated to cause **herpes zoster (shingles).** In shingles, vesicles similar to chickenpox occur in areas such as the waist, upper chest, and face, affecting nerve branches of the cutaneous sensory nerves.

   Reye's syndrome is an occasional, severe complication of chickenpox, influenza, and sometimes other viral diseases. Brain damage or death may result from brain swelling, which prevents blood circulation. Use of aspirin may increase chances for the Reye's syndrome complication.

Herpes Simplex.   The herpes simplex virus is able to remain latent for long periods of time following an initial infection, which usually occurs in infancy and affects the oral mucous membrane. Infections recur in the form of **cold sores** or **fever blisters** when triggered by stress at a later time. Cold sores are often mistaken for **canker sores,** which are similar in appearance but appear in different areas such as the tongue, cheeks, and inner surface of the lips. Their cause is unknown but may be related to stress. Transmission by skin contact among wrestlers is common (**herpes gladiatorum**). **Herpetic whitlow** is infection of the finger caused by contact with HSV-1 lesions. These infections are due to **herpes simplex type 1 virus** (HSV-1). Another infection caused by a similar virus, **herpes simplex type 2 virus** (HSV-2), is transmitted primarily by sexual contact and will be discussed in Chapter 26. Either virus may occasionally cause **herpes encephalitis.** The official names of these viruses are human herpesvirus 1 and 2.

Measles (Rubeola).   The **measles** virus is spread by the respiratory route. Measles is extremely dangerous; complications such as encephalitis, which may result in brain damage, occur once in every 1000 cases, and once in about 3000 cases there is a fatality. **Koplik's spots** on the oral mucosa are diagnostically useful.

A rare neurological complication occurring after recovery from measles is **subacute sclerosing panencephalitis.**

Rubella. Rubella (*German measles*) is caused by a virus and is much milder than rubeola, often being subclinical. There is a light fever and a macular rash of small red spots. Transmission is by the respiratory route. Infections during the first trimester of pregnancy can lead to the **congenital rubella syndrome,** causing serious defects such as deafness, eye cataracts, hearing defects, and mental retardation. Diagnostic tests are mainly used to determine the immune status of pregnant women.

Other Viral Rashes

Fifth Disease (Erythema infectiosum). The name **fifth disease** derives from a 1905 list of skin rash diseases; it was fifth on the list. Caused by human parvovirus B19, symptoms are mild and influenza-like, but with a "slapped-cheek" facial rash.

Roseola. Roseola is a mild, common childhood disease. A high fever is followed in a few days by a body rash lasting for a day or two. The pathogen is human herpesvirus 6 (HHV-6).

Fungal Diseases of the Skin and Nails

A fungal infection of the body is referred to as a **mycosis**. The ability of fungi to resist high osmotic pressure and low moisture makes the skin susceptible to fungal infections.

Cutaneous Mycoses. Fungi that colonize hair, nails, and the outer layer of the epidermis are **dermatophytes. Ringworm,** or **tinea** (called **dermatomycoses**), is caused by members of the genera *Microsporum, Trichophyton,* and *Epidermophyton.* **Tinea capitis,** or ringworm of the scalp, is spread by contact among children and by contact with various fomites, dogs, and cats. Ringworm of the groin (jock itch) is **tinea cruris,** and ringworm of the feet (athlete's foot) is **tinea pedis.** Hair and nails, high in keratin, are areas for which fungal diseases, such as **tinea unguium** (onychomycosis), have an affinity. Topical agents for tinea infections are miconazole and clotrimazole. When nails are infected, itraconazole or terbinatine are recommended. An oral antibiotic, griseofulvin, localizes in keratinized tissue and is an effective treatment.

Subcutaneous Mycoses. Subcutaneous mycoses are caused by fungi inoculated into a wound. The most commonly encountered is **sporotrichosis,** caused by *Sporothrix schenckii.* The nodule formed by the growing fungus may spread along the lymphatic vessels.

Candidiasis. *Candida albicans* is a yeastlike fungus that causes infections of the mucous membranes of the genitourinary tract (**vaginitis**) and oral mucosa (**thrush**). These infections, collectively called **candidiasis,** often appear when bacterial populations are suppressed by antibiotics that do not affect fungi. For topical use, clotrimazole or miconazole are recommended.

Parasitic Infestation of the Skin

Scabies. The tiny mite *Sarcoptes scabiei* burrows under the skin to lay its eggs. This sets up an inflammation (**scabies**) that itches intensely. Treatment is by topical application of permethrin.

Pediculosis (Lice). Infestation by lice is called **pediculosis.** It is caused by a subspecies of *Pediculus humanus, Pediculus humanus capitis.* Treatment is by insecticides, but resistance is common. An alternative treatment is to comb out the nits (egg cases attached to hairs).

MICROBIAL DISEASES OF THE EYE

Inflammation of the Eye Membranes: Conjunctivitis

Conjunctivitis is an inflammation of the **conjunctiva,** the mucous membrane that lines the eyelid and covers the outer surface of the eyeball. It is often called *red eye* or *pinkeye.*

Bacterial Diseases of the Eye

Ophthalmia Neonatorum.   **Ophthalmia neonatorum** is a dangerous eye inflammation transmitted to the newborn during passage through the birth canal. Newborn infants have their eyes washed with 1% silver nitrate or antibiotics to prevent this infection.

Inclusion Conjunctivitis.   **Inclusion conjunctivitis** involves only the conjunctiva, not the cornea. This disease may be spread (to newborns) by an infection in the birth canal or by unchlorinated water. Tetracycline is used topically in treatment.

Trachoma.   **Trachoma** is the greatest single cause of blindness in the world. It is an infection of the epithelial cells of the eye caused by *Chlamydia trachomatis*. Scar tissue forms on the cornea. It is transmitted by contact with fomites.

Other Infectious Diseases of the Eye

Herpetic Keratitis.   **Herpetic keratitis** is caused by herpes simplex type 1 virus, with an epidemiology similar to that of cold sores. Corneal ulcers occur and may lead to blindness. Trifluridine is the drug of choice against this viral infection.

Acanthamoeba Keratitis.   A freshwater ameba, *Acanthamoeba*, causes a keratitis (*Acanthamoeba* keratitis) in the eyes of contact lens wearers. The damage is severe and may require a corneal transplant to correct.

SELF-TESTS

In the matching section, there is only one answer to each question; however, the lettered options (a, b, c, etc.) may be used more than once or not at all.

I. Matching

___ 1. The inner layer of the skin, composed of connective tissue.

___ 2. The lining of the inner eyelid and the surface of the eyeball.

___ 3. Some of these specialized epithelial cells are ciliated.

___ 4. The outermost epidermal layer; consists largely of dead cells containing the protein keratin.

___ 5. Extracellular material to which epithelial cells of mucous membrane are attached.

___ 6. A skin rash that arises from disease conditions.

a. Mucous membrane

b. Stratum corneum

c. Dermis

d. Conjunctiva

e. Basement membrane

f. Exanthem

g. Enanthem

II. Matching (Skin Lesions)

___ 1. Vesicles.

___ 2. Papules.

___ 3. Bullae.

___ 4. Macules.

a. Flat, reddened

b. Small, fluid-filled

c. Fluid-filled lesions larger than about 1 cm

d. Raised lesions

III. Matching

___ 1. *Streptococcus pyogenes.*

___ 2. *Staphylococcus aureus.*

___ 3. Tinea.

___ 4. *Propionibacterium acnes.*

___ 5. *Papillomavirus.*

a. Scalded skin syndrome

b. Acne

c. Erysipelas

d. Warts

e. Ringworm

IV. Matching

___ 1. Variola.

___ 2. Varicella.

___ 3. Herpes zoster.

___ 4. Rubeola.

___ 5. Rubella.

___ 6. Shingles.

a. Impetigo

b. Smallpox

c. Measles

d. Chickenpox

e. German measles

V. Matching

___ 1. Ringworm.

___ 2. Tinea pedis.

___ 3. Dermatophytes.

___ 4. Thrush.

___ 5. Sporotrichosis.

___ 6. Treated with isotretinoin (Accutane).

a. Cutaneous mycoses

b. Superficial mycoses

c. Subcutaneous mycoses

d. Candidiasis

e. Nodular cystic acne

f. Comedonal acne

VI. Matching

____ 1. Swimmer's ear, usually caused by pseudomonads.

____ 2. Boils.

____ 3. Idoxuridine is an effective chemotherapeutic treatment.

____ 4. Chlamydia-caused disease.

____ 5. *Mycobacterium* spp. are the pathogens involved.

a. Herpetic keratitis

b. Trachoma

c. Otitis externa

d. Furuncles

e. Buruli ulcer

VII. Matching

____ 1. Treatment of cystic acne.

____ 2. The location of M protein of streptococci.

____ 3. Causes birth defects.

____ 4. Prevention of ophthalmia neonatorum.

____ 5. Scabies.

a. Teratogenic

b. Fibrils on cell surface

c. Isotretinoin

d. Mite

e. Silver nitrate

VIII. Matching

____ 1. Ringworm of the scalp.

____ 2. First disease deliberately eliminated on Earth.

____ 3. Koplik's spots are diagnostic.

____ 4. Athlete's foot.

____ 5. Occasional complication of chickenpox and influenza.

____ 6. Fungal infection of the nails.

a. Reyes syndrome

b. Measles (rubeola)

c. Smallpox

d. Tinea pedis

e. Tinea capitis

f. Tinea unguium

Fill in the Blanks

1. The eyes are washed by tears, and the enzyme _____ in tears destroys many bacteria.

2. If a boil undertakes a more extensive invasion of the surrounding tissue, it is termed a

   _____.

3. *Streptococcus pyogenes* is an example of group A _____ hemolytic streptococci.

4. The blue-green pus caused by opportunistic infections of burn patients is due to *Pseudomonas aeruginosa* forming water-soluble _____ pigment.

5. Benzoyl peroxide is useful in treating _____.

6. A fungal infection of the body is referred to as a(n) _____.

7. The presence of viruses in the bloodstream is termed _____.

8. The single most prevalent infectious cause of blindness in the world is the chlamydial disease

_____.

9. Herpetic keratitis is caused by herpes simplex type _____ virus.

10. An eyelash follicle infection is called a(n) _____.

11. *Propionibacterium acne* metabolizes _____, forming free fatty acids that cause the inflammation leading to acne scarring.

12. Congenital rubella syndrome is caused by rubella infection during the first

_____ of pregnancy.

13. The _____ test is used to distinguish pathogenic *Staphylococcus aureus* from other species of *Staphylococcus*.

14. Scarlet fever rash is caused by a(n) _____ -type toxin.

15. A protozoan disease of the eye is caused by the aquatic ameba _____.

16. Pediculosis is caused by *Pediculus humanus* _____.

17. An infection of the finger caused by HSV-1 is called herpetic _____.

Critical Thinking

1. A surgical patient acquires a nosocomial infection. The primary organism isolated from the surgical wound is coagulase positive and resistant to penicillin. What is the probable etiologic agent? Why is this organism a common cause of nosocomial infections?

2. Discuss the role of *Propionibacterium* in acne. Does *Propionibacterium* provide any benefits to its host?

3. Why are the symptoms of shingles different from those of chickenpox even though they are caused by the same virus?

4. A 43-year-old patient consulted his physician concerning mild inflammation of his eyes. The patient was instructed to avoid wearing his contact lenses for one week. The symptoms worsened over the next few days. After eliminating the possibility of a bacterial agent, the physician ordered a corneal scraping. What etiologic agent does the physician suspect? What evidence of this agent is the physician looking for in the scraping?

5. Why were efforts to eliminate smallpox successful?

ANSWERS

Matching

I.	1. c	2. d	3. a	4. b	5. e	6. f
II.	1. b	2. d	3. c	4. a		
III.	1. c	2. a	3. e	4. b	5. d	
IV.	1. b	2. d	3. d	4. c	5. e	6. d
V.	1. a	2. a	3. a	4. d	5. b	6. e
VI.	1. c	2. d	3. a	4. b	5. e	
VII.	1. c	2. b	3. a	4. e	5. d	
VIII.	1. e	2. c	3. b	4. d	5. a	6. f

Fill in the Blanks

1. lysozyme   2. carbuncle   3. beta-   4. pyocyanin   5. acne   6. mycosis   7. viremia   8. trachoma 9. 1   10. sty   11. sebum   12. trimester   13. coagulase   14. erythrogenic   15. *Acanthamoeba* 16. *capitis*   17. whitlow

Critical Thinking

1. The etiologic agent is probably *Staphylococcus aureus*. It is a common cause of nosocomial infections because it is a part of the normal microbiota of most people. That means hospital staff, visitors, and even the patient are possible sources of infection.

2. *Propionibacterium acnes* is part of the normal microbiota of the skin. This organism is able to metabolize sebum and produces acid by-products that discourage the growth of potential pathogens.

   When sebum channels become blocked and rupture, bacteria—especially *Propionibacterium acnes*— become involved. The bacterium metabolizes sebum into free fatty acids that cause an inflammatory response. This leads to tissue damage and possible scarring.

3. The symptoms of shingles are different from those of chickenpox because shingles is a different expression of the virus. After having had chickenpox, the patient has partial immunity to the virus.

4. The physician suspects *Acanthamoeba* keratitis and is looking for trophozoites or cysts in the corneal scraping.

5. The efforts to eliminate smallpox were successful primarily because of the lack of animal reservoirs. Once an effective vaccine was available, a concerted effort was coordinated by the World Health Organization that resulted in elimination of the disease.

22 Microbial Diseases of the Nervous System

STRUCTURE AND FUNCTION OF THE NERVOUS SYSTEM

The **central nervous system (CNS)** consists of the **brain** and **spinal cord.** It is the control center that picks up sensory information from the environment and, after interpreting it, sends out impulses to coordinate body activities. The **peripheral nervous system** consists of all the nerves branching from the brain and spinal cord. These nerves are the communication lines between the CNS and the body, and the external environment. The skull protects the brain and the backbone protects the spinal cord; both the brain and spinal cord are covered by the **meninges** (Figure 22.1). The meninges consist of the *dura mater* (outermost layer), the *arachnoid mater* (middle layer), and the *pia mater* (innermost layer). **Cerebrospinal fluid** circulates in the space between the pia mater and the arachnoid layers (*subarachnoid space*). The **blood–brain barrier** consists of capillaries that permit certain substances to pass from the blood to the brain but that

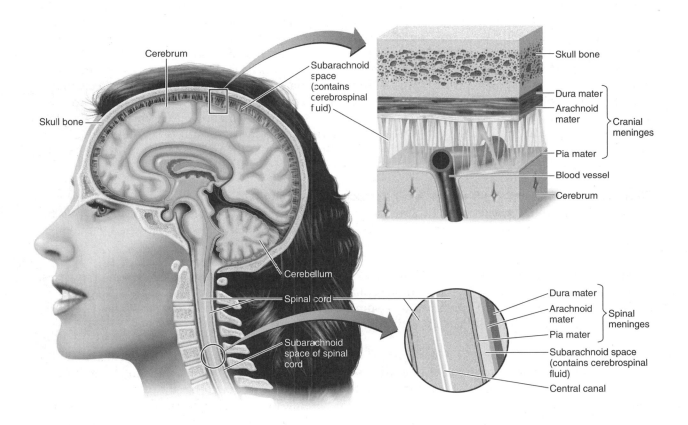

Figure 22.1 The meninges and cerebrospinal fluid. The meninges, whether cranial or spinal, consist of three layers: dura mater, arachnoid mater, and pia mater. Between the arachnoid mater and the pia mater is the subarachnoid space, in which cerebrospinal fluid circulates. Notice that the CSF is vulnerable to contamination by microbes carried in the blood that are able to penetrate the blood–brain barrier at the walls of the blood vessels.

prevent others from passing. Microorganisms can gain access to the CNS by trauma (accidental, or medical procedures such as spinal taps). Also, microorganisms may enter by movement along peripheral nerves or by the bloodstream and lymphatic system. An inflammation of the meninges is **meningitis;** an inflammation of the brain is **encephalitis.** If both the brain and the meninges are affected, the inflammation is called **meningoencephalitis.** Antibiotics often are unable to cross the blood–brain barrier.

BACTERIAL DISEASES OF THE NERVOUS SYSTEM

Bacterial Meningitis

The three major types of bacterial meningitis are meningococcal meningitis, caused by *Neisseria meningitidis;* pneumococcal meningitis, caused by *Streptococcus pneumoniae;* and *Haemophilus influenzae* meningitis. Nearly 50 other bacteria have been reported to cause meningitis, as can fungi, viruses, and protozoa. Meningitis patients complain of headache and have symptoms of nausea and vomiting. This may proceed to convulsions, coma, and even death.

***Haemophilus influenzae* Meningitis.**   *Haemophilus influenzae* type b, the main serological type of medical importance, is an aerobic, gram-negative bacterium. Strains that lack a capsule are called *nontypable. H. influenzae* is often a member of the normal throat flora. It is the most common cause of bacterial meningitis among children under 4 years of age. The incidence has been decreasing since the introduction of a vaccine (*Hib vaccine*).

***Neisseria* Meningitis (Meningococcal Meningitis).**   **Meningococcal meningitis** is caused by the bacterium *Neisseria meningitidis* (the **meningococcus**), an aerobic, gram-negative bacterium. It is a frequent inhabitant of the throat behind the nose, and a throat infection can lead to bacteremia followed by meningitis. Symptoms are thought to be due to endotoxins produced by the bacteria, which can also sometimes enter the bloodstream and cause serious tissue damage. The disease primarily affects the very young, with the highest incidence in the first year. The meningococcus occurs in six capsular serotypes associated with invasive disease (A, B, C, W-135, S, and Y). Serotypes B and C predominate in the United States and other industrialized countries. Sub-Saharan Africa regularly has outbreaks of serotype A. The available vaccine is directed at the polysaccharide capsules of A, C, Y, and W-135, which are the most commonly encountered.

***Streptococcus pneumoniae* Meningitis (Pneumococcal Meningitis).**   **Pneumococcal meningitis** is caused by *Streptococcus pneumoniae,* a common inhabitant of the nasopharyngeal system. About half the cases are among children aged 1 month to 4 years. The mortality rate is high. A conjugated vaccine was recently introduced.

Diagnosis and Treatment of the Most Common Types of Bacterial Meningitis.   Chemotherapy must be prompt; broad-spectrum third-generation cephalosporins are the first choices. Diagnosis requires a sample of cerebrospinal fluid obtained by a spinal tap. Gram stains and cultures can be made. Serological tests such as rapid latex agglutination tests are used.

Listeriosis.   *Listeria monocytogenes,* the gram-positive rod causing **listeriosis,** is widely distributed, mainly in animal feces. Listeriosis is a mild disease in most adults but is more serious in the immunosuppressed, including cancer patients. When infecting a pregnant woman it often causes abortion or serious damage or death to the fetus. The organism proliferates within phagocytic cells and grows at refrigeration temperatures. It is a major concern in the food industry.

Tetanus.   The cause of **tetanus** is an obligately anaerobic, endospore-forming, gram-positive rod, *Clostridium tetani.* Growth of *C. tetani* in wounds releases **tetanospasmin,** a neurotoxin that blocks the nerve pathway that signals muscle relaxation. Spasms result as opposing muscle sets contract (spastic paralysis). Contraction of the jaw muscles prevents opening of the mouth (**lockjaw**). In extreme cases, the body bows backward (**opisthotonos**). Death from respiratory failure occurs in many cases. Most people in the United States have been immunized by the tetanus toxoid included in the DTaP (diphtheria, tetanus, acellular pertussis) vaccine. A booster of toxoid may be given when a dangerous wound is received; this booster causes the body to renew its immunity level in a rapid anamnestic response. If the patient has had no previous immunity, toxoid may not cause a rapid enough appearance of antibodies, and a temporary immunity can be provided by *tetanus immune globulin (TIG)* pooled from immunized humans.

Botulism. The cause of **botulism** is an obligately anaerobic, endospore-forming, gram-positive bacterium, *Clostridium botulinum,* found in soil and freshwater sediments. The exotoxin produced is a neurotoxin that blocks the transmission of nerve impulses across the synapses, causing progressive *flaccid paralysis.* Symptoms typically appear in a day or two. Most botulism results from attempts at preservation by heat that fail to eliminate the *C. botulinum* endospore but provide anaerobic conditions for its growth. There are several toxin types. **Type A toxin** is found mostly in the western United States. It is proteolytic and probably the most virulent. **Type B toxin** is less virulent, and both proteolytic and nonproteolytic strains occur. Type B is responsible for most outbreaks in the eastern United States and Europe. **Type E toxin** is produced by a strain found in wet soils and sediments and often involves seafood. This organism is nonproteolytic, its endospores are less heat-resistant, and the toxin can be produced at refrigerator temperatures. The botulinal toxin is heat-labile; that is, it is destroyed by boiling. It is not formed in acid foods below pH 4.7. Nitrates are included in some meat products to prevent bacterial growth following germination of endospores. For treatment, antitoxins of trivalent ABE type are available, but they do not affect attached toxins. Toxin can be identified by inoculating mice with the suspected toxin; if the mice who survive the inoculation are protected by, for example, type A antitoxins, then the suspected toxin is considered to be type A. **Wound botulism** can occur from *C. botulinum* growth in wounds, and **infant botulism** may occur from growth in the infant GI tract, which is rarely a factor in adults. Botulinum toxin has commercial uses. Injection of the toxin (Botox) at intervals of a few months eliminates forehead wrinkles.

Leprosy. **Leprosy**, sometimes called **Hansen's disease** for the person who first isolated the organism, is caused by an acid-fast rod, *Mycobacterium leprae.* It is probably the only bacterium that grows primarily in the peripheral nervous system. The microorganisms have not been grown on artificial media but can be grown in armadillos. There are two main forms of leprosy. The **tuberculoid (neural) form** (roughly the same as *paucibacillary* in the WHO system) is characterized by regions of the skin that have lost sensation and are surrounded by a border of nodules. The **lepromin test,** similar to the tuberculin test in design, usually is positive for this form. In the **lepromatous (progressive) form** of leprosy (roughly the same as *multibacillary* in the WHO system), skin cells are infected, and disfiguring nodules form over the body. Mucous membranes of the nose (one sign is a lion-faced appearance) and the hands are particularly affected; the organism prefers cooler regions of the body. Progression to this stage of leprosy indicates a less effective immune system.

Leprosy is not very contagious. Death is usually a result of complications from diseases such as tuberculosis. Laboratory diagnosis is by identification of acid-fast rods in lesions. The sulfone drugs, such as dapsone, are the most effective treatment. Rifampin and a fat-soluble dye, clofazimine, also are used.

VIRAL DISEASES OF THE NERVOUS SYSTEM

Poliomyelitis. **Poliomyelitis** is informally called **polio.** Symptoms usually are a few days of headache, sore throat, and fever; only a few cases are paralytic. The poliovirus occurs in three different serotypes. The primary mode of transmission is ingestion of fecal-contaminated water. In paralytic cases, the virus penetrates capillary walls and enters the CNS, where it displays a high affinity for nerve cells. The motor nerve cells in the upper spinal cord, called the anterior horn cells, are particularly affected. In recent years, persons who had polio as children have been showing signs of muscle weakness called *post-polio syndrome.*

The **Salk vaccine** (*inactivated polio vaccine, IPV*) uses viruses inactivated by formalin. This vaccine requires a series of injections and periodic booster shots. The **Sabin vaccine** (*oral polio vaccine, OPV*) contains three live attenuated strains. The immunity from this vaccine resembles that acquired naturally, but on rare occasions the vaccine itself may cause the disease, probably by a mutation to virulence. Diagnosis is usually based on isolation of the virus and observation of cytopathic effects on cell cultures.

Rabies. The **rabies virus,** a rhabdovirus with a bullet shape, is typically transmitted in the saliva of a biting rabid animal. The virus travels along peripheral nerves to the CNS, where it produces encephalitis. Symptoms include spasms of the muscles of the mouth and pharynx when swallowing liquids. This painful reaction causes an aversion to water, which is the basis of the term **hydrophobia** (fear of water) that is applied to rabies.

Animals with **furious rabies** are restless, snapping at anything. Some animals suffer from **paralytic rabies,** in which there is only minimal excitability, although they may snap irritably if handled.

Prevention, Treatment, and Distribution of Rabies. Laboratory diagnosis is usually made by a fluorescent antibody test. In early years, the Pasteur treatment was standard for prevention of rabies in exposed individuals. The vaccination procedure consisted of injections of dried spinal cords of rabbits infected with rabies virus. Currently, treatment of exposed people begins with the administration of human *rabies immune globulin (RIG)*, followed by active immunization. This is called **post-exposure prophylaxis (PEP).** The latter is done with human diploid cell vaccines (HDCV) grown in human diploid cell lines. HDCV is administered in five or six injections over a 28-day period. Indications for antirabies treatment are unprovoked bites from dogs, cats, skunks, and similar animals. In the United States, rabies is often transmitted by bats; the bites may be unnoticed, so just finding a bat in the vicinity of small children or sleeping individuals is considered a reason for preventive immunization. It also is possible to contract rabies by inhaling aerosols of the virus.

Related *Lyssavirus* Encephalitis. In recent years, fatal cases of encephalitis clinically indistinguishable from classic rabies have occurred in countries considered free of rabies. These are caused by genotypes of the genus *Lyssavirus* that are closely related to classic rabies virus and include the Australian bat lyssavirus and European bat lyssavirus. There are seven known genotypes of the genus *Lyssavirus*.

Arboviral Encephalitis. Encephalitis is caused by viruses called arboviruses (<u>ar</u>thropod-<u>bo</u>rne viruses) that are transmitted by a mosquito, is common in the United States. Horses frequently are infected, accounting for terms such as **Eastern equine encephalitis (EEE)** and **Western equine encephalitis (WEE). St. Louis encephalitis (SLE)** and **California encephalitis (CE),** which do not infect horses, also occur. In the Far East, **Japanese B encephalitis** is a similar disease. All cause at least some neurological damage; however, EEE is the most severe in the United States, causing a high incidence of brain damage, deafness, and similar neurological problems. It also has the highest mortality rate. In 1999 an outbreak of West Nile virus occurred in the New York City area and has now become endemic almost nationwide. Diagnosis is usually made by serological tests. Control of mosquitoes is an essential preventive measure.

FUNGAL DISEASE OF THE NERVOUS SYSTEM

Cryptococcus neoformans Meningitis (Cryptococcosis)

Cryptococcus neoformans is a yeastlike fungus widely distributed in soil, especially in soil enriched by pigeon droppings. The disease **cryptococcosis** is thought to be transmitted by inhalation of dried infected pigeon droppings. Other, similar species, such as *C. grubii* and *C. gattii*, are also known to cause cryptococcosis in certain areas. Often the disease does not proceed beyond an infection of the lungs, but it may spread by the bloodstream to the brain and meninges. Laboratory diagnosis is by latex agglutination tests of cerebrospinal fluid. The drugs of choice for treatment are amphotericin B and flucytosine.

PROTOZOAN DISEASES OF THE NERVOUS SYSTEM

African Trypanosomiasis. **African trypanosomiasis** is a disease caused by protozoa that affects the nervous system. The flagellates *Trypanosoma brucei gambiense* and *Trypanosoma brucei rhodesiense* enter the body by a tsetse fly bite. Later they move into the cerebrospinal fluid, where the resulting symptoms are responsible for the informal name of sleeping sickness. Animals are reservoirs for the parasite.

Human-to-human transmission by bites of the insect vector is more likely with *T. b. gambiense. T. b. rhodesiense* causes a disease that is more acute, running its course much more quickly. Suramin and pentamidine are used in chemotherapy. A newer drug, eflornithine, blocks an enzyme needed for proliferation. It is very effective against *T. b. gambiense* but variable against *T. b. rhodesiense.*

Amebic Meningoencephalitis.   A severe neurological disease, **primary amebic meningoencephalitis (PAM)** is caused by *Naegleria fowleri.* It is most common in children who swim in ponds or streams, initially infecting the nasal mucosa. The fatality rate is nearly 100%. A similar disease, **granulomatous amebic encephalitis (GAE),** is slower to develop and often affects organs other than the brain. This organism is also found in freshwater environments. The disease is caused by a species of *Acanthamoeba*, but is not the same one that causes *Acanthamoeba* keratitis, an eye disease.

NERVOUS SYSTEM DISEASES CAUSED BY PRIONS

There are a number of fatal diseases of the human central nervous system caused by prions. **Sheep scrapie** is a typical prion disease. **Creutzfeldt-Jakob disease** and **kuru** are similar diseases in humans. A current international public health concern is **bovine spongiform encephalopathy (BSE),** or *mad cow disease,* which affects many herds of cattle in the British Isles. A prion-caused neurological disease, **chronic wasting disease,** occurs in elk and deer in the United States. In Great Britain and a few other locales, a disease similar to Creutzfeldt-Jakob disease has appeared in humans. However, it differs in several significant ways, especially because it affects a younger population, and is called **variant Creutzfeldt-Jakob disease (vCJD).** A connection with BSE is feared. The agents causing these diseases are unknown; they do not have any detectable nucleic acid, and one hypothesis is that they are composed entirely of protein (**prions**). Chapter 13 has a further discussion of this subject.

DISEASE CAUSED BY UNIDENTIFIED AGENTS

Chronic fatigue syndrom (CFS), or *myalgic encephalomyelitis,* is a debilitating condition of unknown etiology. It is fairly common in the United States, mainly affecting women. Current thinking is that the cause may be overstimulation of the immune system in response to some infection—probably viral.

SELF-TESTS

In the matching section, there is only one answer to each question; however, the lettered options (a, b, c, etc.) may be used more than once or not at all.

I. Matching

_____ 1. A membrane layer covering the brain and spinal cord.

_____ 2. A prion-caused disease.

_____ 3. Opisthotonos.

_____ 4. Hansen's disease.

_____ 5. Human diploid cell vaccine is used in treatment.

_____ 6. Myalgic encephalitis

a. Meninges

b. Tetanus

c. Rabies

d. Kuru

e. Chronic fatigue syndrome

f. Leprosy

II. Matching

____ 1. Innermost layer of the meninges.

____ 2. Outermost layer of the meninges.

____ 3. Middle layer of the meninges.

a. Dura mater

b. Ventricles

c. Subarachnoid space

d. Arachnoid

e. Pia mater

III. Matching

____ 1. Formerly treated by the Pasteur treatment.

____ 2. Treated by human diploid cell vaccine after exposure.

____ 3. Caused by a bullet-shaped rhabdovirus.

____ 4. Also known as hydrophobia.

____ 5. Thought to be transmitted by inhalation of the pathogen in dried pigeon droppings.

____ 6. Caused by the bacterium *Neisseria meningitidis*.

____ 7. Protozoan disease.

a. Rabies

b. Meningococcal meningitis

c. *Haemophilus influenzae* meningitis

d. Cryptococcosis

e. Poliomyelitis

f. Pneumococcal meningitis

g. African trypanosomiasis

IV. Matching

____ 1. A prion-caused disease.

____ 2. A mosquito-borne virus.

____ 3. The drugs of choice for treatment are amphotericin B and flucytosine.

____ 4. Opposing muscles contract, causing spastic paralysis.

____ 5. Pathogen grows at refrigerator temperatures.

a. Creutzfeldt-Jakob disease

b. Meningococcal meningitis

c. Listeriosis

d. *Cryptococcus neoformans* meningitis

e. Tetanus

f. California encephalitis

V. Matching

___ 1. Uses live viruses.

___ 2. On rare occasions, the vaccine has caused the disease by a mutation to virulence.

a. Salk polio vaccine

b. Sabin polio vaccine

VI. Matching

___ 1. An amebic protozoan found in ponds and streams that causes a primary amebic meningoencephalitis.

___ 2. Spread by the bite of a tsetse fly.

___ 3. An important cause of bacterial meningitis.

___ 4. The cause of African sleeping sickness.

a. *Naegleria fowleri*

b. *Trypanosoma brucei gambiense*

c. *Cryptococcus neoformans*

d. *Streptococcus pneumoniae*

VII. Matching

___ 1. Probably the most virulent; the most common type in western United States.

___ 2. Outbreaks often involve seafoods; nonproteolytic.

___ 3. Toxin can be produced at refrigerator temperatures.

a. Type A botulism

b. Type B botulism

c. Type E botulism

Fill in the Blanks

1. An infection of the brain is called _____.

2. An infection of the meninges is called _____.

3. The brain and the spinal cord comprise the _____ nervous system.

4. The nerves branching from the brain and spinal cord comprise the _____ nervous system.

5. The _____ consists of capillaries that permit certain substances, mostly lipid-soluble, to pass from the blood to the brain but prevent other substances from passing.

6. The bacterium that causes _____ can be successfully grown in armadillos.

7. Of the several types of arthropod-borne encephalitis that occur in the United States, the most severe in its effects is _____.

8. The T in the DTaP vaccine stands for _____.

9. The vaccine called Hib prevents meningitis caused by _____.

10. The _____ fluid circulates in spaces within the brain (subarachnoid space).

11. A chemical sometimes added to foods to prevent the growth of *Clostridium botulinum* is

    _____.

12. People with a less effective cell-mediated immune system are more likely to develop the

    _____ form of leprosy.

13. The lepromatous (progressing) form of leprosy is roughly the same as _____

    in the WHO system of naming.

Critical Thinking

1. The neurotoxin causing botulism results in a flaccid paralysis, whereas the neurotoxin causing
   tetanus causes muscle spasms. Explain the difference.

2. Folklore associates the transmission of tetanus with rusty nails. Explain a rational basis for this belief.

3. The Salk vaccine for polio, which uses a killed virus, is somewhat less effective in preventing the
   disease, requires booster doses, and requires an unpleasant injection. However, it is now being
   recommended in place of the live, orally administered Sabin vaccine. Why?

4. People who have been infected with rabies may be able to develop immunity from a vaccination
   given after exposure to the virus. Why is this possible?

ANSWERS

Matching

   I.   1. a   2. d   3. b   4. f   5. c   6. e
  II.   1. e   2. a   3. d
 III.   1. a   2. a   3. a   4. a   5. d   6. b   7. g
 IV.   1. a   2. f   3. d   4. e   5. c
  V.   1. b   2. b
 VI.   1. a   2. b   3. d   4. b
VII.   1. a   2. c   3. c

Fill in the Blanks

1. encephalitis   2. meningitis   3. central   4. peripheral   5. blood–brain barrier   6. leprosy
7. Eastern equine   8. tetanus   9. *Haemophilus influenzae* type b   10. cerebrospinal   11. nitrate
12. lepromatous (progressive)   13. multibacillary

Critical Thinking

1. Botulism toxin prevents the movement of a signal along the nerve, and the muscles never receive signals to relax or contract. In normal operation, a nerve impulse initiates contraction of the muscle. At the same time, an opposing muscle receives a signal to relax so as not to oppose the contraction. The tetanus toxin blocks the relaxation pathway so that both opposing sets of muscles contract.

2. The endospore-forming bacterium that produces the tetanus neurotoxin is common in soil, where a rusty nail might be found. If the nail is rusty, it is more likely to be contaminated with these bacteria than a shiny, new nail. The bacterium is an obligate anaerobe and would grow well in a deep puncture wound such as that caused by a nail.

3. The Sabin vaccine is easier and more pleasant to administer and provides generally better immunity because it mimics an active infection. However, on rare occasions, it reverts to virulence and causes polio. It has been decided recently that the advantages do not outweigh this risk.

4. The rabies virus has an unusually long incubation period. This makes it possible for a patient to respond to vaccination and develop immunity to the virus during the incubation period.

23 Microbial Diseases of the Cardiovascular and Lymphatic Systems

STRUCTURE AND FUNCTION OF THE CARDIOVASCULAR AND LYMPHATIC SYSTEMS

The **cardiovascular system** consists of the heart, blood, and blood vessels. *Blood* is a mixture of *plasma* and formed elements such as *red blood cells (erythrocytes)* and *white blood cells (leukocytes)*.

As the blood circulates, some plasma leaves blood capillaries and enters spaces between tissue cells, the *interstitial spaces* (Figure 23.1). There it is called *interstitial fluid*. As the interstitial fluid moves around tissue cells, it is picked up by *lymph capillaries*, where it is now called *lymph*. The lymph capillaries are permeable to microorganisms; from the lymph capillaries the lymph is transported to larger vessels, the *lymphatics*. Eventually all lymph is returned to the blood. At certain points in the lymphatic system there are *lymph nodes*, through which lymph flows. Microorganisms in the lymph encounter macrophages in the lymph nodes. These organs constitute the **lymphatic system.** Swollen lymph nodes, an indication of infection, are called **buboes.**

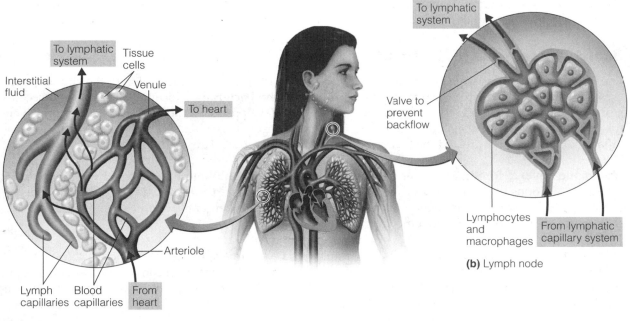

(a) Capillary system in lung

Figure 23.1   Relationship between the cardiovascular and lymphatic systems. (a) From the blood capillaries, some blood plasma filters into the surrounding tissue and enters the lymph capillaries. This fluid, now called lymph, returns to the heart through the lymphatic circulatory system (darker gray), which channels the lymph to a vein. (b) All lymph returning to the heart must pass through at least one lymph node. (See also Figure 16.5 in the text.)

BACTERIAL DISEASES OF THE CARDIOVASCULAR AND LYMPHATIC SYSTEMS

Sepsis and Septic Shock

Septicemia results from the proliferation of bacterial pathogens, with associated toxins, in the bloodstream. **Sepsis** is a *systemic inflammatory response syndrome (SIRS)* caused by a focus of infection that releases mediators of inflammation into the bloodstream. SIRS must exhibit at least two of the following conditions: fever, rapid heart or respiratory rates, or a high count of white blood cells. Sepsis is often accompanied by **lymphangitis,** inflamed lymph vessels visible as red streaks under the skin. Symptions of sepsis are fever, chills, and accelerated breathing and heart rate. When accompanied by a drop in blood pressure and dysfunction of at least one organ, it is called **severe sepsis.** In the final stage of sepsis, when low blood pressure can no longer be controlled by addition of fluids, it is called **septic shock.** This is usually caused by gram-negative bacteria. Their lysis releases endotoxins that cause the drop in blood pressure (*shock*), etc.

Gram-Negative Sepsis. Septic shock is also known as *gram-negative sepsis* or *endotoxic shock.* The best treatment is a genetically modified version of human activated protein C (not to be confused with C-reactive protein), drotrecogin alfa (Xigris), an anticoagulant.

Gram-Positive Sepsis. *Gram-positive sepsis,* usually caused by staphylococci and streptococci, also is a cause of blood-related disease. Enterococci such as *Enterococcus faecium* and *Enterococcus faecalis* are leading causes of nosocomial blood infections. They have a natural resistance to penicillin and have acquired resistance to other antibiotics, including vancomycin. Group B streptococci (*Streptococcus agalactiae*) are a common cause of *neonatal sepsis.*

Puerperal Sepsis. An infection of the uterus that often leads to septicemia is called **puerperal sepsis,** also known as **puerperal fever** or **childbirth fever.** Group A beta-hemolytic streptococci such as *Streptococcus pyogenes* are the most common cause.

Bacterial Infections of the Heart

The innermost of the layers surrounding the heart is endocardium. An inflammation of the endocardium is called **endocarditis. Subacute** (meaning it develops slowly) **bacterial endocarditis** is usually caused by alpha-hemolytic streptococci. **Acute bacterial endocarditis** develops much more rapidly and progressively. It is usually caused by *Staphylococcus aureus.* It can affect heart valves and is often fatal. **Pericarditis,** inflammation of the sac around the heart, the pericardium, is usually caused by streptococci.

Rheumatic Fever. **Rheumatic fever** is generally considered an autoimmune reaction by the body to repeated infections by *Streptococcus pyogenes.* Inflammation causes arthritis and damages heart valves. Some people develop a complication of rheumatic fever called **Sydenham's chorea** (Saint Vitus' dance), characterized by involuntary movements.

Tularemia. **Tularemia** is a zoonotic disease (transmitted by animals) of the lymph nodes caused by *Francisella tularensis,* a gram-negative, facultatively anaerobic rod. Hunters in contact with small animals such as rabbits (*rabbit fever*) are often infected, but arthropod bites (*deer fly fever,* for example) may also be involved. The infection results in a small ulcer at the infection site and, later, lymph node involvement. The microorganism grows intracellularly in phagocytes, a problem in chemotherapy.

Brucellosis (Undulant Fever). Several species of the genus *Brucella* cause **brucellosis,** also known as **undulant fever** because of the periodic spiking fever. *Brucella abortus* and *Brucella melitensis* are most commonly transmitted by the ingestion of the unpasteurized milk of cattle and goats, respectively. *Brucella suis* is transmitted mostly by contact with swine carcasses, and this is the most common mode of transmission in the United States today. Brucellosis caused by *B. melitensis* is the most severe. *Brucella* tend to multiply in the uterus of susceptible animals, favored by the presence of *mesoerythritol* in the fetus and in surrounding membranes.

Anthrax.   Infections by *Bacillus anthracis*, the cause of **anthrax**, are initiated by endospores. In the blood they are taken up by macrophages, where they multiply and kill the macrophages. *B. anthracis* produces two exotoxins, which share a third toxic component, the *protective antigen*, that binds the toxins to target cells to permit their entry. One toxin, the *edema toxin*, causes local edema (swelling) and interferes with phagocytosis. The other toxin, *lethal toxin*, targets and kills macrophages. The capsule of *B. anthracis* is composed of amino acid residues that do not stimulate immunity.

Anthrax occurs in three forms. **Cutaneous anthrax** causes a skin lesion that progresses to a papule, then a vesicle, and then becomes a depressed ulcerated area with a black eschar (scab). A relatively rare form of anthrax is **gastrointestinal anthrax,** which causes ulcers in the gastrointestinal tract and high mortality. The most dangerous is **inhalational (pulmonary) anthrax.** Endospores are inhaled into the lungs, and then the bloodstream is infected. Death is usually the result of septic shock.

Gangrene.   Anaerobic conditions occur in tissue if a wound interrupts the blood supply (*ischemia*). In turn, ischemia leads to death (*necrosis*) of tissue, or **gangrene.** Substances from dead cells provide nutrients for bacteria such as the endospore-forming anaerobes of the genus *Clostridium*. Gas gangrene, especially in muscle tissue, occurs when *Clostridium perfringens* ferments carbohydrates in wound tissue and produces carbon dioxide and hydrogen gases. This microorganism produces exotoxins and enzymes that further interfere with blood supply and otherwise favor the spread of the infection. This condition spreads the area of necrosis; ultimately severe toxemia and death may ensue. Treatments for gas gangrene include *debridement* (surgical removal of tissue) or amputation. **Hyperbaric chambers,** in which the patient is subjected to an oxygen-rich atmosphere that interferes with the growth of the obligately anaerobic clostridia, are particularly useful for abdominal gangrene. Antibiotics such as penicillin are effective.

Systemic Diseases Caused by Bites and Scratches

Animal bites inflicted on humans often become infected. *Pasteurella multocida*, which normally is more likely to cause sepsis in animals, can cause infections in humans. The infection may be localized at the site of the wound or may progress to pneumonia and life-threatening sepsis. Anaerobes such as *Clostridium*, *Bacteroides*, and *Fusobacterium* can also infect deep bites. Infections resulting from human bites are often troublesome.

Cat-Scratch Disease.   Cat-scratch disease, usually resulting from minor scratches, may cause swollen lymph nodes and prolonged fever. The pathogen is a gram-negative bacterium, *Bartonella henselae*.

Rat-Bite Fever.   Two bacterial pathogens are involved as causes of **rat-bite fever,** which occur as two distinctly different versions. The more common pathogen is *Streptobacillus moniliformis*, which causes *streptobacillary rat-bite fever*. (When the pathogen is ingested, the disease is *Haverhill fever*.) The other pathogen is *Spirillum minus*, which causes spirillar fever and is more likely to occur from wild rodent bites. (In Asia, where most cases occur, it is called *spirillar fever*, or *sodoku*.) Bites from handling laboratory rats might cause either disease.

Vector-Transmitted Diseases

Plague.   In the Middle Ages, **plague** was known as the Black Death because of the characteristic blackish areas on the skin caused by hemorrhages. The name **bubonic plague** derives from the buboes that form in the lymph nodes in the groin and armpit regions. The disease is caused by a gram-negative rod, *Yersinia pestis*, that is able to survive and even increase in number inside phagocytic cells. Normally the disease is transmitted by the rat flea from rat reservoirs, but in the western United States the disease is endemic in wild rodents.

A particularly dangerous development is **pneumonic plague**, which arises as a complication in cases of **septicemic plague** when the bacteria are carried by the blood to the lungs. Because of airborne droplet infection, pneumonic plague is highly infective. Untreated plague has a mortality rate of 50–75%, and for pneumonic plague the rate approaches 100%. Antibiotics such as streptomycin and tetracycline are useful for prophylactic protection of exposed people and for treatment.

Relapsing Fever.   All members of the genus *Borrelia* (a spirochete) cause **relapsing fever.** The disease is transmitted by soft ticks that feed on rodents. A high fever is accompanied by jaundice and rose-colored skin spots. The fever subsides after 3 to 5 days, but several relapses, each caused by a different antigenic type of the pathogen, may occur.

Lyme Disease (Lyme Borreliosis).   **Lyme disease** is caused by the bite of a tick, *Ixodes pacificus* on the Pacific coast and mostly *Ixodes scapularis* in the rest of the United States. A skin lesion spreads from the site of the bite, clearing in the center. Later complications are arthritis and, occasionally, heart and neurological abnormalities. A spirochete, *Borrelia burgdorferi,* is the pathogen. The disease can be successfully treated with antibiotics.

Ehrlichiosis and Anaplasmosis.   Human monocytotropic ehrlichiosis (HME) is caused by *Ehrlichia chafeensis,* a rickettsia-like, obligately intracellular bacterium. Within cells it forms aggregates of bacteria (*morulae*). The vector is the Lone Star tick, and the reservoir is the white-tailed deer. A similar tickborne disease, **human granulocytic anaplasmosis (HGA),** was formerly called *human granulocytic ehrlichiosis.* The pathogen is an obligately intracellular bacterium, *Anaplasma phagocytophilium.* The vector is a tick, *Ixodes scapularis.*

Typhus.   Several related diseases are caused by rickettsias that are spread by arthropod vectors such as mites, ticks, mosquitoes, fleas, and lice.

Epidemic Typhus.   **Epidemic typhus,** caused by *Rickettsia prowazekii,* is louseborne. The pathogen is excreted in the louse feces and rubbed into the wound when the bitten host scratches the bite. Symptoms are a high and prolonged fever, stupor, and a rash of red spots caused by subcutaneous hemorrhaging. Mortality rates can be high. A vaccine is available.

Endemic Typhus.   The related, but less severe disease, **endemic murine typhus** occurs sporadically rather than in epidemics. The most common hosts are rodents such as rats and squirrels. The pathogen, *Rickettsia typhi,* is transmitted by rat fleas.

Spotted Fevers.   The best-known spotted fever in the United States is **tickborne typhus,** or **Rocky Mountain spotted fever,** which is most prevalent in the southeastern United States and Appalachia. The rickettsial pathogens are transmitted to humans by ticks. In the east, dog ticks are mainly responsible, and in the Rocky Mountains, wood ticks. The rickettsias are often transmitted among ticks by *transovarian passage,* infecting tick eggs as they are produced. Clinically, the disease resembles the typhus fevers.

VIRAL DISEASES OF THE CARDIOVASCULAR AND LYMPHATIC SYSTEMS

Burkitt's Lymphoma.   The Epstein-Barr virus (a herpesvirus; the official name is *human herpesvirus 4*) causes two forms of cancer: **Burkitt's lymphoma** and **nasopharyngeal carcinoma.**

Infectious Mononucleosis.   **Infectious mononucleosis,** or *mono,* is caused by the Epstein-Barr virus. The disease is characterized by enlarged and tender lymph nodes, enlarged spleen, fever, sore throat, headache, nausea, and general weakness. It probably is most commonly spread by saliva. As a result of the infection, mononuclear white blood cells proliferate in a manner resembling leukemia. It is rarely fatal. Diagnosis involves detection of nonspecific heterophile antibodies and immunofluorescent microscope techniques against the Epstein-Barr virus itself. Recovery results in a good immunity, but patients may continue to shed the virus because of latent infection.

Cytomegalovirus Infections

The cytomegalovirus (CMV) is a very large herpesvirus that remains latent in white blood cells, where it is not much affected by the immune system. The virus is spread by contact such as kissing and contact with body fluids, as well as by blood transfusions and transplanted tissues. Infected cells cause the appearance of inclusion bodies known as "owl's eyes"and are useful in diagnosis. Almost all of us eventually become infected. The infected cell swells (*cytomegaly;* that is, cell enlargement) and is the basis for the viral name. The official name is *human herpesvirus 5.* **Cytomegalic inclusion disease** is a disease of newborns responsible for certain congenital abnormalities that include mental retardation or hearing loss. The cause is a primary (not recurrent) infection by CMV during pregnancy.

The infection by CMV causes few or no symptoms in otherwise healthy people. However, it can cause a serious eye infection, **cytomegalovirus retinitis.** The virus can also cause a devastating infection of the fetus when a primary (not recurrent) infection occurs in a pregnant woman.

Chikungunya Fever.   A tropical disease that is now causing concern in some temperate areas is **chikungunya fever**; it is often referred to simply as *chik.* The symptoms are a high fever and crippling joint pains that tend to persist, sometimes for months. The vector is the *Aedes* mosquito, primarily *A. aegypti.* It is also spread by *A. albopictus,* which is an aggressive daytime biter and can survive in colder climates.

Classic Viral Hemorrhagic Fevers

Most hemorrhagic fevers are zoonotic diseases. Those that are medically familiar are "classic" hemorrhagic fevers. **Yellow fever,** transmitted by the bite of the mosquito *Aedes aegypti,* is first among this group. Mortality is high, mostly from liver damage. It is endemic in much of Africa and tropical Central and South America. Monkeys are animal reservoirs of the virus. **Dengue** is also transmitted by *Aedes aegypti.* It is endemic in the Caribbean and is characterized by fever and severe muscle and joint pain. It is seldom fatal, but the symptoms are painful, giving the disease the alternative name of *breakbone fever.* Dengue also occurs as **dengue hemorrhagic fever.** This is often rapidly fatal.

Emerging Viral Hemorrhagic Fevers

Certain other hemorrhagic diseases are considered new or **emerging hemorrhagic fevers.** All of them result in profuse bleeding, both internally and externally—contact with this blood spreads the disease. Mortality is extremely high. **Lassa fever,** an African disease, is caused by an arenavirus shed in rodent urine. Similar viruses cause similar diseases in South America called **Argentine** and **Bolivian hemorrhagic fevers.** In the United States, an arenavirus, **Whitewater Arroya virus,** has a reservoir in Californian wood rats.

Filoviruses, the **Marburg virus** (or **Green Monkey Virus**) and the **Ebola hemorrhagic fever** are the cause of two other severe hemorrhagic fevers. Their natural reservoir in Africa is unknown but is thought to be in small mammals.

Hantavirus Pulmonary Syndrome

Best known in the southwestern United States, **hantavirus** infections are widespread in the world. It is a frequently fatal pulmonary infection, causing the lungs to fill with fluids. Elsewhere in the world, hantaviruses are best known as **hemorrhagic fever with renal syndrome.** They are transmitted by inhalation of hantaviruses (a bunyavirus) in dried excretions from infected rodents.

PROTOZOAN DISEASES OF THE CARDIOVASCULAR AND LYMPHATIC SYSTEMS

Chagas' Disease (American Trypanosomiasis).   An example of a protozoan disease of the cardiovascular system is **Chagas' disease (American trypanosomiasis).** The causative agent is *Trypanosoma cruzi,* a flagellated protozoan. The disease occurs in a small area of the southern United States and throughout Mexico, Central America, and South America. The arthropod vector is the reduviid bug. The disease is transmitted to humans when insect bites are contaminated by the insect's feces. Reservoirs for *T. cruzi* are various wild animals. Most damage is caused by inflammatory reactions after transport by the blood to the liver, spleen, heart, and so on. One symptom is loss of involuntary muscular contractions in the esophagus and gastrointestinal tract. These organs become grossly enlarged—*megaesophagus* and *megacolon.* The disease is most dangerous to children, in whom it is most likely to damage the heart.

Toxoplasmosis.   The protozoan *Toxoplasma gondii,* which causes the disease of the blood and lymph vessels called **toxoplasmosis,** forms spores (oocysts). Cats are essential for reproduction of the organisms, because the sexual phase appears to occur only in their intestinal tract. Shed in cat feces, the protozoa infect a new host, usually by ingestion. *Oocysts* contain *sporozoites* that invade host cells and form *tachyzoites,* which reproduce. In the chronic phase, the parasite is in *tissue cysts* containing *bradyzoites.* Loss of immune function can result in reactivation. Eating undercooked meat of an animal infected in this manner or inhalation of dried cat feces may cause human infections. In pregnant women, infection of the fetus across the placenta can cause drastic damage to the fetus. A serological test using fluorescent antibody or an ELISA can be used in diagnosis; however, recently developed PCR tests have revolutionized prenatal diagnosis.

Malaria.   **Malaria** is characterized by chills and fever and, often, vomiting and severe headaches. Symptoms typically occur in cycles of 1 to 3 days. It is caused by the spore-forming protozoa of the genus *Plasmodium.* Malaria is found anywhere the protozoan parasite is present in human hosts and wherever the *Anopheles* mosquito is found. (It can also be spread by blood transfusions or contaminated syringe needles.) It is a serious and increasing problem in tropical Asia, Africa, Central America, and South America.

There are four major forms of malaria: *Plasmodium falciparum* (most dangerous and geographically widespread), *Plasmodium vivax* (also widely distributed), *Plasmodium malariae,* and *Plasmodium ovale* (both of which have a lower infection rate and are geographically restricted, milder diseases). The infecting plasmodium, in the form of a *sporozoite,* enters liver cells, where it undergoes schizogony (asexual reproduction). Later, *merozoites* are released into the bloodstream to infect red blood cells. When more merozoites are released from rupturing red blood cells, paroxysms of chills and fever result. Anemia from loss of red blood cells and excessive enlargement of the liver and spleen are added complications. Some merozoites become male and female *gametocytes* and are picked up by feeding mosquitoes. These then pass through a sexual cycle in the mosquito to produce sporozoites.

Laboratory diagnosis is made by identifying the protozoa in blood smears. Sufferers from sickle cell disease and symptomless carriers of the gene (sickle cell trait), tend to be resistant to malaria. Quinine derivatives such as primaquine, mefloquine, and chloroquine are used in chemotherapy and prevention. Alternatives are malarone and derivatives of artemisinin. The WHO recommends artemisinin combination therapies (ACT) for treatment.

Leishmaniasis

Leishmaniasis is a widespread and complex disease that exhibits several clinical forms. The protozoan pathogens are of about 20 species, often divided into three groups for simplicity. Transmission is by several species of sand fly. *Leishmania donovani* causes **visceral leishmaniasis,** called *kala azar* in Asia, Africa, and southeast Asia, where parasites invade internal organs. This is a debilitating disease that, untreated, is fatal within a couple of years. *Leishmania tropica* causes a **cutaneous leishmaniasis** sometimes called

Oriental sore. Infection causes a lesion that ulcerates and eventually leaves a prominent scar. *Leishmania braziliensis* causes **mucocutaneous leishmaniasis,** so called because it affects mucous membranes as well as skin. It causes disfiguring destruction of tissues of the nose, mouth, and upper throat. It is often called *American leishmaniasis* because it is common in tropical areas of this hemisphere.

Treatments are drugs containing antimony or sodium stibogluconate. A new oral drug, miltefosine, promises to be more effective.

Babesiosis.   **Babesiosis** is tickborne and caused by the protozoan *Babesia microti.* It is endemic in the same areas as are Lyme disease and human granulocytic ehrlichiosis, and coinfection often occurs. Usually subclinical, in immunocompromised patients it is often serious, resembling malaria.

HELMINTHIC DISEASES OF THE CARDIOVASCULAR AND LYMPHATIC SYSTEMS

Schistosomiasis.   **Schistosomiasis** is caused by a flatworm helminth, a fluke. The disease is a major world health problem. Waters become contaminated with ova excreted with human wastes. The motile larval form of *Schistosoma,* the *miracidium,* is released from the ova and enters a species of snail. Eventually, the pathogen emerges from the snail as a fork-tailed *cercaria.* This penetrates the skin of the human host and is carried by the bloodstream to the liver or urinary bladder (depending on the species of fluke), where it matures into an adult form, producing a new supply of ova. Defensive reactions by the body to the presence of ova cause tissue damage (*granulomas*). The disease causes damage, such as abscesses and ulcers, to the liver and to other organs, such as the lungs or urinary system.

Schistosoma haematobium causes urinary schistosomiasis because it causes an inflammation of the urinary bladder wall. Similarly, *S. japonicum* and *S. mansoni* cause intestinal inflammation. *S. mansoni* is endemic in South America and the Caribbean, including Puerto Rico.

Diagnosis is mainly by detection of flukes or ova in fecal or urine specimens. Praziquantel and oxamniquine are used in chemotherapy.

Swimmer's Itch.   **Swimmer's itch** sometimes troubles swimmers in lakes in the northern United States. This is an allergic reaction to cercaria of a similar wildfowl parasite that does not mature in humans.

SELF-TESTS

In the matching section, there is only one answer to each question; however, the lettered options (a, b, c, etc.) may be used more than once or not at all.

I. Matching

____ 1. The largest vessels carrying oxygenated blood from the heart.

____ 2. The walls are only one cell thick; they aid in the exchange of materials.

____ 3. The venous equivalent of arterioles.

a. Capillaries

b. Venules

c. Arterioles

d. Veins

e. Arteries

II. Matching

____ 1. Blood cells important in phagocytosis and antibody production.

____ 2. A lymphoid organ.

____ 3. Small, oval structures in the lymphatic system; sites of considerable defensive activity by the body.

____ 4. Plasma that bathes tissue cells after their passage through capillary walls.

____ 5. Blood cells that carry oxygen.

a. Platelets

b. Erythrocytes

c. Leukocytes

d. Plasma

e. Lymph

f. Interstitial fluid

g. Lymph nodes

h. Tonsils

III. Matching

____ 1. A toxic, inflammatory condition arising from the spread of bacteria or bacterial toxins from a focus of infection.

____ 2. Of autoimmune origin due to group M proteins of streptococci.

____ 3. Swollen lymph nodes.

____ 4. Heart infection that develops rapidly, damages valves; usually caused by *Staphylococcus aureus.*

____ 5. Drop in blood pressure due to gram-negative sepsis.

____ 6. Characterized by red streaks on skin from the site of infection.

a. Lymphangitis

b. Sepsis

c. Pericarditis

d. Acute bacterial endocarditis

e. Buboes

f. Rheumatic fever

g. Septic shock

IV. Matching

___ 1. Probably transmitted by saliva.

___ 2. Childbirth fever.

___ 3. Often transmitted by contact with small animals such as rabbits.

___ 4. Undulant fever, at one time transmitted by ingestion of contaminated milk, is now mostly transmitted by contact with animal carcasses.

___ 5. Caused by a spore-forming rod that is often present in the soil.

___ 6. The cat is essential in the reproductive cycle and the transmission of the causative organisms.

___ 7. Caused by a protozoan that forms oocysts.

___ 8. Heterophil antibodies are used in diagnosis.

___ 9. Caused by the Epstein-Barr virus.

___ 10. Transmitted by sand flies.

___ 11. Kala azar.

___ 12. Caused by CMV.

a. Brucellosis

b. Toxoplasmosis

c. Anthrax

d. Rocky Mountain spotted fever

e. Tularemia

f. Puerperal sepsis

g. Infectious mononucleosis

h. Leishmaniasis

i. Cytomegalic inclusion disease

V. Matching

___ 1. The bite of a tick transmits a spirochete, *Borrelia burgdorferi*.

___ 2. A swimming stage called a cercaria is an essential part of the life cycle of the pathogen.

___ 3. A rickettsial disease transmitted by dog ticks or wood ticks.

___ 4. A rickettsial disease transmitted by a louse.

___ 5. A rickettsial disease transmitted by a rat flea.

___ 6. Chagas' disease.

___ 7. A spore-forming protozoan is the cause.

___ 8. Saint Vitus' dance.

a. Gangrene

b. Lyme disease

c. Epidemic typhus

d. Rocky Mountain spotted fever

e. Endemic murine typhus

f. Yellow fever

g. Plague

h. Sydenham's chorea

i. American trypanosomiasis

j. Malaria

k. Schistosomiasis

VI. Matching

____ 1. Tickborne protozoan disease.

____ 2. Mosquito-transmitted hemorrhagic fever.

____ 3. Pulmonary infection transmitted by inhalation of dried rodent urine and feces.

____ 4. Tickborne bacterial disease.

____ 5. Transmitted by contact with infected blood; filoform virus.

a. Babesiosis

b. Hantavirus

c. Ebola hemorrhagic fever

d. Lyme disease

e. Yellow fever

Fill in the Blanks

1. The surgical removal of tissue, short of amputation, is called _____.

2. The fluid portion of the blood is called _____.

3. A general name for a white blood cell is _____.

4. When a gram-negative bacterium lyses, it releases part of its cell walls as harmful

_____.

5. Group M proteins are associated with the bacterial genus _____.

6. *Brucella suis* is most likely to infect people coming into contact with animals such as

_____.

7. Sydenham's chorea is a complication of _____.

8. Burkitt's lymphoma is caused by the same virus that causes _____.

9. The official name of the Epstein-Barr virus is human herpesvirus _____.

10. Infections caused by obligate anaerobes such as *Clostridium perfringens* are sometimes treated by putting the patients in _____ chambers.

11. Many years ago, Semmelweis showed how proper hygiene and disinfection of hands and instruments could prevent _____ in maternity wards.

12. When malaria is transmitted by the bite of a mosquito, the parasite stage that is injected into the host is a(n) _____.

13. The most dangerous type of malaria is caused by *Plasmodium* _____.

14. Snails are essential to the life cycle of the disease organism causing _____.

15. Cats are essential to the life cycle of the disease organism causing _____.

16. The bite of the *Anopheles* mosquito causes _____.

17. The form of schistosomiasis known as urinary schistosomiasis is caused by

S. _____.

Critical Thinking

1. At one time the most common cause of sepsis was gram-positive bacteria such as *Staphylococcus aureus*. Although such infections have recently increased because of invasive medical procedures, today the most common and the most dangerous form is gram-negative sepsis. Speculate on why gram-positive sepsis is not as common today as gram-negative sepsis. Also, why is gram-negative sepsis so dangerous—relatively more dangerous than gram-positive?

2. After examining a patient with acute bacterial endocarditis, her physician concludes that the probable cause was a chronic tooth infection. What is the probable etiologic agent? How can dental infections lead to endocarditis?

3. Why is it so important to treat cases of strep throat?

4. What is a hyperbaric chamber, and what role does it play in the treatment of gas gangrene?

ANSWERS

Matching

I.   1. e   2. a   3. b
II. 1. c   2. h   3. g   4. f   5. b
III. 1. b   2. f   3. e   4. d   5. g   6. a
IV. 1. g   2. f   3. e   4. a   5. c   6. b   7. b   8. g   9. g   10. h   11. h   12. i
V.   1. b   2. k   3. d   4. c   5. e   6. i   7. j   8. h
VI. 1. a   2. e   3. b   4. d   5. c

Fill in the Blanks

1. debridement   2. plasma   3. leukocyte   4. endotoxins   5. streptococci   6. swine   7. rheumatic fever   8. infectious mononucleosis   9. 4   10. hyperbaric   11. puerperal fever   12. sporozoite 13. *falciparum*   14. schistosomiasis   15. toxoplasmosis   16. malaria   17. *haematobium*

Critical Thinking

1. Gram-positive bacteria are now much more readily controlled by antibiotics than are gram-negative bacteria. The availability of antibiotics is the primary reason that we see less of the gram-positive type of sepsis. Gram-negative bacteria release endotoxins when they die. Antibiotics that kill them can even temporarily worsen the condition.

2. The causative agent was probably *Staphylococcus aureus, Streptococcus pyogenes,* or another organism associated with dental infections. Microorganisms that are released into the blood during tooth infections can find their way to the heart, resulting in endocarditis.

3. Untreated cases of strep throat may lead to rheumatic fever. Approximately 3% of untreated cases of strep throat lead to rheumatic fever.

4. A hyperbaric chamber is a chamber that contains a pressurized, oxygen-rich atmosphere. The oxygen saturates the infected tissues and prevents the growth of obligate anaerobes such as *Clostridium*. This procedure is used to treat abdominal gas gangrene.

24 Microbial Diseases of the Respiratory System

STRUCTURE AND FUNCTION OF THE RESPIRATORY SYSTEM

The **upper respiratory system** consists of the nose and throat, including the middle ear and *auditory tubes* (Figure 24.1). Mucus traps dust and microorganisms, and cilia assist in moving these to the throat for elimination. The **lower respiratory system** consists of the *larynx, trachea, bronchial tubes,* and *lungs* (Figure 24.2). *Alveoli* are air sacs in lung tissue where oxygen and carbon dioxide are exchanged, and the double-layered membrane around the lungs is the *pleura*. Macrophages in the alveoli destroy many pathogens, and IgA antibodies in mucus, saliva, and tears aid in resistance.

NORMAL MICROBIOTA OF THE RESPIRATORY SYSTEM

A number of potentially pathogenic organisms are part of the microbiota of the upper respiratory system, but they seldom cause illness. The lower respiratory system is almost sterile.

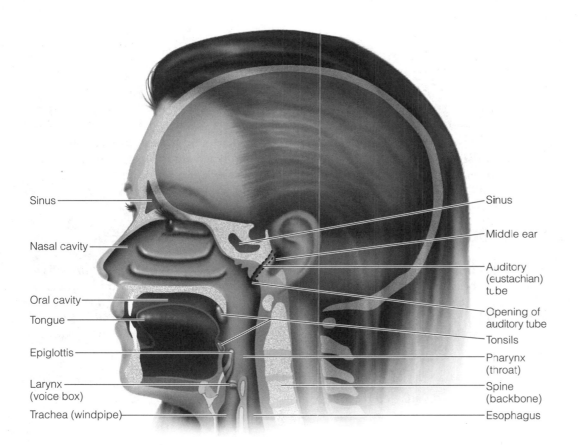

Sinus

Nasal cavity

Oral cavity

Tongue

Epiglottis

Larynx
(voice box)

Trachea (windpipe)

Sinus

Middle ear

Auditory
(eustachian)
tube

Opening of
auditory tube

Tonsils

Pharynx
(throat)

Spine
(backbone)

Esophagus

Figure 24.1   Structures of the upper respiratory system.

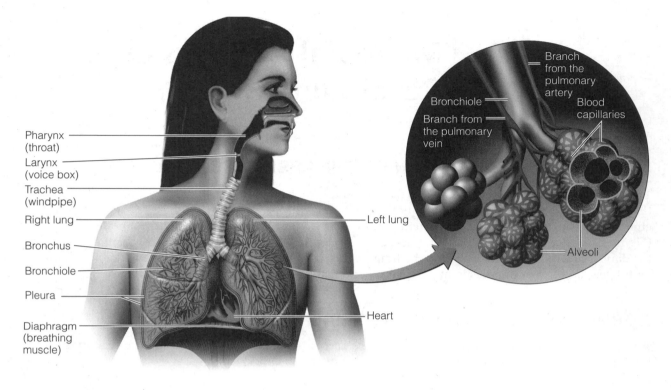

Figure 24.2 **Structures of the lower respiratory system.**

MICROBIAL DISEASES OF
___THE UPPER RESPIRATORY SYSTEM___

Among the diseases of the upper respiratory system are **pharyngitis,** a sore throat; **laryngitis,** an infected larynx (see Figure 24.1); **tonsillitis,** inflamed tonsils; and **sinusitis,** an infected sinus. **Epiglottitis,** an inflammation of a flaplike structure of cartilage that prevents material from entering the larynx, is a possibly life-threatening disease. It usually is caused by *Haemophilus influenzae* type b.

BACTERIAL DISEASES OF THE UPPER RESPIRATORY SYSTEM

Streptococcal Pharyngitis (Strep Throat).   *Streptococcus pyogenes*, the group A streptococcus (GAS), is an important cause of **streptococcal pharyngitis.** Symptoms are inflammation and fever; complications are tonsillitis or otitis media. GAS pathogenicity is enhanced by virulence factors such as *streptokinases* (which lyse fibrin clots and allow spreading) and *streptolysins* (which are cytotoxic to tissue cells, red blood cells, and protective leukocytes). Diagnosis makes use of latex agglutination tests that may require as little as 10 minutes to perform. Treatment is usually with penicillin or erythromycin.

Scarlet Fever.   Some strains of *S. pyogenes* have been lysogenized by a phage and produce an *erythrogenic toxin* that causes **scarlet fever.** The skin affected by the toxin turns pinkish red, and a high fever occurs. Affected skin may peel.

Diphtheria.   *Corynebacterium diphtheriae*, a gram-positive, non–spore-forming pleomorphic rod, causes **diphtheria.** Dividing cells often fold into V- and Y-shaped figures. Many well people are symptomless carriers.

DTaP vaccine (T = tetanus, aP = acellular pertussis) stimulates immunity to the diphtheria toxin. Exotoxin, whose production is controlled by a gene acquired from a lysogenic phage, is carried by the bloodstream and interferes with protein synthesis in the host's cells. A membrane composed of fibrin and dead human and bacterial cells (diphtheria, meaning "leather") may form in the throat and block air passage.

Laboratory diagnosis can be made by bacterial isolation on selective and differential media. Antitoxins are essential in treatment but must be administered before toxin enters the cells. Antibiotics are also useful but are not a substitute for antitoxin.

Cutaneous diphtheria, characterized by skin lesions, is fairly common in tropical countries. In the United States, it affects mainly lower socioeconomic groups.

Otitis Media.   Infections of the upper respiratory system may cause **otitis media,** or middle ear infection, leading to earache. The symptoms are due to pus formation and pressure against the eardrum. *Staphylococcus aureus, Streptococcus pneumoniae, Haemophilus influenzae,* and *Moraxella catarrhalis* are the most commonly identified causes of this disease. A vaccine directed at prevention of pneumococcal pneumonia (*S. pneumoniae*) lessens the incidence of otitis media in children.

VIRAL DISEASE OF THE UPPER RESPIRATORY SYSTEM

The Common Cold.   The **common cold** is caused mainly by viruses classified as coronaviruses or rhinoviruses. There are more than 200 viruses that cause colds, so immunity to any one is of limited use. Immunity develops in isolated populations as a result of IgA antibodies, and immunities accumulate with age. Laryngitis and otitis media are common complications. Symptoms can be alleviated, but there is no cure.

MICROBIAL DISEASES OF THE LOWER RESPIRATORY SYSTEM

A lower respiratory tract infection affecting the bronchi (see Figure 24.2) is **bronchitis** or **bronchiolitis.** A severe complication of bronchitis is **pneumonia.**

BACTERIAL DISEASES OF THE LOWER RESPIRATORY SYSTEM

Pertussis (Whooping Cough).   Infection by *Bordetella pertussis,* a gram-negative coccobacillus, can cause **pertussis (whooping cough).** Ciliary action is blocked by accumulations of dense masses of these bacteria in the trachea and bronchi. Eventually, *tracheal cytotoxin* causes loss of ciliated cells. The initial, **catarrhal,** stage resembles a common cold. The **paroxysmal** stage is characterized by attempts to clear accumulations by coughing. The bacteria may be cultured from a throat swab inserted through the nose on a wire while the patient coughs. Recovery results in good immunity. An acellular vaccine (DTaP) is available.

Tuberculosis.   *Mycobacterium tuberculosis,* a slow-growing rod that sometimes forms filaments, is the cause of **tuberculosis.** The organism is *acid-fast,* meaning it retains the carbol fuchsin dye of the Ziehl–Neelsen staining technique. The stain characteristically reflects the large lipid content of the cell wall, which also may make it resistant to drying, sunlight, and chemical disinfectants.

Tuberculosis is usually acquired by inhaling the tubercle bacillus. In the alveoli (see Figure 24.2), the bacilli are phagocytized by macrophages. These may destroy the pathogen; if they do not, the bacilli multiply within the macrophages. New macrophages are attracted to the site, and a **tubercle** (a small lump that is characteristic of tuberculosis) forms. Hypersensitivity reactions against bacillus-laden macrophages kill many of them, and a **caseous center** in the tubercle, surrounded by macrophages and lymphocytes, forms. As the caseous lesions heal, they become calcified and are called **Ghon complexes,** which show up on X-ray images. If the disease is not arrested, the caseous area enlarges—called **liquefaction**—eventually forming an air-filled **tuberculous cavity.** The bacilli now multiply rapidly, eventually leading to release of the microbes into the circulatory system. The disseminated infection, affecting many organs, is called **miliary tuberculosis.**

Mycobacterium bovis is the cause of **bovine tuberculosis,** which is usually spread by unpasteurized milk. This disease mainly affects bones and the lymphatic system. A vaccine, **BCG,** is a live attenuated culture of *M. bovis.*

Skin tests such as the Mantoux test involve injecting a purified protein derivative (PPD) of the tuberculosis bacterium. T cells sensitized by exposure to tuberculosis cause a delayed hypersensitivity reaction at the injection site. Some rapid diagnostic tests have become available. One is QuantiFERON-TB Gold (QFT-G), which detects interferon gamma (IFN-γ). Another is T-SPOT/TB, which enumerates the T cells that produce IFN-γ. A new polymerase chain reaction test, Xpert MTB/RIF, can diagnose tuberculosis by detecting *M. tuberculosis.*

The intracellular growth of tuberculosis shields it from antibiotics. Drugs such as rifampin, isoniazid, and ethambutol are used as treatment, usually in combinations to avoid resistance. Antibiotic resistance in *M. tuberculosis* is increasing. Resistance to the first-line drugs, isoniazid and rifampin, is termed *multi-drug-resistant (MDR)* tuberculosis; additional resistance to at least three of the six main classes of second-line drugs is called *extensively drug-resistant (XDR)* tuberculosis.

Tuberculosis, which the body combats by cell-mediated immunity, is a frequent complication of AIDS. Other mycobacterial diseases affect AIDS patients. Most are a group of related organisms called the *M. avium-intracellulare* complex.

Bacterial Pneumonias

The term *pneumonia* is applied to many pulmonary infections.

Pneumococcal Pneumonia.   **Pneumococcal pneumonia,** caused by the pneumococcus *Streptococcus pneumoniae,* is the classic form of pneumonia. It typically causes an infection of the lobes of the lung, *lobar pneumonia.* The organism can be distinguished from other similar organisms by inhibition of its growth near a disk of optochin, or by bile solubility. Serological typing is useful. A vaccine is used mostly to immunize elderly or debilitated individuals. Penicillin is the drug of choice in treatment.

Haemophilus influenzae Pneumonia.   Clinically, *Haemophilus influenzae* pneumonia is similar to pneumonia caused by *S. pneumoniae.* A Gram stain often helps in differentiation; *H. influenzae* is gram-negative.

Mycoplasmal Pneumonia (Primary Atypical Pneumonia).   Pneumonias typically have a bacterial etiology. If a bacterial agent is not isolated, the pneumonia is considered *atypical.* The most common atypical type (**primary atypical pneumonia**) is caused by a bacterium with no cell wall, *Mycoplasma pneumoniae,* which does not grow on routine isolation media. The bacterium can, however, be grown and isolated on horse serum and yeast extract medium, on which the minute colonies may have a "fried-egg"appearance. The disease is relatively mild and is sometimes called **walking pneumonia.** Tetracyclines are often effective in treatment.

Legionellosis.   **Legionellosis (Legionnaires' disease),** characterized by high fever and cough, is caused by a gram-negative rod, *Legionella pneumophila.* The bacterium may grow in the water of air-conditioning cooling towers or in the hot-water lines of hospitals. Erythromycin and rifampin are the drugs of choice for treatment. *L. pneumophila* also causes a mild, self-limiting respiratory disease called **Pontiac fever.**

Psittacosis (Ornithosis).   Psittacine birds, such as parakeets and other parrots, carry this disease, hence the name **psittacosis,** or **ornithosis.** *Chlamydophila psittaci,* a gram-negative obligately intracellular bacterium, is the cause. It forms **elementary bodies,** which are the airborne infective agent. Within the cell they develop into larger **reticulate bodies** that reproduce by fission; eventually they become elementary bodies and leave the cell. The dried droppings of sick birds are the usual source of these.

Chlamydial pneumonia.   In recent years there have been outbreaks of pneumonia caused by *Chlamydophila pneumoniae.* It resembles mycoplasmal pneumonia and is transmitted person to person.

Q Fever.   *Coxiella burnetii,* an obligately intracellular bacterium, causes **Q fever.** It is a parasite of cattle ticks that eventually contaminates dairy products. An occasional complication is endocarditis, especially in cases of chronic Q fever.

Melioidosis.   **Melioidosis,** caused by the gram-negative *Burkholderia pseudomallei,* closely resembles glanders, a disease affecting horses. The pathogen is widely distributed in moist soils in southeast Asia and northern Australia. Sporadic cases occur elsewhere. Most commonly seen as pneumonia, it also causes abscesses, severe sepsis, and even encephalitis. Mortality rates are high.

VIRAL DISEASES OF THE LOWER RESPIRATORY SYSTEM

Viral Pneumonia.   **Viral pneumonia** is seldom confirmed by culture but is often assumed if mycoplasmal and bacterial pneumonia are ruled out. **Respiratory syncytial virus** is the most common cause, especially in children. It has a significant rate of mortality.

Influenza (Flu).   The **influenza** virus has an envelope with two types of projections: *hemagglutinin (HA) spikes* and *neuraminidase (NA) spikes.* There are fifteen subtypes of HA and nine of NA. Antibodies against influenza are directed mainly at these spikes. Variation in their composition alters the antigenic type of the virus and helps it evade established resistance in the human population. *Type A* of the virus is more widespread and severe, and *type B* is milder and geographically limited. **Antigenic drift** is the result of minor variations of antigenic makeup; **antigenic shifts** are major changes involving recombination (*reassortment*) of the eight RNA segments of the genome. Because both avian and mammalian influenza viruses grow in swine, these animals are where reassortment occurs. Vaccines are available to high-risk groups; they are usually multivalent, directed at the types then in circulation. These latter cause the designation—the Hong Kong virus of 1968, for example—to change and enable the virus to evade almost all established immunity. The ultimate goal is a flu vaccine that will protect against all flu strains. Perhaps it could use a *conserved protein* as a target antigen. These might be in the membrane of the infected cells or in the stalk of the viral hemagglutinin. Amantadine, zanamivir (*Relenza*), or oseltamivir phosphate (*Tamiflu*), if taken very early, will significantly reduce symptoms.

FUNGAL DISEASES OF THE LOWER RESPIRATORY SYSTEM

Histoplasmosis.   **Histoplasmosis** is caused by the dimorphic fungus *Histoplasma capsulatum,* and some cases may resemble tuberculosis. It is, however, much milder. It is found mainly in the Mississippi and Ohio River valleys. Bird droppings in the soil encourage growth of the fungus, which is spread by airborne conidia. Amphotericin B or itraconazole are used in chemotherapy.

Coccidioidomycosis.   A dimorphic fungus, *Coccidioides immitis,* whose sources are found in the soil of the American Southwest, is the cause of **coccidioidomycosis.** It is transmitted by inhalation of *arthrospores.* Amphotericin B and itraconazole are used to treat the disease, but most cases are subclinical; about 1% are progressive, resembling tuberculosis.

***Pneumocystis* Pneumonia.**   *Pneumocystis* **pneumonia** is caused by *Pneumocystis jirovecii* (formerly *P. carinii*). It has been uncertain whether the organism is a protozoan or a fungus. Recently, microbiologists have generally concurred that it is related to certain yeasts. The parasite may be transmitted by direct human contact; it causes disease mostly in immunosuppressed individuals, especially AIDS patients. Pneumocystis infection causes the alveoli to become filled with a frothy exudate; untreated infections are usually fatal. Trimethoprim-sulfamethoxazole is effective for treatment.

Blastomycosis (North American Blastomycosis). Another dimorphic fungus, *Blastomyces dermatitidis*, causes **North American blastomycosis.** It is found most often in the Mississippi River valley. Although most cases are subclinical, a rapidly progressive infection with cutaneous ulcers and abscess formation can occur. Amphotericin B or itraconazole is used in treatment.

Other Fungi Involved in Respiratory Disease

Other fungi, such as *Aspergillus fumigatis* and other species of *Aspergillus*, may cause **aspergillosis.** Compost piles are likely growth sites. Other mold genera also may cause pulmonary infections in susceptible individuals.

SELF-TESTS

In the matching section, there is only one answer to each question; however, the lettered options (a, b, c, etc.) may be used more than once or not at all.

I. Matching

___ 1. Mycoplasmal pneumonia.

___ 2. The causative organism is suspected to grow in the water-cooling towers of air-conditioning systems.

___ 3. Swine may be "mixing vessels" in which virulent mutations arise by reassortment.

___ 4. Production of the harmful exotoxin involved in this disease is controlled by a lysogenic phage.

___ 5. Pharyngitis.

___ 6. DTaP vaccine is useful in prevention.

___ 7. Resembles glanders, a disease affecting horses.

a. Diphtheria

b. Legionellosis

c. Primary atypical pneumonia

d. Influenza

e. Streptococcal sore throat

f. Melioidosis

II. Matching

____ 1. The infectious agent causing the disease is an elementary body formed intracellularly by a bacterial pathogen.

____ 2. Caused by *Coxiella burnetii*.

____ 3. Caused by a dimorphic fungus growing in soil contaminated by bird droppings.

____ 4. Caused by a dimorphic fungus widely distributed in airborne arthrospores in the American Southwest.

____ 5. Caused by rickettsial organisms parasitic of cattle ticks.

____ 6. A fungal disease found with considerable frequency in the Ohio and Mississippi River valleys.

____ 7. Also called ornithosis.

____ 8. Caused mainly by viruses classified as coronaviruses or rhinoviruses.

a. Common cold

b. Histoplasmosis

c. Coccidioidomycosis

d. Q fever

e. Psittacosis

III. Matching

____ 1. Caused by a lysogenized bacterium.

____ 2. Pertussis.

____ 3. The DTaP vaccine helps prevent this disease.

____ 4. Infection of the middle ear.

____ 5. Ghon complexes are one possible characteristic.

____ 6. The Mantoux test is useful in screening for this disease.

____ 7. Caused by an acid-fast organism.

____ 8. Ethambutol, isoniazid, and rifampin are useful in treatment.

____ 9. A disease of the respiratory tract near the larynx caused by *Haemophilus influenzae*.

a. Scarlet fever

b. Otitis media

c. Whooping cough

d. Tuberculosis

e. Acute epiglottitis

IV. Matching

___ 1. Double-layered membrane around the lungs.

___ 2. Disseminated, late form of tuberculosis.

___ 3. Air sacs in lung tissue where oxygen and carbon dioxide are exchanged.

___ 4. An early stage of pertussis.

___ 5. A nonrespiratory form of diphtheria.

a. Alveoli

b. Pleura

c. Miliary

d. Catarrhal

e. Cutaneous

Fill in the Blanks

1. The disease agent causing _____ is characterized by designations HA and NA.

2. Among diseases of the upper respiratory tract are pharyngitis, laryngitis, tonsillitis, sinusitis, and epiglottitis. Of these, only _____ is likely to be life-threatening.

3. *Mycobacterium avium-intracellulare* complex is a frequent opportunistic infection characteristic of the disease _____.

4. Immunity to the common cold is based mainly on antibodies of the _____ type.

5. The vaccine used to help prevent tuberculosis is called the _____ vaccine.

6. The Ziehl–Neelsen staining technique is valuable in diagnosis of _____.

7. Tuberculosis bacteria entering the lungs may be ingested and destroyed by alveolar _____.

8. Minor year-to-year variation in the antigenic makeup of the influenza virus is called antigenic _____.

9. Major changes in the antigenic makeup of the influenza virus that allow it to evade almost all previous immunity are called antigenic _____.

10. Laboratory diagnosis of _____ can be made by bacterial culture from a swab passed through the nose on a wire while the patient coughs.

11. Diagnosis for streptococcal sore throat may involve streaking a throat swab onto blood agar to observe the type of _____.

12. The organism *Coxiella burnetii* is the cause of the disease _____.

13. The terms *liquefaction* and *caseous center* are associated with a progressive case of _____.

14. A virulence factor in group A streptococci that enzymatically dissolves fibrin clots and promotes the spread of the pathogen is called _____.

Critical Thinking

1. Discuss the new approach that has been suggested to develop a vaccine to prevent the common cold.

2. What are the consequences of *Bordetella pertussis* growing on cilia of the trachea?

3. Why do certain ethnic groups such as American Indians, Asians, and Hispanics tend to have higher rates of tuberculosis than do people of northern European descent? (*Hint:* See Chapter 17.)

4. What are HA spikes and NA spikes, and how do they relate to reproduction of the influenza virus?

5. Why does the incidence of histoplasmosis and coccidioidomycosis sometimes increase after earthquakes in California?

ANSWERS

Matching

I. 1. c   2. b   3. d   4. a   5. e   6. a   7. f
II. 1. e   2. d   3. b   4. c   5. d   6. b   7. e   8. a
III. 1. a   2. c   3. c   4. b   5. d   6. d   7. d   8. d   9. e
IV. 1. b   2. c   3. a   4. d   5. e

Fill in the Blanks

1. influenza   2. epiglottitis   3. AIDS   4. IgA   5. BCG   6. tuberculosis   7. macrophages   8. drift
9. shift   10. pertussis   11. hemolysis   12. Q fever   13. tuberculosis   14. streptokinase

Critical Thinking

1. Because there are so many agents that can cause colds, it would be difficult or even impossible to make a vaccine effective against them all. Research has shown that most of these organisms use as few as two receptors to attach to the host's mucosa. For this reason, researchers have considered developing a vaccine that will elicit production of an antibody to block those receptors.

2. Although *Bordetella* does not invade tissue, it will grow on cilia of the trachea, impeding their action and resulting in loss of the ciliated cells and an accumulation of mucus. The patient coughs desperately to eliminate the mucus. In small children, the coughing may be so forceful as to break the ribs.

3. People of northern European descent have a certain level of innate immunity to tuberculosis because many generations have been exposed to the disease. American Indians, Asians, and Hispanics lack this innate immunity.

4. HA spikes and NA spikes are projections located on the outside of an influenza virus. HA spikes (hemagglutinin) allow the virus to recognize and attach to host cells before infecting them. NA spikes (neuraminidase) help the virus to separate from the host cell after intracellular reproduction.

5. Earthquakes produce dust that may harbor these organisms. This situation is especially true in years with an especially wet winter followed by an especially dry summer.

25 Microbial Diseases of the Digestive System

STRUCTURE AND FUNCTION OF THE DIGESTIVE SYSTEM

The first group of organs of the essentially tubelike **gastrointestinal (GI) tract,** or **alimentary canal,** is the *mouth, pharynx* (throat), *esophagus* (food tube), *stomach, small intestine,* and *large intestine* (Figure 25.1). **Accessory structures** consist of the *teeth, tongue, salivary glands, liver, gallbladder,* and *pancreas.* Except for the teeth and tongue, these lie outside the GI tract and produce secretions released into the tract. The action of secretions such as bile, pancreatic enzymes, saliva, and stomach and intestinal enzymes converts ingested foods into their end-products of digestion. By the time the liquefied food leaves the small intestine as *feces,* the absorption of sugars, fatty acids, and amino acids produced by digestion is almost complete. Structures such as lymph nodes and Peyer's patches are collectively called gut-associated lymphoid tissue (GALT).

NORMAL MICROBIOTA OF THE DIGESTIVE SYSTEM

In the mouth, saliva contains millions of bacteria per milliliter. The stomach and small intestine have relatively few because of their acidity. The large intestine has enormous numbers of bacteria, mostly anaerobes or facultative anaerobes. Among the antimicrobial defenses of the intestinal tract are *Paneth cells,* millions of which line the intestinal wall and phagocytize bacteria. They also produce antibacterial proteins called *defensins* and the antibacterial enzyme *lysozyme.*

BACTERIAL DISEASES OF THE MOUTH

Dental Caries (Tooth Decay)

The hard surface of the teeth accumulates masses of microorganisms and their products. This is called **dental plaque** and is an important factor in **dental caries,** or tooth decay. The most important *cariogenic* (caries-causing) bacterium is *Streptococcus mutans.* This, and certain other bacteria, convert sucrose sugar into a gummy polysaccharide called *dextran,* which localizes the bacteria and their acid production, forming a hole in the *enamel* of the tooth (Figure 25.2). Some antimicrobial protection is provided by lysozyme in saliva and the tissue exudate called *crevicular fluid.* Old plaque deposits are called *dental calculus* or *tartar.* Once the enamel has been penetrated, other bacteria are involved in decay of the *dentin* and the *pulp.*

Periodontal Disease

Many people who avoid tooth decay later lose their teeth to **periodontal** (surrounding the tooth) **disease.** As gums recede with age, they expose the *cementum.*

Gingivitis. The most common periodontal disease is **gingivitis,** infection of the *gingivae,* or gums, a condition characterized by bleeding of the gums while the person brushes the teeth.

Periodontitis. A chronic form of gingivitis is **periodontitis.** The gums are inflamed and bleed. Pus often forms in pockets surrounding the teeth. Tooth loss arises from destruction of bone and tissue supporting the teeth. *Porphyromonas* species are important factors.

   Acute necrotizing ulcerative gingivitis (Vincent's disease or **trench mouth)** is a serious mouth infection. Chewing causes pain, and the breath is foul. *Prevotella intermedia* is an important pathogen.

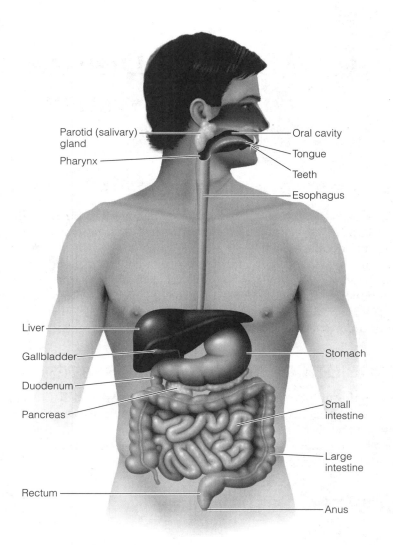

Parotid (salivary) gland
Pharynx
Oral cavity
Tongue
Teeth
Esophagus
Liver
Gallbladder
Duodenum
Pancreas
Stomach
Small intestine
Large intestine
Rectum
Anus

Figure 25.1   The human digestive system.

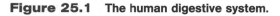

BACTERIAL DISEASES OF THE LOWER DIGESTIVE SYSTEM

Intoxication is a disease resulting from the ingestion of toxins preformed in food by microbial growth. **Infection** is a disease resulting from microbial growth in the tissues of the host. Severe **diarrhea** accompanied by blood and mucus is termed **dysentery.** The general term **gastroenteritis** describes inflammation of the stomach and intestinal mucosa. Typical symptoms are *abdominal cramps, nausea,* and *vomiting.* Oral rehydration therapy to replace water and certain chemicals is an effective treatment.

Staphylococcal Food Poisoning (Staphylococcal Enterotoxicosis)

Staphylococcus aureus produces an exotoxin (which, because it affects the digestive system, is called an **enterotoxin**) that causes nausea, vomiting, and diarrhea about 1 to 6 hours after ingestion. Serological type A enterotoxin is responsible for most of these cases. The enterotoxin is heat-resistant and withstands boiling for 30 minutes. *Coagulase-positive* staphylococci produce **coagulase,** an enzyme that coagulates blood plasma and is usually associated with production of the toxin. Staphylococci are resistant to drying, radiation, and osmotic pressures. Custard, cream pies, and ham are foods with a high incidence of this poisoning, largely because of the presence of sugar concentrations or curing agents, which suppresses nonstaphylococcal competition.

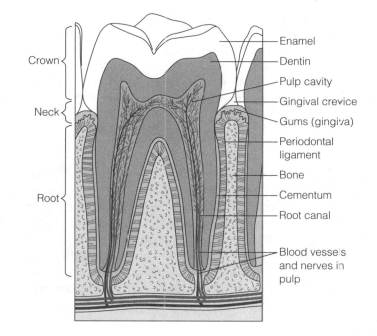

Figure 25.2 A healthy human tooth.

Food can be contaminated by staphylococci on fingers or in skin lesions. If allowed incubation time for toxin formation (**temperature abuse**), illness can result. Pathogenic staphylococci ferment mannitol, produce hemolysis, are coagulase-positive, form yellow colonies, and cause no obvious spoilage when growing on foods. The source of contamination can sometimes be determined by *phage typing* the bacteria.

Shigellosis (Bacillary Dysentery). Bacteria of the genus *Shigella—S. sonnei, S. dysenteriae, S. flexneri,* and *S. boydii*—all cause **shigellosis (bacillary dysentery).** *S. sonnei,* the most common species in the United States, causes a relatively mild disease, whereas *S. dysenteriae,* not common in the United States, causes a much more severe disease. It is known as the Shiga bacillus (its toxin is known as **Shiga toxin**).

Salmonellosis (*Salmonella* Gastroenteritis). Salmonella are not assigned to species but rather are differentiated by hundreds of serovars. They are considered members of two species, primarily *Salmonella enterica* (see Chapter 11). All serovars of *Salmonella* are considered pathogenic to some degree. The Kauffmann-White serological scheme to identify serovars assigns numbers and letters to different antigens: O (somatic or body), Vi (capsular), and H (flagellar). This scheme is used by most clinical laboratories. **Salmonellosis** results from ingestion of the organisms and growth of the cells in the intestinal tract. An incubation time of about 12 to 36 hours is normal. Symptoms are moderate fever, nausea, and diarrhea; there is a low mortality rate. Salmonellae are common inhabitants of the intestinal tract of many animals; poultry and egg products are particularly frequent sources of food contamination.

Typhoid Fever. *Salmonella typhi* is a very virulent serovar of *Salmonella* and the cause of **typhoid fever.** Incubation is about 2 weeks. Typical symptoms are high fever and malaise. Diarrhea appears only late in the disease—the second or third week. In severe cases there may be perforation of the intestinal wall and dissemination of the organism in the body. An occasional characteristic of the disease is **chronic carriers**, which continue to spread the disease in feces after recovery. This pathogen is not found in animals but is transmitted in the feces of humans. Antimicrobials such as third-generation cephalosporins are effective treatments.

Cholera. *Vibrio cholerae,* the cause of **cholera,** is spread by the fecal–oral route. The organism produces an enterotoxin (**cholera toxin**) in the intestines that causes rapid loss of body fluids and electrolytes, as well as violent vomiting. The serogroup O:1, biotype El Tor or eltor, of *V. cholerae* causes the epidemic form of the disease.

Treatment of any form of cholera is based mainly on replacement of lost fluids and electrolytes; antibiotics are not very effective. Untreated, the death rate can be as high as 50%; it is about 1% when properly treated.

Noncholera Vibrios

Cholera is a distinct disease resulting from the cholera toxin produced by *V. cholerae O:1* and *O:139.* Other serotypes of *V. cholerae* cause gastroenteritis, but it is usually milder. These latter organisms are indigenous in most Pacific and Gulf coastal waters. They are more likely to be invasive with symptoms such as bloody stools and fever. *Vibrio parahaemolyticus* causes a gastroenteritis similar to cholera but not as severe. Because it is present in coastal waters, outbreaks are usually associated with seafood. It is the most common gastroenteritis in Japan. The organism has a requirement for sodium, and growth media must contain 2–4% sodium chloride. *V. vulnificans* is mostly ingested with raw oysters from polluted estuaries. It is very dangerous to people who are immunocompromised or suffer from liver disease.

Escherichia coli Gastroenteritis

One of the most numerous microorganisms in the human intestinal tract, *Escherichia coli* is normally harmless, but some strains can cause *E. coli* **gastroenteritis.** Taxonomists consider *E. coli* to be indistinguishable from the genus *Shigella.*

Escherichia coli gastroenteritis is caused by toxin-secreting pathogenic strains adapted to invasion of intestinal epithelial cells. Eight or more pathogenic varieties (pathovars) of *E. coli* have been characterized. We will discuss five such pathovars.

Enteropathogenic *E. coli* (EPEC). Major cause of diarrhea in developing countries and potentially fatal to infants. Form pedestals beneath them at site of attachment.

Enteroinvasive *E. coli* (EIEC). Probably synonymous with *Shigella* (has same pathogenic mechanisms). Enter submucosa of intestinal tract through M cells in same manner and cause a *Shigella*-like dysentery.

Enteroaggregative *E. coli* (EAEC). Found only in humans; growth habit causes a "stacked-brick" configuration on tissue culture cells.

Enterohemorrhagic *E. coli* (EHEC). These also form pedestals, as with EPEC. Sometimes termed *Shiga-toxin producing E. coli* (STEC). Most common serotype is O157:h7; less well known strains include O121 and O104:H21.

Enterotoxigenic *E. coli* (ETEC). Secretes enterotoxins that causes diarrhea; often fatal for children under age 5. One enterotoxin resembles cholera toxin. Not invasive; remains in intestinal lumen.

Traveler's diarrhea. Most common bacterial cause is ETEC; the second most frequent isolate is EAEC. Other likely causes are *Salmonella, Shigella,* and *Campylobacter,* as well as various unidentified pathogens.

***Campylobacter* Gastroenteritis.** *Campylobacter* are gram-negative, spirally curved rods that are part of the intestinal flora of animals such as sheep and cattle. Recently they have been identified as one of the most common causes of diarrheal illness in the United States. Improved methods of isolation that recognize *Campylobacter*'s microaerophilic growth requirements have greatly increased isolations. *Campylobacter* **gastroenteritis** is characterized by fever, cramping abdominal pain, and diarrhea. An unusual, occasional complication is the neurological disease Guillain-Barré syndrome, a usually temporary paralysis.

Helicobacter Peptic Ulcer Disease. The organism *Helicobacter pylori* is the probable cause of stomach (peptic) ulcers. The microbe produces urease, which changes urea into alkaline ammonia, which neutralizes stomach acid. This permits its growth in the stomach mucosa, where it leads to inflammation and eventually ulceration. Effective treatments include antibiotics, including metronidazole. Pepto-Bismol also is often part of the drug regimen.

Yersinia Gastroenteritis. The symptoms of *Yersenia* **gastroenteritis,** also called **yersiniosis,** are diarrhea, fever, headache, and abdominal pain severe enough to cause frequent misdiagnosis as appendicitis. It is caused by *Yersinia enterocolitica* and *Y. pseudotuberculosis,* gram-negative inhabitants of the intestines of many domestic animals. Both organisms are capable of growth at refrigerator temperatures.

***Clostridium perfringens* Gastroenteritis.** A very common, if often unrecognized, form of food poisoning in the United States is diarrhea caused by *Clostridium perfringens* (*C. perfringens* **gastroenteritis**). This endospore-forming anaerobe is the same organism responsible for gas gangrene; it produces a number of damaging toxins. Cooked meat or stews that contain meat are a common source of the illness. Cooking lowers the oxygen level for clostridial growth, and the spores survive the heating. An exotoxin is produced as the organism grows in the intestinal tract, causing abdominal pain and diarrhea. Most cases are not clinically diagnosed. Symptoms appear in 8 to 12 hours.

***Clostridium difficile*–Associated Diarrhea.** *Clostridium difficile*–**associated diarrhea** is a common ailment that occurs mostly in hospitals and nursing homes. It is usually precipitated by extended use of broad-spectrum antibiotics that eliminate competing intestinal bacteria and permit the rapid proliferation of the toxin-producing *C. difficile.* Oral rehydration is important, and metronidazole, which targets the metabolism of anaerobes, is often a part of therapy.

***Bacillus cereus* Gastroenteritis.** *Bacillus cereus* is a gram-positive, spore-forming bacterium that causes *B. cereus* **gastroenteritis** by producing toxins that cause either a diarrheal episode resembling a *C. perfringens* gastroenteritis, or nausea and vomiting.

VIRAL DISEASES OF THE DIGESTIVE SYSTEM

Mumps. The parotid glands, just below and in front of the ear, are the particular target of the mumps virus. **Mumps** is characterized by inflammation and swelling of the parotid glands, accompanied by fever and pain. *Orchitis,* an inflammation of the testes, also may occur. Incubation time is about 16 to 18 days. Introduction of the measles–mumps–rubella (MMR) vaccine has caused a sharp decline in cases.

Hepatitis

Hepatitis A. Hepatitis A is a disease primarily affecting the liver. Caused by the **hepatitis A virus (HAV),** it is primarily transmitted by the *fecal–oral* route. The virus is shed in the feces. Symptoms are loss of appetite (*anorexia*), malaise, nausea, diarrhea, abdominal discomfort, fever, and chills lasting for 2 days to 3 weeks. *Jaundice,* a skin yellowing caused by liver infection, sometimes appears. Mortality is low, and the disease is often subclinical. The incubation period is 2 to 6 weeks, which complicates epidemiology. There is no chronic form. A vaccine is available.

Hepatitis B. Hepatitis B is caused by the **hepatitis B virus (HBV).** This is a completely different virus from HAV, as can be seen in Table 25.1, which summarizes several different types of hepatitis. HBV is most commonly transmitted by blood but can also be transmitted by many other body fluids, such as saliva, breast milk, and semen. Three distinct structures with the hepatitis B surface antigen (HB$_s$Ag) can be found in patients' blood. **Dane particles** probably represent intact hepatitis B virions, whereas **filamentous** and **spherical particles** are probably unassembled components of Dane particles (Figure 25.3). Many cases are asymptomatic, but the acute disease is generally more severe than hepatitis A. About 10% of patients enter a *chronic phase* and carry the virus indefinitely. Liver cancer is often a subsequent complication of chronic infections. A vaccine is available.

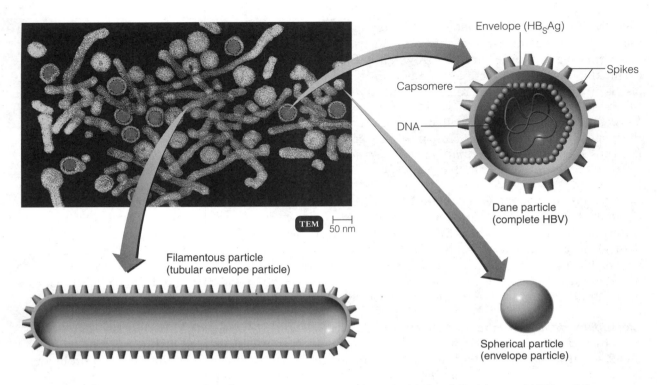

Figure 25.3 Hepatitis B virus (HBV). Micrograph and illustrations depict the distinct types of HBV particles discussed in the text.

Drugs available for retreatment include alpha interferon and peginterferon as well as nucleoside analogs such as lamivudine and others.

Hepatitis C. Screening of donated blood for HBV antibodies became so effective that hepatitis of this type became rare. However, another hepatitis named non-A, non-B appeared, which has since been named **hepatitis C,** caused by the **hepatitis C virus (HCV).** A serological test has been developed, in turn, to detect this form. Many cases are subclinical or mild, but the disease is much more likely than HBV to become chronic and cause liver damage. The treatment of choice is a combination of peginterferon alfa-2A and ribavirin.

Hepatitis D (Delta Hepatitis). A hepatitis virus discovered in 1977 was originally called the **delta antigen;** in people infected with HBV, this led to a high incidence of severe liver damage and mortality. Hepatitis D can occur as an acute (*coinfection form*) or a chronic (*superinfection form*) hepatitis. The severe symptoms are associated with superinfection. The **hepatitis D virus (HDV)** is a single strand of RNA with a capsid; it cannot cause infection. Only when it obtains an external envelope of HBsAg from coinfection with HBV is it infectious.

Hepatitis E. Another form of non-A, non-B hepatitis is spread by fecal–oral transmission, much like hepatitis A, which it clinically resembles. The agent is **hepatitis E virus (HEV).** It does not cause chronic liver disease but is responsible for a very high mortality rate in pregnant women.

Other Types of Hepatitis. There is evidence of blood-transmitted viruses that cause hepatitis known as **hepatitis F (HFV)** and **hepatitis G (HGV).** HGV is found worldwide and in the United States is more prevalent than HCV. HGV is closely related to HCV and is sometimes called GB virus C. It does not cause a significant disease condition.

See Table 25.1 for a summary of the characteristics of viral hepatitis.

Table 25.1 Characteristics of Viral Hepatitis

Disease	Pathogen	Symptoms	Incubation Period	Method of Transmission	Diagnostic Test	Treatment	Vaccine
Hepatitis A	Hepatitis A virus, Picornaviridae	Mostly sub-clinical; fever, headache; malaise, jaundice in severe cases; no chronic disease	2–6 weeks	Ingestion	IgM antibodies	Immunoglobulin	Inactivated virus. Post-exposure immune globulin
Hepatitis B	Hepatitis B virus, Hepadnaviridae	Frequently sub-clinical; similar to HAV, but no headache; more likely to progress to severe liver damage; chronic disease occurs	4–26 weeks	Parenteral; sexual contact	IgM antibodies	Interferon alfa and nucleoside analogues	Genetically modified vaccine produced in yeast
Hepatitis C	Hepatitis C virus, Flaviviridae	Similar to HBV, more likely to become chronic	2–22 weeks	Parenteral	PCR for viral RNA	Peginterferon and ribavirin	None
Hepatitis D	Hepatitis D virus, Deltaviridae	Severe liver damage; high mortality rate; chronic disease may occur	6–26 weeks	Parenteral; requires coinfection with hepatitis B	IgM antibodies	None	HBV vaccine is protective
Hepatitis E	Hepatitis E virus, Caliciviridae	Similar to HAV, but pregnant women may have high mortality; no chronic disease	2–6 weeks	Ingestion	IgM antibodies, PCR for viral RNA	None	HAV vaccine is protective

Viral Gastroenteritis

About 90% of viral gastroenteritis is caused by either rotaviruses or Norwalk-like viruses; the latter are collectively called noroviruses.

Rotavirus is probably the most common cause of viral gastroenteritis. More than 90% of children in the United States are infected by the age of 3—and in some cases the parents are infected as well. After an incubation period of 2 or 3 days, the patient suffers from a low fever, diarrhea, and vomiting, which persists for about a week.

Noroviruses were originally called *Norwalk viruses*, which are members of the caliciviruses. Humans become infected by fecal–oral transmission of food and water. The viruses are unusually persistent on environmental surfaces. After an incubation period of 18 to 48 hours, the patient suffers from 2 or 3 days of vomiting and diarrhea.

FUNGAL DISEASES OF THE DIGESTIVE SYSTEM

Some fungi produce toxins called *mycotoxins.*

Ergot poisoning, or **ergotism,** is caused by the toxin ergot, which is produced by the fungus *Claviceps purpurea*, which grows on grains.

Aflatoxin is produced by the mold *Aspergillus flavus*, which grows on grains and so on. It is toxic to livestock and is suspected to be a human carcinogen.

PROTOZOAN DISEASES OF THE DIGESTIVE SYSTEM

Giardiasis. A flagellated protozoan, *Giardia lamblia* (frequently known as *G. intestinalis* and occasionally as *G. duodenalis*), causes the prolonged diarrheal disease in humans called **giardiasis.** The disease is also characterized by malaise, nausea, weakness, weight loss, and abdominal cramps. Treatment is by metronidazole and quinacrine hydrochloride. Nitazoxanide (Alina) has activity similar to metronidazole and is also now recommended for cryptosporidiosis (see below). The general population has many carriers who may excrete the cyst stage, which is relatively resistant to the chlorine used to disinfect drinking water.

Cryptosporidiosis. **Cryptosporidiosis** is caused by *Cryptosporidium hominis* (formerly *C. parvum* genotype 1) which rarely infects animals, and by *C. parvum* (previously *C. parvum* genotype 2) which infects both humans and livestock. This disease is transmitted by ingestion of oocysts and causes diarrhea. It is particularly likely to affect immunosuppressed humans. The oocysts are resistant to chlorination and must be removed from water supply systems by filtration. Coinfections with giardiasis are common.

Cyclospora Diarrheal Infection. *Cyclospora* diarrheal infection is caused by a protozoan, *Cyclospora cayetanensis.* There have been several outbreaks in the United States from ingestion of oocysts on uncooked foods such as berries.

Amebic Dysentery. The cysts of *Entamoeba histolytica* are ingested with contaminated food and water and can cause severe **amebic dysentery,** or **amebiasis.** Severe infections can cause perforation of the intestinal wall and abscesses. For treatment, metronidazole plus iodoquinal are the drugs of choice.

HELMINTHIC DISEASES OF THE DIGESTIVE SYSTEM

Tapeworms

Cysticerci, larval forms of the **tapeworm** that become encysted in muscles of the intermediate host, are ingested by humans and develop into adult tapeworms. They attach to the intestinal wall of the human host and shed eggs with the host's feces. *Taenia saginata* (the beef tapeworm), *Taenia solium* (the pork tapeworm), and *Diphyllobothrium latum* (the fish tapeworm) are the best-known tapeworms in the United States. Tapeworms may be 6 or 7 meters in length. Niclosamide is used in treating tapeworm infections.

The pork tapeworm, *T. solium,* differs from the beef tapeworm by producing the larval stage in the human host. **Taeniasis** is an infestation of the human intestine by the adult tapeworm; it is a benign and asymptomatic condition, but the host expels eggs continuously. **Cysticercosis,** an infection with the larval stage, can develop when humans ingest eggs, which can lodge in many tissues (usually the brain and muscle). Larvae may also lodge in the eye, **ophthalmic cysticercosis,** or in the central nervous system, especially the brain, **neurocysticercosis.**

Hydatid Disease. *Echinococcus granulosus,* a tapeworm only a few millimeters in length, may infect humans by way of contact with dog feces. The dog becomes infected by eating the flesh of infected sheep. The tapeworm egg migrates in the human to various tissues and develops into a large **hydatid cyst.** These cysts are damaging because of their size, and, if they rupture, they can cause anaphylactic shock.

Nematodes

Pinworms. The most familiar nematode infection in humans is by *Enterobius vermicularis* (**pinworm**). The tiny worm migrates out of the anus of the human host and lays its eggs, causing local itching. Drugs such as pyrantel pamoate or mebendazole are effective treatments.

Hookworms. **Hookworm** infestations were once common in the southeastern states, mostly caused by *Necator americanus.* The hookworm attaches to the intestinal wall and feeds on blood and tissue. Blood loss can lead to craving for starch or clay soils (*pica*), which is a symptom of iron deficiency anemia. Mebendazole is an effective treatment.

Ascariasis.   One of the most widespread helminthic diseases is **ascariasis,** caused by *Ascaris lumbricoides.* The worms can be up to about a foot in length but cause few symptoms, because they live on partially digested food in the intestinal tract. Ingested eggs hatch into small, wormlike larvae that pass from the bloodstream into the lungs. From the lungs they migrate into the throat and are swallowed, developing into adults in the intestinal tract. Mebendazole is an effective treatment.

Whipworm (*Trichuris trichiura*).   The common name, whipworm, is derived from the morphology. There is a thin, hair-like body and a thicker posterior end, resembling a coiled whip with a handle. The egg is distinctive, resembling a tea tray with handles. The ingested egg hatches and enters the crypts of Lieberkühn in the large intestine. There they grow and eventually locate themselves with the posterior end of the worm extending into the intestinal lumen and the hair-like anterior remaining buried in the mucosa. They feed on cell contents and blood.

Trichinellosis.   The small roundworm *Trichinella spiralis* causes **trichinellosis.** The disease is acquired by ingestion of undercooked meat, usually pork, that contains the encysted larvae. In the intestine of a human the ingested cyst matures into the adult form, producing larvae that invade tissue, especially muscles of the diaphragm and eye. Mebendazole is an effective drug.

SELF-TESTS

In the matching section, there is only one answer to each question; however, the lettered options (a, b, c, etc.) may be used more than once or not at all.

I. Matching

_____ 1. Produced from sucrose by *Streptococcus mutans.*

_____ 2. Pertaining to the inflammation and degeneration of gums and bones supporting the teeth.

_____ 3. An enzyme produced by pathogenic strains of *Staphylococcus aureus.*

_____ 4. A disease resulting from ingestion of a toxin.

_____ 5. A disease resulting from microbial growth in tissues of the host.

_____ 6. Loss of appetite.

_____ 7. Vincent's disease.

_____ 8. Chronic gingivitis.

a. Periodontal disease

b. Coagulase

c. Intoxication

d. Infection

e. Dental plaque

f. Anorexia

g. Periodontitis

h. Acute necrotizing ulcerative gingivitis

II. Matching

_____ 1. Affects the parotid glands.

_____ 2. Causes diarrhea; same organism that causes gas gangrene.

_____ 3. Caused by a small roundworm that encysts its larval form in muscle tissue.

_____ 4. MMR vaccine used.

_____ 5. A virus commonly causing gastroenteritis.

_____ 6. Caused by *Salmonella typhi.*

_____ 7. A disease caused by growth of *Salmonella* bacteria in the intestinal tract; incubation time of 12 to 36 hours is normal.

_____ 8. Orchitis is a possible complication.

a. Hydatid disease

b. Salmonellosis

c. Norovirus

d. Trichinellosis

e. Typhoid fever

f. *Clostridium perfringens* gastroenteritis

g. Mumps

III. Matching

_____ 1. Shigellosis.

_____ 2. *Streptococcus mutans.*

_____ 3. The most common food poisoning in Japan; sodium-requiring bacteria.

_____ 4. Prolonged diarrhea caused by a flagellated protozoan.

_____ 5. *Entamoeba histolytica.*

_____ 6. Human enteric caliciviruses.

_____ 7. Coagulase-positive organisms.

_____ 8. *Trichuris trichiura*

a. Dental caries

b. Staphylococcal food poisoning

c. Bacillary dysentery

d. *Vibrio parahaemolyticus* gastroenteritis

e. Norwalk-like agents

f. Giardiasis

g. Amebic dysentery

h. Whipworm

IV. Matching

____ 1. Viral hepatitis, usually foodborne.

____ 2. A term associated with the serological classification of *Salmonella*.

____ 3. An infectious viral particle that causes disease only from coinfection with the hepatitis B virus.

____ 4. Mycotoxin.

____ 5. Associated with tapeworm infestation.

____ 6. Formerly non-A, non-B hepatitis.

____ 7. Found in intestinal tract; phagocytic; produce defensins and lysozyme.

a. Aflatoxin

b. Hepatitis A

c. Hepatitis B

d. Paneth cells

e. HDV

f. Kauffmann-White

g. Bacteriocinogen

h. Hydatid cysts

i. Hepatitis C

V. Matching

____ 1. Not a tapeworm.

____ 2. Causes neurocysticercosis in humans.

____ 3. Pork tapeworm.

____ 4. Fish tapeworm.

____ 5. Beef tapeworm.

____ 6. Cause of hydatid disease.

____ 7. A tapeworm for which humans are not the definitive host.

a. *Taenia solium*

b. *Taenia saginata*

c. *Echinococcus granulosis*

d. *Diphyllobothrium latum*

e. *Ascaris lumbricoides*

VI. Matching

____ 1. Growth characterized by a "stacked-brick" configuration.

____ 2. The cause of hemolytic uremic syndrome.

____ 3. Not an invasive strain of *E. coli*, but probably the most common cause of traveler's diarrhea.

____ 4. Enteroinvasive *E. coli*.

a. ETEC

b. STEC

c. EAEC

d. EIEC

Fill in the Blanks

1. The epidemic form of cholera is caused by serogroup _____ of *Vibrio cholerae.*

2. The term for skin yellowing due to liver infection is _____.

3. The term for the larval form of tapeworms that becomes encysted in the muscles of the intermediate host is _____.

4. The toxin produced by the mold *Aspergillus flavus* is _____.

5. The toxin produced by the fungus *Claviceps purpurea* is _____.

6. _____ immunoglobulins in saliva tend to prevent microbial attachment.

7. Dental plaque is formed mostly from sticky polymers of _____ produced from metabolism of sucrose.

8. The enterotoxin responsible for most cases of staphylococcal food poisoning is serological type

_____.

9. The Dane particle probably represents an intact virion of the hepatitis type

_____.

10. Pathogenic staphylococci will grow in the presence of 7.5% sodium chloride and ferment the sugar

_____.

11. _____ disease is a collective term for a number of conditions characterized by inflammation and degeneration of the gums, supporting bone, and so on.

12. Gut-associated _____ tissue is abbreviated as GALT.

Critical Thinking

1. Is food contaminated with *Staphylococcus aureus* safe to eat after reheating? Why?

2. *Salmonella enterica,* the cause of salmonellosis, infects both humans and many animals; *Salmonella typhi,* the cause of typhoid fever, infects only humans. Which would be more affected by the development of modern water treatment and sewage treatment practices by cities?

3. Infection of humans by the pork tapeworm, *Taenia solium,* is potentially very dangerous, whereas infection by the beef tapeworm, *Taenia saginata,* is relatively benign. Why?

4. There are a great many antibiotics that affect bacteria and relatively few drugs that affect protozoa and helminths. Why?

5. Why must a person be infected with HBV in order to become infected with HDV?

ANSWERS

Matching

I.	1. e	2. a	3. b	4. c	5. d	6. f	7. h	8. g
II.	1. g	2. f	3. d	4. g	5. c	6. e	7. b	8. g
III.	1. c	2. a	3. d	4. f	5. g	6. e	7. b	8. h
IV.	1. b	2. f	3. e	4. a	5. h	6. i	7. d	
V.	1. e	2. a	3. a	4. d	5. b	6. c	7. c	
VI.	1. c	2. b	3. a	4. d				

Fill in the Blanks

1. O:1   2. jaundice   3. cysticercus   4. aflatoxin   5. ergot   6. IgA   7. dextran   8. A   9. B
10. mannitol   11. Periodontal   12. lymphoid

Critical Thinking

1. Food contaminated with *S. aureus* is not safe to eat after reheating, because the enterotoxin that *S. aureus* produces is little affected by heat. The toxin causes the symptoms of staphylococcal food poisoning.

2. Typhoid fever has been made into a rare disease in the United States by modern sanitation. Salmonellosis has not been much *affected,* because foods become contaminated during animal processing. Such contamination is not significantly influenced by water or sewage treatment.

3. The pork tapeworm produces a larval stage in humans as well as animals. It is the presence of this larval form in the brain that causes neurocysticercosis. The beef tapeworm lives out its life in the intestinal tract without producing a larval stage in humans. Humans have become well adapted to the presence of the adult helminth.

4. Protozoa and helminths closely resemble humans at the cellular level, that is, the way in which the cells metabolize and reproduce. Therefore, it is more difficult to find drugs that selectively inhibit protozoa and helminths without affecting the human host.

5. The HDV antigen is not capable of causing infection without first being covered by the HBV envelope.

26 Microbial Diseases of the Urinary and Reproductive Systems

The **urinary system** consists of organs that regulate the chemical composition of the blood and excrete waste products. The **reproductive system** consists of organs that produce gametes for reproduction of the species or support the developing embryo.

STRUCTURE AND FUNCTION OF THE URINARY SYSTEM

The urinary system consists of two kidneys, two ureters, a urinary bladder, and the urethra (Figure 26.1). The **urine** passes down the **ureters** from the **kidneys,** where wastes have been removed from the blood, to the **bladder.** Elimination eventually occurs through the **urethra.** The urethra in the male also conveys seminal fluid. Valves in ureters prevent backflow to the kidneys from the bladder, an aid in preventing infection. Also, the acidity of urine and the flushing action of excretion help prevent infection.

STRUCTURE AND FUNCTION OF THE REPRODUCTIVE SYSTEM

The **female reproductive system** consists of two **ovaries,** two **uterine (fallopian) tubes,** the **uterus,** the **vagina,** and the external genitals. The ovum (egg) is released from the ovary (a process called ovulation) and passes down the uterine tube. If it is fertilized in the uterine tube, it becomes implanted in the wall of

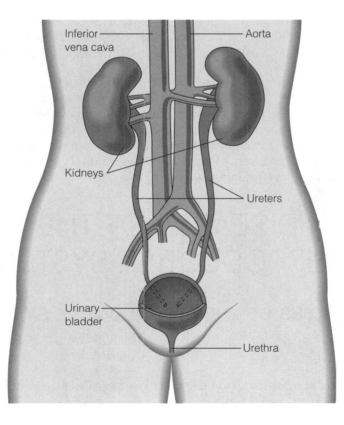

Figure 26.1 Organs of the human urinary system, shown here in the female.

the uterus, where it develops into an embryo and then into a fetus. The vagina is a copulatory canal and part of the birth canal. The external genitals are the clitoris, labia, and glands for lubricating secretions. The **male reproductive system** consists of two **testes,** which make sperm and male sex hormones; ducts to transport seminal fluid and sperm; and the **penis.**

NORMAL MICROBIOTA OF THE URINARY AND REPRODUCTIVE SYSTEMS

The urine in the bladder and the organs of the upper urinary tract are normally sterile. In infancy, the female vagina is populated by acid-forming lactobacilli, but the acidity becomes more neutral in childhood, and the microbiota then contain a variety of corynebacteria, cocci, and bacilli. At puberty, acid-forming lactobacilli again become the dominant flora. At menopause, the microbiota return to those of childhood, and the pH becomes neutral.

___ DISEASES OF THE URINARY SYSTEM ___

BACTERIAL DISEASES OF THE URINARY SYSTEM

Infections of the urinary system are mostly opportunistic. Inflammation of the urethra is called *urethritis;* of the bladder, *cystitis;* and of the ureters, *ureteritis.* Microbial populations of as few as 1000/ml of any single bacterial type, or as few as 100/ml of coliforms such as *E. coli,* are considered an indication of infection.

Cystitis.   **Cystitis** is an inflammation of the urinary bladder and is more common in females. Most cases are due to infection by *E. coli* or *Staphylococcus saprophyticus.* Symptoms include *dysuria* (difficult, urgent urination) and *pyuria* (leukocytes in the urine). A positive urine test for *leukocyte esterase,* an enzyme produced by neutrophils, indicates an active infection.

Pyelonephritis.   **Pyelonephritis** is an inflammation of the kidneys. Scar tissue can be formed that impairs kidney function. It is generally a complication of infections elsewhere in the body; in females, these usually are lower urinary tract infections. *E. coli* is responsible for about 75% of the cases.

Leptospirosis.   **Leptospirosis** is caused by a spirochete, *Leptospira interrogans.* The disease is characterized by chills, fever, and headache. Localization of the pathogens in the kidneys may result in jaundice and possible kidney failure. Leptospirosis is mainly a disease of animals, which excrete the organism with their urine. Domestic dogs are commonly immunized against it. The disease is transmitted to humans and animals by contact with urine-contaminated water. The organism can pass through minor abrasions in mucous membranes. An emerging form of leptospirosis, **pulmonary hemorrhagic syndrome**, has appeared globally. It causes massive bleeding of the lungs.

DISEASES OF THE _____ REPRODUCTIVE SYSTEM _____

BACTERIAL DISEASES OF THE REPRODUCTIVE SYSTEM

Diseases of the reproductive system transmitted by sexual activity are called **sexually transmitted infections (STIs).**

Gonorrhea.   One of the most common reportable communicable diseases in the United States is **gonorrhea,** which is caused by a gram-negative diplococcus, *Neisseria gonorrhoeae.* To be infective, the bacteria must attach to the epithelial wall of the urethra by means of fimbriae. The organism may invade spaces separating mucosal cells found in the oral-pharyngeal area and cervix, as well as the eyes and rectum.

Men are likely to become aware of gonorrheal infection by painful urination and discharge of pus from the urethra. Complications such as sterility resulting from partial blockage of reproductive ducts by scarring may occur.

Women usually are unaware of the infection, but pain can accompany the fairly common occurrence of a spread of the infection to the uterus and uterine tubes.

Complications of gonorrhea may involve the joints, heart (**gonorrheal endocarditis**), meninges (**gonorrheal meningitis**), eyes, and pharynx. **Gonorrheal arthritis** is caused by the growth of the gonococcus in fluids in the joints. **Ophthalmia neonatorum** is a condition resulting from infection of the eyes of an infant as it passes through the birth canal. Laws usually require administration of antibiotics at birth to prevent this. Gonorrheal infections also can be transferred by hand contact in adults. Gonorrhea can affect any area of sexual contact. **Pharyngeal gonorrhea** often resembles the symptoms of a sore throat, and **anal gonorrhea** can be painful, with the discharge of pus.

There is no immunity to reinfection. Currently, recommended treatment is with cephalosporin antibiotics. Tetracycline often is administered in treatment because many gonorrhea patients also have concurrent chlamydial infections. Diagnosis can be made by identifying the organism in the infectious discharges of both men and women. The organisms are observed as gram-negative cocci in pairs contained within phagocytic leukocytes. Serological diagnostic tests and DNA probes are also used for diagnosis.

Nongonococcal Urethritis (NGU).   Inflammations of the urethra not caused by *Neisseria gonorrhoeae* are called **nongonococcal urethritis (NGU)** or **nonspecific urethritis.** NGU may be the most common sexually transmitted infection in the United States, but it is not reportable. Most of these infections are caused by chlamydias, obligate intracellular parasites not easily cultured. Tetracycline is an effective treatment. Other bacteria, such as the mycoplasmas *Ureaplasma urealyticum* or *Mycoplasma hominis,* may cause NGU.

Pelvic Inflammatory Disease.   **Pelvic inflammatory disease (PID)** is a collective term for any extensive bacterial infection of the female pelvic organs. PID is often caused by *N. gonorrhoeae;* however, coinfection with chlamydial bacteria is frequent, and they are also an important agent of this disease. Infection of the uterine tubes (**salpingitis**) is the most serious kind of PID. Scarring can block passage of ova from the ovary to the uterus. This can result in sterility or **ectopic** (or tubal) **pregnancy** (the fetus develops in the uterine tube rather than in the uterus).

Syphilis.   The number of **syphilis** cases in the United States has remained fairly stable compared to gonorrhea. The causative organism is a gram-negative spirochete, *Treponema pallidum.* (Separate strains cause the tropical skin disease **yaws.**) Syphilis is transmitted mainly by sexual contact. The incubation period averages about 3 weeks. There are several identifiable stages of the disease.

Primary Stage Syphilis.   The initial symptom of the **primary stage** is a **chancre,** or sore, which usually appears at the site of the infection. The fluid formed in the chancre is highly infectious, and spirochetes can be seen with darkfield microscopy. In a few weeks this lesion disappears; many women are entirely unaware of it. Serological tests are positive in about 80% of patients in the primary stage. Bacteria enter the blood and the lymphatic system and become distributed in the body.

Secondary Stage Syphilis.   Several weeks after the primary stage, the **secondary stage,** characterized by rashes, appears. The rash occurs on the skin and mucous membranes, and lesions are very infectious. Nonsexual transmission—kissing, dentistry, and so on—is most likely at this stage. Serological tests become almost uniformly positive. In a few weeks the symptoms subside. The damage done at this stage and the later tertiary stage is caused by inflammatory response to circulating immune complexes that lodge at various body sites.

Latent Period.   Following the secondary stage, the **latent period** is entered. The majority of cases do not progress beyond the secondary stage, even without treatment, perhaps because of developing immunity.

Tertiary Stage Syphilis. In fewer than half the cases, the disease reappears after an interval of years in a **tertiary stage.** Tertiary, or late stage syphilis, can be classified by affected tissues or types of lesion. **Gummatous syphilis** is characterized by *gummas,* rubbery masses that cause local tissue destruction. **Cardiovascular syphilis** most seriously results in a weakening of the aorta. **Neurosyphilis** affects the central nervous system with a variety of symptoms: *paresis,* a sign of dementia; and *tabes dorsalis,* loss of coordination of voluntary movement. It is now rare for syphilis to be allowed to progress to the tertiary stage.

Congenital Syphilis. When syphilis is transmitted from mother to fetus, causing **congenital syphilis,** the infant's mental development is often impaired.

Diagnosis and Treatment of Syphilis. The **VDRL (Venereal Disease Research Laboratory)** slide test and other rapid screening tests for diagnosis, such as the **rapid plasma reagin (RPR)** card test, detect antibodies, not against the spirochete itself, but reagin-type antibodies. Simple **rapid diagnostic tests (RDTs)** are serological tests that can be done by a fingerprick blood sample. None of these serological tests detect the pathogen directly, and all are classified as *nontreponemal serological tests.* Confirmation of a positive screening test can be made by *treponemal-type serological tests* such as the **fluorescent treponemal antibody absorption (FTA-ABS)** tests.

Only treponemal-type tests are used for confirmatory testing, and they are almost always done in a central laboratory. Benzathine penicillin, a long-lasting formulation, has continued to be effective in the treatment of syphilis, especially in the primary stage.

Lymphogranuloma Venereum. *Chlamydia trachomatis,* the probable major cause of NGU and the cause of trachoma, also is responsible for **lymphogranuloma venereum.** The disease is found in much of the tropical world and in the southeastern United States. Infection causes the regional lymph nodes to become enlarged and tender. Scarring of the lymph drainage ducts occasionally leads to massive enlargement of the male genitals and rectal narrowing in females.

Chancroid. **Chancroid (soft chancre)** is mainly a tropical disease, but cases in the United States are reported—often associated with drug users. In chancroid, a painful ulcer forms on the genitals and spreads to adjacent lymph nodes. The causative organism is *Haemophilus ducreyi.*

Bacterial Vaginosis

Vaginitis, infections of the vagina, are most often caused by the fungus *Candida albicans,* the protozoan *Trichomonas vaginalis,* or the bacterium *Gardnerella vaginalis. C. albicans* usually is an opportunist, whereas *T. vaginalis* and *G. vaginalis* usually are sexually transmitted.

G. vaginalis infection, once called nonspecific vaginitis, involves interaction with anaerobic vaginal bacteria. There is no sign of inflammation, so the term **bacterial vaginosis** is preferred to vaginitis. Diagnosis is based on a fishy odor, a vaginal pH above 4.5, and "clue cells," sloughed-off vaginal epithelial cells covered with gram-negative rods, in the discharge. Metronidazole, which eliminates the anaerobic bacteria, is used for treatment.

VIRAL DISEASES OF THE REPRODUCTIVE SYSTEM

Genital Herpes

Herpes simplex virus type 2, and sometimes type 1, causes a sexually transmitted disease called **genital herpes.** (The official names are human herpesvirus-1 and human herpesvirus-2.) The lesions cause a burning sensation, and vesicles appear. The lesions are infectious; there is a serious danger of infection to an infant at birth—**neonatal herpes.** Such an infection can be fatal or cause serious neurological damage.

The virus may enter a latent stage in nerve cells and reappear at intervals, much like cold sores. There is no cure at this time for genital herpes, although oral and topical administrations of acyclovir and related antivirals alleviate symptoms and decrease the frequency of recurrence in many cases. Valacyclovir, taken daily, reduces sexual transmission significantly.

Genital Warts.   Warts are caused by viruses, and sexual transmission of **genital warts** (or *condyloma acuminata*) is common. The papillomaviruses, especially serotype 16, that cause warts also are sometimes factors in cervical cancer. Vaccines (Gardasil and Cervarix) against cancer-causing serotypes are available for young girls.

AIDS.   AIDS is caused by a virus that is often sexually transmitted; it is discussed in Chapter 19 because it primarily affects the immune system.

FUNGAL DISEASE OF THE REPRODUCTIVE SYSTEM

Candidiasis.   *Candida albicans* is a yeastlike fungus that grows on mucous membranes. Infection of the mucous membrane of the vagina is **vulvovaginal candidiasis.** There is irritation; severe itching; a thick, yellow, cheesy discharge; and a yeasty odor. Predisposing conditions include pregnancy, diabetes, certain tumors, and treatment with immunosuppressive drugs or broad-spectrum antibiotics. Diagnosis is made by microscopic identification of the fungus, and treatment consists of topical application of clotrimazole and miconazole, or an oral dose of fluconazole.

PROTOZOAN DISEASE OF THE REPRODUCTIVE SYSTEM

Trichomoniasis.   The protozoan *Trichomonas vaginalis* is a fairly normal inhabitant of the female vagina and the male urethra. **Trichomoniasis** results when the acidity of the vagina is disturbed and the protozoa outgrow the normal microbiota. Men rarely have any symptoms; women exhibit a purulent discharge with a disagreeable odor, accompanied by irritation and itching. Diagnosis is made by microscopic examination of the discharge. Metronidazole is used in treatment.

The Torch Panel of Tests

A panel of tests directed at detecting diseases likely to cause birth defects in newborns in termed the TORCH panel from the acronym, Toxoplasmosis; Other (such as syphilis, hepatitis B, enterovirus, Epstein-Barr virus, varicella virus); Rubella: Cytomegalovirus, Herpes simplex virus.

SELF-TESTS

In the matching section, there is only one answer to each question; however, the lettered options (a, b, c, etc.) may be used more than once or not at all.

I. Matching

___ 1. One of the most common reportable communicable bacterial diseases in the United States.

___ 2. The VDRL test is useful in diagnosis.

___ 3. Caused only by a protozoan.

___ 4. Most of the cases probably are caused by chlamydias.

___ 5. In diagnosis, the causative organism of this disease is seen as pairs of gram-negative cocci contained within phagocytic leukocytes.

___ 6. *Treponema pallidum.*

___ 7. Caused by a spirochete.

___ 8. Metronidazole has been an effective treatment.

___ 9. Caused by a virus.

a. Nongonococcal urethritis

b. Gonorrhea

c. Genital herpes

d. Syphilis

e. Trichomoniasis

II. Matching (Stages of Syphilis)

___ 1. Characteristic symptoms are mainly rashes; serological tests at this stage are almost uniformly positive.

___ 2. The disease is not normally infectious during this stage, but it can be transmitted from mother to fetus.

___ 3. The usual symptom characteristic of this stage is the chancre.

___ 4. Lesions in this stage can cause extensive damage; symptoms may be due to immune reactions.

a. Primary

b. Latent

c. Tertiary

d. Secondary

III. Matching

___ 1. The site of ectopic pregnancies.

___ 2. Where the ovum is implanted to develop into a fetus.

___ 3. The channel for eventual elimination of urine.

___ 4. Carries urine from the kidneys to the urinary bladder.

___ 5. The site of cystitis.

a. Ureter

b. Bladder

c. Urethra

d. Uterus

e. Uterine tube

f. Urine

IV. Matching

___ 1. Caused by *Chlamydia trachomatis*.

___ 2. Caused by *Haemophilus ducreyi*.

___ 3. Pathogen often discharged in animal urine.

___ 4. Caused by a spirochete.

a. Glomerulonephritis

b. Leptospirosis

c. Chancroid

d. Lymphogranuloma venereum

Fill in the Blanks

1. Inflammation of the urethra is called _____.

2. Inflammation of the ureters is called _____.

3. Cases of gonorrhea are often treated with tetracycline antibiotics that are directed at a probable coinfection with _____.

4. Slide tests to screen for syphilis detect _____ -type antibodies.

5. Herpes simplex virus type _____ is the most common cause of genital herpes.

6. The uterine tubes are also known as the _____ tubes.

7. Pyelonephritis is an inflammation of the _____.

8. Domestic dogs are commonly immunized against a disease, _____, which is spread by contact with urine-contaminated water.

9. To be infective, the organism that causes gonorrhea attaches to the epithelial walls of the urethra by means of _____ .

10. Syphilis transmitted by the mother to the fetus is known as _____ syphilis.

11. Rubbery lesions of the tertiary stage of syphilis, called _____ , sometimes cause extensive tissue damage.

12. Another name for chancroid is _____ .

13. A drug useful in alleviating symptoms of genital herpes is _____ .

14. It is customary to treat syphilis with long-acting benzathine _____ .

15. A urine test that is used to diagnose cystitis detects an enzyme produced by neutrophils, which is called the leukocyte _____ test.

16. The VDRL test for syphilis is an example of a _____ serological test. (treponemal or nontreponemal)

17. The "T" in the TORCH panel of tests is for _____ .

Critical Thinking

1. Explain why bacterial infections of the urinary system are more common in females than in males.

2. How is the widespread use of oral contraceptives related to the increased incidence of gonorrhea and other STIs?

3. List and discuss three diseases caused by *Chlamydia trachomatis.*

4. Why is trichomoniasis more common in females than in males?

ANSWERS

Matching

I.   1. b   2. d   3. e   4. a   5. b   6. d   7. d   8. e   9. c
II. 1. d   2. b   3. a   4. c
III. 1. e   2. d   3. c   4. a   5. b
IV. 1. d   2. c   3. b   4. b

Fill in the Blanks

1. urethritis   2. ureteritis   3. chlamydia   4. reagin   5. 2   6. fallopian   7. kidneys   8. leptospirosis
9. fimbriae   10. congenital   11. gummas   12. soft chancre   13. acyclovir   14. penicillin   15. esterase
16. nontreponemal   17. toxoplasmosis

Critical Thinking

1. Women have a shorter urethra than do men; the urethra in women is closer to the anus than it is in men; the female urethra has an abundance of normal microbiota.

2. Oral contraceptives have contributed to the increased incidence of gonorrhea and other STIs for two major reasons. First, oral contraceptives increase the moisture content and raise the pH of the vagina. This increases the susceptibility of mucosal cells to infection. Second, oral contraceptives have in many cases replaced condoms and spermicides, both of which help prevent disease transmission.

3. Three diseases caused by *Chlamydia trachomatis* include the following:
   - Lymphogranuloma venereum is a disease seen most often in tropical countries. After 7 to 12 days, a small lesion appears at the site of infection, ruptures, and heals. One week to 2 months later, the organisms invade the lymphatic system, cause scarring, and may result in enlargement of the genitals in males or rectal narrowing in females.
   - Trachoma is the single greatest cause of blindness and is common in the arid regions of Africa and Asia and in the southwestern United States. It is transmitted by hands, fomites, and flies.
   - Nongonococcal urethritis (NGU) refers to any inflammation of the urethra not caused by gonorrhea. The most common cause is *Chlamydia*. Signs and symptoms include painful urination and a watery discharge.

4. Trichomoniasis is more common in females than in males because *Trichomonas vaginalis* may be part of the normal microbiota of women.

27 Environmental Microbiology

MICROBIAL DIVERSITY AND HABITATS

Microbes that live in extreme conditions of temperature, acidity, alkalinity, or salinity are called **extremophiles.** Their enzymes are called **extremozymes.** Many are members of the domain Archaea.

Symbiosis

Symbiosis is the interaction between coexisting organisms or populations. **Mycorrhizae** are fungi symbiotically associated with plant roots. **Endomycorrhizae** (*vesicular-arbuscular mycorrhizae*) form large spores. Hyphae from these penetrate plant roots and form **vesicles** (probably storage structures) and **arbuscules** (bushlike structures that release nutrients to plants as they decay). **Ectomycorrhizae** do not form vesicles or arbuscules but rather form an *external mantle* over the smaller roots, usually of trees such as pine or oak. **Truffles** (a gourmet food) are ectomycorrhizae.

SOIL MICROBIOLOGY AND BIOGEOCHEMICAL CYCLES

Billions of organisms—microscopic as well as comparatively huge insects and earthworms—form the soil community. Bacteria are present in millions per gram.

Biogeochemical cycles are largely dependent on the activity of soil and water microbes.

The Carbon Cycle

The first step in the **carbon cycle** is the utilization of atmospheric carbon dioxide (about 0.03% of the atmosphere) in photosynthesis to produce organic matter (Figure 27.1). Eventually this organic matter, from green plants or creatures living directly or indirectly off plants, is changed by decay into carbon dioxide and released again to the atmosphere. Much carbon dioxide also is released by the burning of fossil fuels. There is considerable concern that accumulations of carbon dioxide in the atmosphere will cause world temperatures to increase—**global warming.**

The Nitrogen Cycle

Nitrogen is needed by all organisms for the synthesis of protein, nucleic acids, and other nitrogen-containing compounds (Figure 27.2). Molecular nitrogen (N_2) constitutes about 80% of the Earth's atmosphere.

As nitrogen-containing organic matter enters the soil (much of the nitrogen is in protein form), it undergoes decay by microorganisms. Decomposition of proteins releases amino acids, called **deamination.** Ammonia (NH_3) is formed by the removal of amino groups from amino acids, called **ammonification.**

In **nitrification,** ammonia is oxidized to yield energy by autotrophic bacteria (Figure 27.2). During nitrification, *Nitrosomonas* bacteria convert ammonia to nitrites (NO_2), and *Nitrobacter* bacteria oxidize the nitrites to nitrates (NO_3), the form preferred by plants as a nitrogen source.

Denitrification is an anaerobic process by which nitrates are used as an electron acceptor in place of oxygen (an example of anaerobic respiration). *Pseudomonas* species appear to be the most important of several groups involved. The result is, by a series of steps, formation of nitrogen gas, which returns to the atmosphere. This process can be an important means of nitrogen loss in soil.

Nitrogen fixation is the ability to use nitrogen directly from the atmosphere, an accomplishment of only a few bacteria or cyanobacteria.

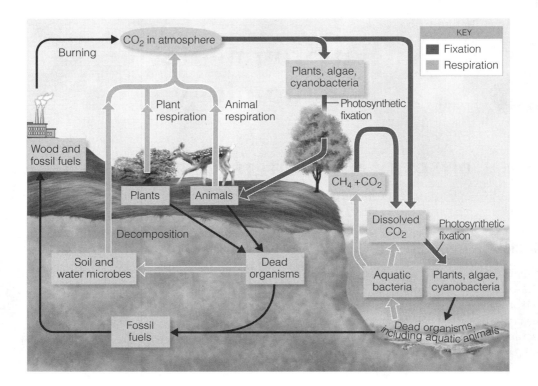

Figure 27.1   The carbon cycle.

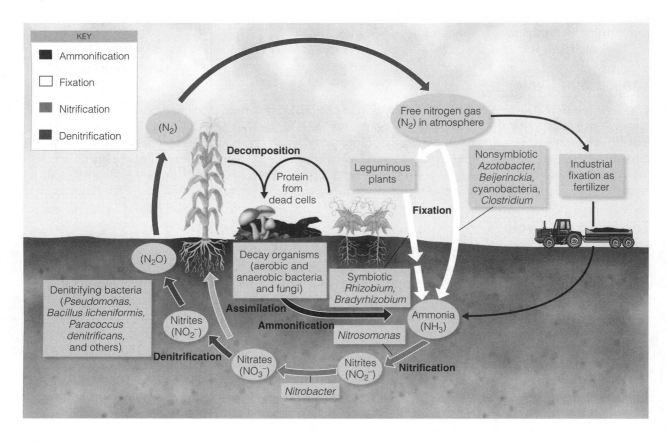

Figure 27.2   The nitrogen cycle. In general, nitrogen in the atmosphere goes through fixation, nitrification, and denitrification. Nitrates assimilated into plants and animals after nitrification go through decomposition, ammonification, and then nitrification again.

Free-living nitrogen-fixing bacteria are found in particularly high concentrations in the *rhizosphere,* which is the region roughly 2 millimeters from the plant root. The rhizosphere is comparatively rich in nutrients and is especially important in the ecology of grasslands. Among these nitrogen-fixing bacteria are the aerobic *Azotobacter* and *Beijerinckia.* Some anaerobic species such as *Clostridium pasteurianum* and a number of species of the photosynthetic cyanobacteria fix nitrogen. The cyanobacteria are important in rice paddies of Asia, where they form a nitrogen-fixing symbiosis with *Azolla* water plants. The **nitrogenase** (nitrogen-fixing) enzyme system operates only anaerobically and must be protected from oxygen; cyanobacteria carry it in **heterocysts.** In agricultural soil, most of the heterotrophic nonsymbiotic nitrogen-fixing bacteria usually lack sufficient carbohydrates needed for energy to reduce significant amounts of nitrogen to ammonia. They seldom fix the amounts that can be observed under laboratory conditions. Nevertheless, they can make important contributions to the nitrogen economy of areas such as grasslands, forests, and the arctic tundra.

Symbiotic nitrogen-fixing bacteria are agriculturally more important. The genera *Rhizobium* and *Bradyrhizobium* infect the roots of leguminous plants such as soybeans, beans, peas, peanuts, alfalfa, and clover. These bacteria are specific for a particular leguminous species. They attach to a root hair and cause formation of an **infection thread** passing into the root itself. *Rhizobium* and *Bradyrhizobium* entering root cells by the thread become enlarged (**bacteroids**). The root cells proliferate to form a **root nodule** packed with bacteroids. Nitrogen is fixed there, with the plant furnishing anaerobic conditions and growth nutrients and the bacteria fixing the nitrogen (Figure 27.3). Similar examples occur in nonleguminous plants such as alder trees, for which an actinomycete (*Frankia*) is the microorganism. **Lichens,** a mutualistic association between a fungus and an alga or cyanobacterium, may be a source of nitrogen when one symbiont is a nitrogen-fixing cyanobacterium.

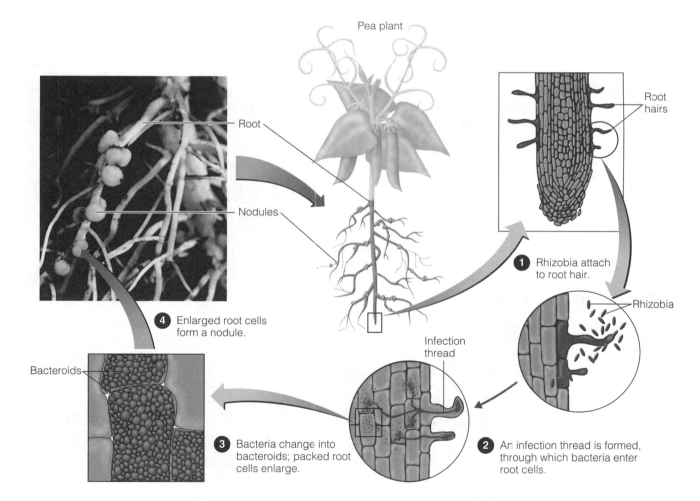

Figure 27.3   Formation of a root nodule. Members of the nitrogen-fixing genera *Rhizobium* and *Bradyrhizobium* form these nodules on legumes. This mutualistic association is beneficial to both the plant and the bacteria.

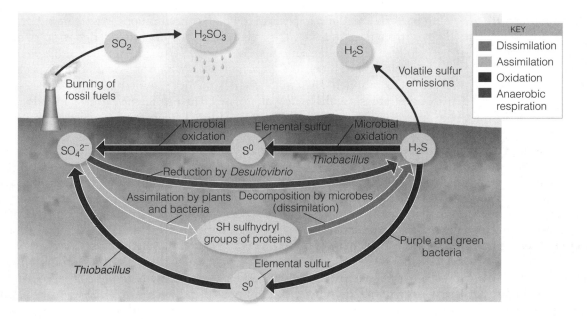

Figure 27.4 **The sulfur cycle.** Note the imortance of aerobic and anaerobic conditions.

The Sulfur Cycle

The sulfur cycle, like the nitrogen cycle, represents numerous oxidation states of the elements (Figure 27.4). Hydrogen sulfide (H_2S) represents a source of energy for autotrophic bacteria. Microbes that use it for energy convert H_2S into elemental sulfur (S^0) granules and fully oxidized sulfates (SO_4^{2-}). Sulfates are incorporated into sulfur-containing amino acids of proteins. As proteins are decomposed (**dissimilation**) the sulfur is released as H_2S to continue the cycle. Members of the green and purple sulfur bacteria also oxidize H_2S, but these organisms are using light for energy; H_2S is used to reduce CO_2, much like H_2O is used in plant photosynthesis.

Life without Sunshine

Ecosystems exist in caves and near deep-ocean hydrothermal vents that maintain entire communities that are not based on photosynthesis. The **primary producers** are chemoautotrophic bacteria rather than photoautotrophic plants or microbes. They use the energy of H_2S and the carbon of CO_2. Other autotrophic bacteria, called **endoliths,** live deep in subsurface rock formations. Hydrogen (H_2) produced by chemical and radioactive reactions in the rocks serves as the primary energy source.

The Phosphorus Cycle

Phosphorus exists primarily as phosphate ions and undergoes little change in oxidation state. The phosphorus cycle mostly involves changes from soluble to insoluble forms, often in relation to pH changes. This cycle has no volatile component that enters the atmosphere, and phosphorus tends to reach a dead end in the seas. Fossil accumulations from old seabeds are mined.

THE DEGRADATION OF SYNTHETIC CHEMICALS IN SOIL AND WATER

Pesticides and other synthetic chemicals, such as plastics, that do not occur in nature (**xenobiotics**) enter the soil in large amounts. Many of these, compared to natural organic matter, are degraded very slowly if at all. DDT is an example of a pesticide that is resistant to biodegradation (*recalcitrant*). Small changes in the molecule can be important. For example, the addition of a single chlorine atom to the ring structure of the herbicide 2,4-D to form 2,4,5-T creates a chemical that is much more recalcitrant.

Bioremediation

The use of microbes to detoxify or degrade pollutants is called **bioremediation.** Oil spills can be effectively treated with "fertilizer" that supplies bacteria with nitrogen and phosphorus missing in petroleum. The addition of specialized pollutant-degrading microbes is called **bioaugmentation.**

Solid Municipal Waste

Solid municipal waste (garbage) is usually placed into landfills. Under anaerobic conditions in landfills, methane is produced.

Organic matter, especially plant remains, can be accumulated into piles and **composted.** Bacteria, many of them thermophilic, change these wastes into a product that uses much less space and can be used on agricultural land.

AQUATIC MICROBIOLOGY AND SEWAGE TREATMENT

Aquatic microbiology refers to the study of microorganisms in lakes, rivers, estuaries, and oceans.

Aquatic Microorganisms

Freshwater Microbiota.   High nutrient levels in water are generally reflected by high microbial numbers. Microbial populations of freshwater bodies tend to be differentiated by the availability of oxygen and light. The **littoral zone** is the region along the shore with rooted vegetation and light that penetrates to the bottom. The **limnetic zone** consists of the surface away from the shore. The **profundal zone** is the deeper water under the limnetic zone, and the **benthic zone** is the sediment at the bottom. Light is important because photosynthetic algae are the main source of organic matter (energy) for the lake. These are the *primary producers*, located in the limnetic zone. The deeper waters of the profundal zone are low in oxygen and light. There, purple and green sulfur bacteria, which use light filtered past algae at higher levels, are found. In this zone are *Desulfovibrio*, which produce hydrogen sulfide from sulfate, and the methane-producing bacteria.

Seawater Microbiota.   Nearly a third of all life on the planet consists of microbes that live under the seafloor. There they make large amounts of methane gas. In the upper, sunlit waters of the oceans, cyanobacteria, such as *Synechococcus* and *Prochlorococcus,* comprise much of the population. Immense numbers of these photosynthetic microbes fix carbon dioxide to provide the nutrients that marine **phytoplankton** (Greek for "wandering plants") use for growth. Particularly numerous are *Pelagibacter ubique,* a tiny bacterium that was only recently discovered using the FISH technique discussed in Chapter 10. These bacteria serve as particulate food for increasing larger consumers that are eventually prey for fish. In waters below 100 meters, Archaea, such as *Crenarchaeota,* account for much of the biomass of oceans. Many deep-ocean microbes, some symbiotic with fish, emit light (**bioluminescence**).

The Role of Microorganisms in Water Quality

Water Pollution

Transmission of Infectious Diseases.   Many diseases are perpetuated by the **fecal–oral route** of transmission, in which a pathogen shed in human feces contaminates water or food and is later ingested. Examples are typhoid fever, cholera, hepatitis A, and the protozoan *Giardia lamblia.* Many helminthic diseases, such as schistosomiasis, are spread among people coming into contact with contaminated water. The water is usually ingested; a larval form penetrates the skin.

Chemical Pollution.   Many industrial and agricultural chemicals enter water and are resistant to degradation. Some become concentrated in biological organisms, as happens with fat-soluble pesticides and metallic mercury. Mercury is converted by benthic bacteria to a soluble form that is taken up by fish. Eating excessive amounts of such fish can cause neurological damage.

Eutrophication, a term meaning "well nourished," can occur when nitrogen and phosphorus, in particular, enter waters. Algae get carbon from atmospheric carbon dioxide and energy from light. Therefore, algae and non–nitrogen-fixing cyanobacteria require only relatively small amounts of nitrogen and phosphorus for heavy growth (**blooms**). Many cyanobacteria fix their own nitrogen; for them, phosphorus is the only ingredient lacking. The death of these organisms results in their degradation by bacteria, a process that depletes the oxygen in water and hastens the filling of the lake. Oil spills are often effectively treated with certain nutrients and specially adapted bacteria.

Coal-mining wastes often are high in sulfur content, mostly iron sulfides (FeS_2). Bacteria such as *Thiobacillus ferrooxidans* obtain energy from the oxidation of the ferrous ion (Fe^{2+}) and convert the sulfides into sulfates. These enter the streams as sulfuric acid, damaging aquatic life and causing formation of insoluble yellow iron hydroxides. These organisms are also used in mining to leach such minerals as uranium from ores in a soluble form.

Water Purity Tests. Microbiological tests for water safety are aimed at detection of **indicator organisms.** In the United States these are usually **coliform bacteria** (aerobic or facultative anaerobic, gram-negative, non–endospore-forming rod-shaped bacteria that ferment lactose with acid and gas formation within 48 hours at 35°C). Coliforms are normal inhabitants of human intestines, and their presence is considered an indication of human wastes entering the water. *Escherichia coli* is the predominant **fecal coliform.** Some coliforms, common soil and plant inhabitants, are **nonfecal coliforms** and are less important as indicators. The U.S. Environmental Protection Agency (EPA) recommends the use of *Enterococcus* bacteria as safety indicators for waters in oceans and bays, where they decrease in population more uniformly than coliforms.

There are two basic methods for detecting and enumerating coliforms. The first is the **multiple-tube fermentation technique.**

Coliforms may also be detected by the **membrane filter method.** Bacteria are retained on the surface of the membrane filter and will form colonies of a distinctive appearance.

A newer method of detecting *E. coli* makes use of the fact that it normally produces the enzyme β-glucuronidase. When a sample containing *E. coli* is added to a medium containing a substrate, MUG, the β-glucuronidase forms a product that is visible under ultraviolet light. Coliforms other than *E. coli* produce only the enzyme β-galactosidase, which forms a yellow color in media containing the substrate ONPG.

Water Treatment

If turbid, water can be held in a reservoir to allow some particulate matter to settle out. The water then undergoes a **flocculation treatment,** by addition of aluminum potassium sulfate (alum), for example. This chemical forms a *floc,* which carries suspended colloidal material with it as it settles. The next step is passage through beds of sand (**sand filtration**), which removes about 99% of the bacteria and viruses remaining at this stage. Microorganisms are trapped mostly by surface adsorption onto the sand grains. Activated charcoal, which absorbs chemical pollutants, may supplement sand filtration. Water is then **chlorinated** to kill microorganisms before distribution. Ozone is considered a possible replacement or supplement for chlorination. **Ozone** (O_3) is generated by electricity at the site of treatment. Arrays of **ultraviolet** tube lamps can also be used to disinfect water.

Sewage (Wastewater) Treatment

After water is used, it becomes **sewage,** a term that includes household washing wastes, industrial wastes, and rainwater from street drains. Sewage is mostly water with as little as 0.03% particulate matter.

Primary Sewage Treatment. The usual first step in sewage treatment is **primary treatment.** Settling chambers remove sand and so on; skimmers remove floating oil and grease; floating debris is ground up. Following this, solid matter is allowed to settle out in sedimentation tanks. Sewage solids collecting on the bottom (40–60% of suspended solids are removed here) are called **sludge.** Sludge is removed periodically, and the effluent (liquid flowing out) may undergo secondary treatment.

Biochemical Oxygen Demand. **Biochemical oxygen demand (BOD)** is a measure of the biologically degrad able organic matter in water, as determined by measuring the amount of oxygen required by bacteria to metabolize it. A classic method is to follow for 5 days the lowering of oxygen levels in water samples in a sealed bottle. The more oxygen used up as the bacteria utilize the organic matter in the sample, the greater the BOD.

Secondary Sewage Treatment. Primary treatment removes 25–35% of the BOD; the remainder is largely in the form of dissolved organic matter. **Secondary treatment** involves high aeration to encourage aerobic microorganisms to oxidize dissolved organic matter to carbon dioxide and water. In the **activated sludge system,** air is added to aeration tanks. This encourages the growth of organisms that rapidly oxidize organic matter and causes formation of suspended bacteria-containing masses called floc or *sludge granules.* After several hours, the sludge is passed to a tank, where it settles to the bottom. Some of the sludge is recycled to the activated sludge tanks as a "starter"for the next sewage batch (this is the basis for the term *activated sludge*). If the sludge floats, the phenomenon is called **bulking,** a common problem. The growth of filamentous bacteria such as *Sphaerotilus natans* is a common cause of bulking. Activated sludge systems remove 75–90% of the BOD. **Trickling filters** are a secondary treatment method whereby sewage is sprayed over a bed of rocks or formed plastic about 6 feet deep. A biofilm composed of microorganisms that aerobically decompose dissolved organic matter in the passing sewage soon forms on the rocks. Trickling filters remove only 80–85% of the BOD, but they are less troublesome to operate and less sensitive to toxic sewage and overloads. **Rotating biological contactor systems** consist of a series of large rotating vertical disks. These rotate slowly, partially submerged, and build up an active biofilm community on their surface.

Sludge Digestion. Sludge accumulating from primary and secondary treatment can be treated in **anaerobic sludge digesters.** The first stage of activity in this process is a fermentation forming organic acids and carbon dioxide. Microorganisms then produce hydrogen and carbon dioxide from these organic acids. The hydrogen and carbon dioxide are used by anaerobic methane-producing bacteria to form methane. Most methane is derived from the energy-yielding reduction of carbon dioxide by hydrogen gas:

$$CO_2 + 4\,H_2 \longrightarrow CH_4 + 2\,H_2O$$

Other methane producers split acetic acid to yield carbon dioxide and methane:

$$CH_3COOH \longrightarrow CH_4 + CO_2$$

Much of the sewage is, therefore, converted to methane and carbon dioxide. Residual solids are pumped to shallow drying beds or to filters to reduce the volume. This sludge is disposed of in several ways, including use as a soil conditioner. Class A sludge has no detectable pathogens; Class B sludge has reduced levels of pathogens.

Septic Tanks. **Septic tanks** settle out suspended solids in a holding tank. These are periodically pumped out. The effluent flows through a system of perforated piping into a soil drainage field, where it enters the soil and is decomposed by soil microorganisms.

Oxidation Ponds. Many small communities and many industries use **oxidation ponds** (also called **lagoons** or **stabilization ponds**). Many incorporate a two-stage system. The first stage is deep and anaerobic, and sludge settles out. Effluent is then pumped into other ponds that are shallow enough to be aerated by wave action. The growth of algae is encouraged in order to produce oxygen to help maintain aerobic conditions in such high organic matter loads. Bacterial action in decomposing organic matter generates the carbon dioxide used by the algae in photosynthesis in the shallow waters.

Tertiary Treatment. Primary and secondary treatments do not remove all of the biodegradable organic matter. Secondary treatment effluent contains about 50% of the original nitrogen and 70% of the phosphorus. These proportions can have a great impact on a lake's ecosystem (see earlier discussion of eutrophication). **Tertiary treatment** removes almost all of the BOD, nitrogen, and phosphorus. Nitrogen is converted to ammonia and evaporates into the air or is converted to nitrogen gas by denitrifying bacteria. Tertiary treatment is very expensive and is used at this time only in highly developed lake areas where eutrophication must be avoided, even at high costs.

SELF-TESTS

In the matching section, there is only one answer to each question; however, the lettered options (a, b, c, etc.) may be used more than once or not at all.

I. Matching

____ 1. Area within about 2 mm of a plant root.

____ 2. Heavy growth of algae in a body of water.

____ 3. Bacteria found within rocks and shales.

a. Rhizosphere

b. Topsoil

c. Blooms

d. Endoliths

II. Matching

____ 1. *Rhizobium* species are important in this process.

____ 2. *Pseudomonas* species are important in this process.

____ 3. *Azotobacter* species are important in this process.

____ 4. *Nitrobacter* species are important in this process.

____ 5. Forms a mantle over small roots.

____ 6. Forms arbuscules inside plant root cells.

____ 7. Truffles.

____ 8. Removal of amino acids from proteins.

a. Symbiotic nitrogen fixation

b. Free-living nitrogen fixation

c. Nitrification

d. Denitrification

e. Ammonification

f. Endomycorrhizae

g. Ectomycorrhizae

h. Dissimilation

i. Deamination

III. Matching (Aquatic Zones)

____ 1. The zone in the region along the shore with rooted vegetation and light that penetrates to the bottom.

____ 2. The deeper water under the limnetic zone.

____ 3. Bottom sediments.

____ 4. The water surface away from shore.

____ 5. Where the primary producers are mainly located.

____ 6. Where the methane-producing bacteria are found.

a. Limnetic zone

b. Benthic zone

c. Littoral zone

d. Profundal zone

IV. Matching (Sewage Treatments)

____ 1. Removes almost all BOD, nitrogen, and phosphorus, not all by biological means.

____ 2. Methane-producing bacteria are an important element.

____ 3. Mainly designed to encourage growth of aerobic bacteria to oxidize dissolved organic matter to carbon dioxide and water.

____ 4. *Zoogloea* bacteria form floc in one method.

a. Primary

b. Secondary

c. Anaerobic sludge digestion

d. Tertiary

V. Matching

____ 1. A morphological form of *Rhizobium*.

____ 2. An aquatic plant that may have a symbiosis with nitrogen-fixing cyanobacteria.

____ 3. A possible substitute or supplement for chlorination treatment of water.

____ 4. A shrimplike creature found in the ocean.

____ 5. The filamentous organism *Sphaerotilus natans* may be a factor in this.

a. Krill

b. Bacteroids

c. Lichen

d. *Azolla*

e. Phytoplankton

f. Ozone

g. Bulking

VI. Matching

____ 1. A human and animal pathogen found in soil.

____ 2. Fecal coliform.

____ 3. Conversion of ammonia to nitrite.

____ 4. Nonsymbiotic nitrogen fixation involves this bacterium.

____ 5. Species of oceanic cyanobacteria.

a. *Bacillus anthracis*

b. *Nitrosomonas*

c. *Beijerinckia*

d. *Escherichia coli*

e. *Synechococcus*

VII. Matching

____ 1. Use of microbes to detoxify or eliminate pollutants.

____ 2. Excess nutrients that lead to overgrowth, especially by photosynthetic organisms, in lakes and streams.

____ 3. Presumably caused by increasing amounts of carbon dioxide in the atmosphere.

____ 4. Microbes, mostly members of Archaea, that live in environments of exceptional temperatures, acidity, or salt concentrations.

____ 5. Microbes that live on the energy of hydrogen found in deep rock formations.

____ 6. An example would be a plastic bottle.

a. Endoliths

b. Extremophiles

c. Eutrophication

d. Bioremediation

e. Global warming

f. Xenobiotics

Fill in the Blanks

1. The form of nitrogen most likely to be used by plants is _____.

2. The removal of amino groups from amino acids to form ammonia is called _____.

3. Denitrification is a process by which nitrates are used as an electron acceptor in place of

   _____.

4. The cyanobacteria protect their nitrogenase systems from oxygen in a body called a

   _____.

5. *Rhizobium* bacteria infect a plant by attaching to a root hair and causing formation of a(n)

   _____, by which they travel into the root of the plant.

6. Many of the cyanobacteria fix nitrogen, and for them the only requirement for heavy growth in most lake waters is the addition of small amounts of _____.

7. Addition of alum to water during purification is called _____ treatment.

8. Sewage solids that collect on the bottom are called _____.

9. In oxidation ponds, the growth of algae is often encouraged in order to produce

   _____.

10. Putting organic municipal waste into piles that encourage partial degradation by microbial action is called _____.

11. A word derived from the Greek, meaning "wandering plants," is _____.

Critical Thinking

1. Explain the benefits of the filamentous growth habit exhibited by actinomycetes.

2. What are indicator organisms, and how are they used in the evaluation of water quality?

3. Compare and contrast two methods of secondary treatment used in sewage treatment.

4. How do mycorrhizae contribute to plant growth?

ANSWERS

Matching

   I.   1. a   2. c   3. d

   II.   1. a   2. d   3. b   4. c   5. g   6. f   7. g   8. i

   III.   1. c   2. d   3. b   4. a   5. c   6. b

   IV.   1. d   2. c   3. b   4. b

   V.   1. b   2. d   3. f   4. a   5. g

   VI.   1. a   2. d   3. b   4. c   5. e

   VII.   1. d   2. c   3. e   4. b   5. a   6. f

Fill in the Blanks

1. nitrate   2. ammonification   3. oxygen   4. heterocyst   5. infection thread   6. phosphorus
7. flocculation   8. sludge   9. oxygen   10. composting   11. phytoplankton

Critical Thinking

1. Actinomycetes are common inhabitants of the soil. Their filamentous growth habits allow them to bridge the gap between soil particles in dry conditions and also maximize surface area, giving actinomycetes a nutritional advantage over other soil-dwelling microbes.

2. Indicator organisms are organisms that are consistently found in human feces in sufficient numbers to be detected in water when fecal contamination occurs. The indicator organism must also grow at least as well as would pathogenic organisms in feces. In the United States, coliform organisms are used as indicator organisms. Their detection is an indication of fecal contamination of water.

3. The activated sludge system involves the addition of air or pure oxygen to the effluent from primary treatment. This encourages the growth of aerobic organisms that oxidize much of the effluent's organic matter.

   Trickling filters involve the spraying of sewage over a bed of rocks or molded plastic. Aerobic bacteria form a gelatinous film on these materials and act very much the same way as was just discussed.

4. There are two types of mycorrhizae: endomycorrhizae (vesicular-arbuscular mycorrhizae) and ectomycorrhizae. Both function as root hairs of plants. They extend the surface area through which plants can absorb nutrients, especially phosphorus.

28 Applied and Industrial Microbiology

FOOD MICROBIOLOGY

Foods and Disease

Like water, foods can become a source of disease. To minimize the potential for this, a system of sanitation training and inspections is used to inspect dairies, restaurants, and other food-preparation operations. To safeguard food preparation in industrial settings, a system known as **hazard analysis and critical control point (HACCP)** is now widely used. It is designed to identify points at which foods are most likely to be contaminated with harmful microbes, or which might allow their proliferation. This system also requires monitoring of adequate temperatures to kill pathogens or adequate storage temperatures to prevent their reproduction.

Industrial Food Canning

Canned foods today undergo what is called **commercial sterilization** by steam under pressure in a large **retort,** which is the minimum processing necessary to destroy the endospore-forming pathogen *Clostridium botulinum.* To ensure this result, the **12D treatment** is applied, by which a theoretical population of botulism bacteria would be decreased by 12 logarithmic cycles (10^{12} endospores would be reduced to only one).

Spoilage of Canned Food. **Thermophilic anaerobic spoilage** is possible in low-acid canned foods. The can is usually swollen from gas production, and the contents have a lowered pH and a sour odor. The cause is thermophilic clostridia, which survived commercial sterilization because of unusually resistant endospores. In **flat sour spoilage,** caused by thermophilic organisms such as *Bacillus stearothermophilus,* the can is not swollen by gas production. Both of these types of spoilage occur with storage at higher-than-normal temperatures; the thermophilic organisms involved will not grow at normal storage temperatures. Canned-food spoilage by **mesophilic bacteria** is due to can leakage or underprocessing. The former is more likely to result in spoilage by non–endospore-formers, and the latter in spoilage by endospore-formers. As the can cools, a vacuum is formed and contamination by leakage can therefore be caused by sucking cool water past the sealant in the crimped lid. These types of spoilage often result in odors of putrefaction in high-protein foods. Botulism is a possible consideration in mesophilic spoilage.

Some acidic foods, such as tomatoes and fruits, are preserved by heating at temperatures below the boiling point. This type of preservation is possible because organisms such as molds, yeasts, and occasional species of acid-tolerant non–endospore-forming bacteria (the only organisms likely to grow in such foods) are easily killed. One problem with such acidic foods can be the heat-resistant **sclerotia** (specialized resistant bodies) of certain molds, which can survive temperatures of 80°C for a few minutes. Other problems are **heat-resistant ascospores** of the mold *Byssochlamys fulva* and the endospore-forming *Bacillus coagulans,* a bacterium capable of growth at a pH as low as almost 4.0.

Aseptic Packaging

The container used in **aseptic packaging** is usually made of laminated paper or plastic that cannot tolerate conventional heat treatment. Rolls of packaging materials are fed into a machine that sterilizes the material in hot hydrogen peroxide or with high-energy electron beams. The containers are formed while still in the sterile environment and filled with conventionally heat-sterilized foods.

Radiation and Industrial Food Preservation

The use of **ionizing radiation** to preserve food is theoretically possible. Actual sterilization by radiation probably causes too much alteration in the taste of most foods for commercial use. The use of radiation is more likely to resemble pasteurization by heat treatment. Radiation is measured in Grays, frequently in terms of thousands of Grays (kGy). **Microwaves** kill bacteria only by their heating effects on food.

High-Pressure Food Preservation

Prewrapped foods can be submerged in tanks of pressurized water. This process kills many bacteria by disrupting cellular functions and has the advantage of preserving the color and taste of food better than many other methods.

The Role of Microorganisms in Food Production

Cheese.   **Cheeses** come in many types, all requiring the formation of a **curd,** which can then be separated from the liquid fraction, or **whey.** Except for unripened cheeses such as ricotta and cottage, the curd undergoes a microbial ripening process. The curd is milk protein, **casein,** and is usually formed by action of the enzyme **rennin** (chymosin), aided by acid conditions that are provided by inoculation with lactic acid–producing bacteria. These lactic bacteria provide the flavor and aroma.

Cheeses generally are classified by hardness. Romano and Parmesan cheeses are *very hard* cheeses; cheddar and Swiss are *hard;* Limburger, blue, and Roquefort cheeses are *semisoft;* and Camembert is a *soft* cheese. Hard cheddar and Swiss cheeses are ripened by lactic bacteria growing anaerobically in the interior. The longer the incubation time, the more acidity and the sharper the taste. (The holes in Swiss are from carbon dioxide produced by a *Propionibacterium* species of bacteria.) Semisoft cheeses are ripened by bacteria and other organisms growing on the surface. Blue and Roquefort cheeses are ripened by *Penicillium* molds inoculated into the cheese. The open texture allows adequate oxygen to reach the molds. Camembert cheese is made in small packets so that the ripening enzymes of surface molds will diffuse into the cheese.

Other Dairy Products.   **Butter** is made by churning cream until the fat phase forms globules of butter separated from the liquid buttermilk fraction. Lactic acid bacteria are allowed to grow in the cream and provide the flavor from the **diacetyls** that give the typical butter flavor and aroma. **Buttermilk** usually is made by inoculating skim milk with bacteria that form lactic acid and diacetyls.

Yogurt is made from low-fat milk by first evaporating much of the water in a vacuum pan. This process is followed by inoculation with a species of lactic acid–producing *Streptococcus* that grows at elevated temperatures. Stabilizers are added to aid in the formation of the thick texture. **Kefir** and **kumiss** are beverages of eastern Europe. In these, acid-producing bacteria are supplemented with a lactose-fermenting yeast to give the drinks a small alcoholic content.

Nondairy Fermentations.   Yeasts are used in baking; the sugars in bread dough are fermented to ethanol and carbon dioxide. The carbon dioxide makes the bubbles of leavened bread, and the ethanol evaporates in baking. Fermentation also is used in making sauerkraut, pickles, and olives. In Asia, soy sauce is a popular fermented food.

Alcoholic Beverages and Vinegar.   **Beer** and **ale** are products of fermentation of grain starches by yeast. Beer is fermented slowly with yeast strains that remain on the bottom (*bottom yeasts*). Ale is fermented relatively rapidly, at a higher temperature, with yeast strains that usually form clumps that are buoyed to the top by carbon dioxide (*top yeasts*). The grain starch is first converted to glucose and maltose by a process called **malting.** Barley is allowed to sprout, then dried and ground. This results in *malt,* which contains amylases to degrade starch. **Sake,** Japanese rice wine, is made from rice without malting. The mold *Aspergillus* converts the rice starch to fermentable sugars. **Distilled spirits,** such as whiskey, vodka, and rum, are made by fermentating carbohydrates from grains, potatoes, and molasses, respectively. The alcohol is then distilled off.

Wines are typically made from the yeast fermentation of grapes, which usually need no additional sugar, although more can be added to make more alcohol. Malic acid production makes some wines too acid, and bacteria are used to convert the malic acid to the less acidic lactic acid (**malolactic fermentation**). Wine can be spoiled by aerobic bacteria that convert alcohol into acetic acid. This process is used to make **vinegar.** Ethanol is made by the anaerobic fermentation of carbohydrates by yeasts. The ethanol is then aerobically oxidized to acetic acid by *Acetobacter* and *Gluconobacter* bacteria.

INDUSTRIAL MICROBIOLOGY

Industrial microbiology—the use of microbes to make industrial products—has centered on the lactic acid and ethanol fermentations. In the future, as petroleum becomes more scarce, we may see a return to many microbial fermentations. In the coming years there will be a revolution in the application of genetically engineered microorganisms in industry—**biotechnology.**

Fermentation Technology

Vessels for industrial fermentations are called **bioreactors.** They are designed to control aeration, pH, and temperatures. Microbes in industrial fermentation produce either primary metabolites, such as ethanol, or secondary metabolites, such as penicillin. A **primary metabolite** is formed at the same time as the cells, and their production curves are almost parallel. **Secondary metabolites** are not produced until the growth phase (**trophophase**) has entered the stationary phase of the growth cycle. The following period, during which most of the secondary metabolite is produced, is the **idiophase.** To improve strains of industrial microbes, radiation is often used to create useful mutants.

Immobilized Enzymes and Microorganisms. Some industrial processes make use of enzymes immobilized on surfaces while a continuous flow of substrate passes over them. Live or dead cells are also immobilized on such surfaces as small spheres or fibers, to make industrial products.

Industrial Products

Amino Acids. Certain **amino acids,** such as **lysine,** are especially important as nutritional supplements because animals cannot synthesize them. They are present only in low levels in vegetable proteins and are therefore a valuable product. Glutamic acid is an amino acid (L-glutamate) used to make the flavor enhancer **monosodium glutamate.** The microbially synthesized amino acids **phenylalanine** and **aspartic acid** are ingredients in the artificial sweetener NutraSweet.

Citric Acid. **Citric acid** is produced industrially by the mold *Aspergillus niger* from molasses. This happens when the mold is provided with only a limited supply of iron and manganese.

Enzymes. Enzymes are valuable industrial products: **amylases** for syrups and paper sizing, **glucose isomerase** to convert glucose into fructose, **proteases** for baking, **proteolytic** enzymes for meat tenderizers and additives to detergents. Rennin, used to form curds in milk, is a fermentation product of certain fungi or genetically engineered bacteria. A procedure long used in Japan to produce amylases is used to make **koji,** an enzyme preparation used to make fermented soy products. Molds such as *Aspergillus* are grown on rice or a wheat-soybean mixture. Primarily, the amylases in koji change cereal starch into sugars, but koji preparations also contain proteolytic enzymes that convert the protein in soybeans into a more digestible and flavorful form. It is the basis of Japanese diet staples such as *soy sauce* and *miso* (a fermented paste of soybeans with a meaty flavor).

Vitamins. Microbes produce several commercial vitamins. **Vitamin C, vitamin B$_{12}$,** and **riboflavin** are examples.

Pharmaceuticals. **Antibiotics** are an important industrial product. Many are still produced by microbial fermentations. Work continues on selection of more productive mutants and manipulations to increase efficiency. **Vaccines** are also an important industrial product. **Steroids,** such as cortisone, and estrogens and progesterone, used in birth control pills, are products of microbial conversions.

Copper Extraction by Leaching.   The recovery of otherwise unprofitable grades of **uranium** and **copper ore** is made possible by the activity of the bacterium *Thiobacillus ferrooxidans*. Acidic water is sprinkled over a pile of the ore, and the bacteria change, for example, insoluble copper compounds to an oxidized, soluble form that moves out of the ore piles and can be reclaimed as metallic copper.

Microorganisms as Industrial Products.   Baker's yeast is a good example of microbes themselves as an industrial product. Others are the symbiotic nitrogen-fixing bacteria *Rhizobium* and *Bradyrhizobium*, which are used to inoculate leguminous plants. The insect pathogen *Bacillus thuringiensis* is commonly sold in preparations applied to garden plants and trees.

Alternative Energy Sources Using Microorganisms

Conversion of biomass (various organic wastes) into alternative fuel sources is called **bioconversion.** **Methane** is an important source of such a process. It is produced even today in commercial amounts, from large landfills and cattle-feeding lots.

Biofuels

Much interest is centered on the use of **biofuels** to eventually replace fossil petroleum-based fuels used for transport. Because the technology is well established, early interest has been concentrated on **ethanol** from sugarcane and corn. However, because such crops are valuable food products, research into ethanol production is now focusing on the use of cellulosic materials from nonfood crops. Ethanol will probably be replaced by "higher" alcohols with longer carbon chains, especially "branched" alcohols. These have a lower capacity to absorb water (a requirement for pipeline transport) and have higher energy content. The use of algae grown in ponds or enclosed reactors could be used to produce **biodiesel** fuel and possibly jet fuel. **Hydrogen,** especially if produced by splitting water, is also an attractive fuel candidate. It can be used in fuel cells to generate electricity and, when burned, does not produce harmful residues.

SELF-TESTS

In the matching section, there is only one answer to each question; however, the lettered options (a, b, c, etc.) may be used more than once or not at all.

I. Matching

_____ 1. Involved in making a less acidic wine.

_____ 2. Component of NutraSweet.

_____ 3. An enzyme that forms milk curd.

_____ 4. Responsible for the flavors and aroma of dairy products.

_____ 5. Resistant bodies found on a number of molds.

_____ 6. An essential amino acid.

_____ 7. A product made by *Aspergillus niger* from molasses.

_____ 8. Molds such as *Aspergillus* grow on rice or wheat-soybean mixtures to produce amylases and proteases.

a. Sclerotia

b. Malolactic fermentation

c. Phenylalanine

d. Rennin

e. Diacetyls

f. Citric acid

g. Lysine

h. Koji

II. Matching (Canned-Food Spoilage Types)

___ 1. Caused by a thermophilic *Bacillus* species.

___ 2. Swollen can; contents would have low pH and a sour odor.

___ 3. Botulism would be a possible consideration in this type.

___ 4. Probable cause would be can leakage or underprocessing.

a. Flat sour

b. Mesophilic

c. Thermophilic anaerobic

III. Matching (Cheeses)

___ 1. Camembert

___ 2. Swiss

___ 3. Cottage

___ 4. Parmesan

___ 5. Roquefort

a. Hard cheese

b. Unripened cheese

c. Very hard cheese

d. Semisoft cheese

e. Soft cheese

Fill in the Blanks

1. The first step in making _____ is to evaporate much of the water from low-fat milk in a vacuum pan.

2. Blue and Roquefort cheeses are ripened by _____ molds.

3. Vinegar is made by the aerobic metabolism of _____ by acetic-acid bacteria.

4. Glutamic acid is a microbial product used to make the flavor enhancer _____ glutamate.

5. Vessels for highly aerobic industrial fermentation are usually called _____.

6. Canned food undergoes the minimum processing needed to destroy any *Clostridium botulinum* endospores; this is called _____ sterilization.

7. In this type of processing (question 6 above), the population of any botulism bacteria would be decreased by _____ logarithmic cycles. (Give a number.)

8. The liquid fraction remaining in milk after curd formation is called _____.

9. The holes in Swiss cheese are from carbon dioxide produced by a(n) _____ species of bacteria.

10. Packaging materials used in aseptic packaging are often sterilized by passing them through a bath of hot _____.

11. A(n) _____ metabolite, such as penicillin from a mold, is not produced until the growth phase has entered the stationary phase of the growth cycle.

12. The growth phase during which most secondary metabolites are produced is the _____ -phase. (Supply the prefix of the word.)

13. Ale is brewed by a type of yeast informally called a _____ yeast. (top or bottom)

14. Supply the missing word: Hazard analysis and _____ control point.

Critical Thinking

1. Compare and contrast commercial sterilization and true sterilization.

2. What advantages and disadvantages are associated with the use of high-energy electron accelerators or irradiation with gamma rays to preserve food?

3. What possible role could the cyanobacterium *Spirolina* play in the colonization of Mars?

4. What is bioconversion? What are its possible benefits?

ANSWERS

Matching

I.   1. b   2. c   3. d   4. e   5. a   6. g   7. f   8. h
II. 1. a   2. c   3. b   4. b
III. 1. e   2. a   3. b   4. c   5. d

Fill in the Blanks

1. yogurt   2. *Penicillium*   3. ethanol   4. monosodium   5. bioreactors   6. commercial   7. twelve
8. whey   9. *Propionobacterium*   10. hydrogen peroxide   11. secondary   12. idio-   13. top   14. critical

Critical Thinking

1. True sterilization kills or eliminates all of the microbes from food. Commercial sterilization is not as rigorous; it is designed to kill *Clostridium botulinum* endospores. If this is achieved, then any other bacteria that cause food spoilage or are pathogenic will be killed as well. The lower temperatures used in commercial sterilization result in a better-tasting food product.

2. Advantages are that irradiation of food with gamma rays is very effective at eliminating microorganisms, worms, and insect pests; also, penetration of the food is excellent because of the high energy content of gamma radiation. Disadvantages are that the process takes several hours and must be conducted behind protective walls; in addition, the public is apprehensive concerning radiation.

   Advantages are that high-energy electron accelerators are faster than irradiation with gamma-rays; it takes only seconds to treat food. Disadvantages are that this method doesn't penetrate well, so it is effective only for thin food products such as sliced meat.

3. *Spirulina* could be grown and used as a food source for the colonists. Because it gets energy from photosythesis and carbon from carbon dioxide, it would be of minimal transport weight.

4. Bioconversion refers to the conversion of biomass into alternate fuel sources such as methane. This practice would help meet our ever-increasing energy needs while helping to dispose of solid waste.